Emerging Infectious Disease

Scrivener Publishing
100 Cummings Center, Suite 541J
Beverly, MA 01915-6106

Infectious and Rare Diseases

Series Editor: Dr. Rishabha Malviya

The series offers a comprehensive exploration of the scientific landscape surrounding infectious and rare diseases. Within the epidemiological and pathogenic realm, the series delves into meticulous analyses of disease prevalence, distribution patterns, and the intricate molecular and cellular mechanisms underpinning pathogenesis. The diagnostic segment scrutinizes state-of-the-art techniques, including advanced imaging and molecular diagnostic modalities, with a focus on integrating artificial intelligence and machine learning to enhance precision and efficiency in disease diagnosis. Immunotherapy and vaccine development form another crucial aspect, featuring a thorough examination of innovative immunotherapeutic strategies tailored for infectious and rare diseases, along with breakthroughs in vaccine technologies. The series also confronts the escalating challenge of antimicrobial resistance in infectious diseases, probing into novel antimicrobial agents and stewardship programs designed to counter resistance. The scope extends to rare disease research, providing in-depth studies on the genetic and molecular foundations of rare diseases. This includes advancements in personalized medicine, featuring gene therapies and precision treatments

Publishers at Scrivener
Martin Scrivener (martin@scrivenerpublishing.com)
Phillip Carmical (pcarmical@scrivenerpublishing.com)

Emerging Infectious Disease

Unravelling Norovirus and Pandemic Dynamics

Edited by

Sachin Kumar
Department of Pharmacology, Delhi Pharmaceutical Sciences and Research University, New Delhi, India

Shriyansh Srivastava
Department of Pharmacology, Delhi Pharmaceutical Sciences and Research University, New Delhi, India

Rishabha Malviya
Department of Pharmacy, School of Medical and Allied Sciences, Galgotias University, Greater Noida, India

and

Ranjit Sah
BMC South, Boston Medical Centre Health System, MA, USA

Scrivener Publishing

WILEY

This edition first published 2026 by John Wiley & Sons, Inc., 111 River Street, Hoboken, NJ 07030, USA and Scrivener Publishing LLC, 100 Cummings Center, Suite 541J, Beverly, MA 01915, USA

For more information about Scrivener publications please visit www.scrivenerpublishing.com.

Wiley Global Headquarters
111 River Street, Hoboken, NJ 07030, USA

For details of our global editorial offices, customer services, and more information about Wiley products visit us at www.wiley.com.

The manufacturer's authorized representative according to the EU General Product Safety Regulation is Wiley-VCH GmbH, Boschstr. 12, 69469 Weinheim, Germany, e-mail: Product_Safety@wiley.com.

Library of Congress Cataloging-in-Publication Data

ISBN 978-1-394-34695-0

Cover image: Generated with AI using Adobe Firefly
Cover design by Russell Richardson

Set in size of 11pt and Minion Pro by Manila Typesetting Company, Makati, Philippines

Printed in the USA

10 9 8 7 6 5 4 3 2 1

Dedication

I dedicate this book to my greatest inspiration, my father, **Mr. Sanjay Srivastav**, and my most substantial support, my mother, **Mrs. Pushpa Srivastava**, who always believed in me. They sowed the seeds of knowledge in my mind and nurtured them with unwavering love and care. I also dedicate this work to my younger brother, **Romil Srivastav**, and my beloved sister, **Soumya Srivastava**, whose presence brings me constant joy and motivation. Finally, I express heartfelt gratitude to all the teachers who teach passionately and patiently, shaping minds and igniting dreams.

Shriyansh Srivastava

Contents

Preface

The 21st century has seen a surge in emerging infectious diseases, including norovirus, which is responsible for millions of gastroenteritis cases annually. Despite its tremendous impact, norovirus study has been difficult due to its discrete nature, genetic complexity, and limits *in vitro* replication. However, recent advances in virology, epidemiology, and molecular biology have offered more information about norovirus dynamics and pandemic potential.

This book seeks to bridge the gap between basic research and applied techniques for preventing norovirus outbreaks. It offers an in-depth understanding of norovirus through a multidisciplinary approach that includes genetics, environmental variables, zoonotic transmission, and epidemic response. The *One Health* concept emphasizes the link between human, animal, and environmental health.

The book additionally analyses the impact of COVID-19 on norovirus circulation and research, highlighting lessons learned and potential for future research. Understanding norovirus in the broader context of pandemic dynamics is critical for developing effective control measures and preventing future outbreaks.

The editors are grateful to all contributing authors for their significant contributions as well as the support of institutions and research communities. The editors hope that this book will be a valuable resource for people looking to better understand norovirus and its global impacts. As infectious illnesses evolve, we must remain committed to interdisciplinary collaboration, scientific innovation, and proactive public health initiatives. Finally, we extend our thanks to Martin Scrivener and the team at Scrivener Publishing for their efforts in bringing this work to fruition.

The Editors
November 2025

Part 1

INTRODUCTION AND GENETIC INSIGHTS

1

Introduction to Viral Storms: Infectious Disease

Shivaprasad Sirimalla[1], Sanjit Sah[2,3,4], Uday Venkat Mateti[1]* and Allen Pinto[1]

[1]Department of Pharmacy Practice, NGSM Institute of Pharmaceutical Sciences (NGSMIPS), Nitte (Deemed to be University), Mangalore, Karnataka, India
[2]Department of Pediatrics, Dr. D. Y. Patil Medical College, Hospital and Research Centre, Dr. D. Y. Patil Vidyapeeth (Deemed-to-be-University), Pimpri, Pune, Maharashtra, India
[3]Department of Public Health Dentistry, Dr. D.Y. Patil Dental College and Hospital, Dr. D.Y. Patil Vidyapeeth (Deemed-to-be-University), Pune, Maharashtra, India
[4]Department of Medicine, Korea University, Seoul, South Korea

Abstract

Infectious diseases, caused by pathogenic microorganisms including bacteria, viruses, and fungi, continue to present significant challenges to global health. The emergence of highly virulent viruses in the twenty-first century has resulted in pandemics that have profoundly affected humans worldwide. Major pandemics and epidemics such as H1N1 influenza, severe acute respiratory syndrome, human immunodeficiency virus, and coronavirus disease have already afflicted humanity. Several factors contribute to the emergence and spread of these viral outbreaks, including environmental changes, urbanization, global travel, and biological factors such as viral mutations and host immune responses. For many infections, geographical and seasonal patterns are inherently linked to the climate, raising the possibility of using climate parameters as predictive indicators in disease early warning systems (EWS). Norovirus infection, a significant global problem, shows predictable seasonality, with activity peaks during the winter seasons and in cold-temperature places. The primary step to prevent the spread of infectious diseases is the early identification of diseases and following effective interventions

**Corresponding author*: udayvenkatmateti@gmail.com; ORCID: 0000-0001-8149-2067
Shivaprasad Sirimalla: ORCID: 0000-0002-9751-3237
Sanjit Sah: ORCID: 0009-0008-5017-083X
Allen Pinto: ORCID: 0009-0008-3011-8962

Sachin Kumar, Shriyansh Srivastava, Rishabha Malviya and Ranjit Sah (eds.) Emerging Infectious Disease: Unravelling Norovirus and Pandemic Dynamics, (3–38) © 2026 Scrivener Publishing LLC

to control the diseases, resulting in reduction of mortality and morbidity. Using climate parameters and EWS is effective in predicting the diseases and can prevent the huge spread of transmission of a disease. Public health measures such as isolating, quarantining, and controlling the border had helped in past events to control the disease spread and it maintained social structure. Analyzing the past events and current examples provides great understanding of different aspects of viral storms; it may help to combat future pandemics in terms of prevention and management.

Keywords: Infectious diseases, viral pandemics, norovirus, climate change, viral transmission, global health

1.1 Introduction to Infectious Disease

Infectious diseases have been a continuously major health concern in human history, which caused various pandemics that impacted human population in various ways such as the economy. There are various pathogens that cause these infectious diseases such as parasites, fungus, viruses, and bacteria, leading to continuous damage to human health globally. Research in the field of infectious disease explored the infection mechanism, host–pathogen interaction, disease spread, and prevention and control of disease. It is very essential to understand completely the dynamics of an infectious disease that changes in response to environment, social, and biological factors [1–3].

Infectious diseases can happen in a large range of severity such as mild and self-limiting, i.e., common cold. Some other infections can cause severe and life-threatening situations such as sepsis. Cytokine storm is one of the severe manifestations of infectious diseases. It is a huge and uncontrolled immune response of the host after an infection. The host's immune system releases a large amount of cytokines into the blood stream that causes inflammation and damage to the tissue and finally leads to organ failure. This unusual situation happens in severe infections such as COVID-19, influenza, etc., giving importance to the host's immune response in combating infections [4–6].

Infectious diseases significantly impacted global health and accounts for a considerable number of morbidity and mortality. It was much more in low- and middle-income countries which are having limited access to healthcare. Infectious diseases also burden the economic and social domains, saturating the healthcare system, causing financial collapse, and worsening social inequalities [7, 8].

Research has made significant progress and contributions in fighting infectious diseases. In the past, diseases like polio, smallpox, and measles were prevented by the invention of vaccines, which has been the most profound advancement and thereby saving millions of lives. The invention of antibiotics and its wide usage made new revolution for bacterial infection and thereby reduced significant mortality rates. Likewise, invention by the use of antiviral drugs has made transformations on the management of viral infections like HIV/AIDS and hepatitis. The medical advancement has significantly decreased the burden on countries caused by the infectious diseases, and this leads to improved quality of life and improved human lifespan [9, 10].

Nevertheless, the combat against infectious diseases is not yet finished. As the new emergence of infectious diseases arises, such as novel coronavirus like SARS-CoV-2 that causes COVID-19, it is still a problem to the health of mankind. The emergence of new diseases is due to various factors such as interaction of humans with wild animals, leading to zoonotic spillover in which deadly pathogens transfer from animals to humans. This event happens due to several factors like increased human–animal interaction, deforestation, and climate change. In most parts of the world, deadly old pathogens like tuberculosis and measles which are under control are again re-emerging as new mutant forms of viruses along with new infectious diseases [11, 12].

Along with newly emerging viruses, antimicrobial resistance (AMR) is another challenge to fight against infectious diseases. Misuse and overuse of antibiotics, antivirals, and antifungals have happened over a period of time, and these lead to the development of drug-resistant strains of these pathogenic microorganisms. AMR has been the major obstacle in fighting the infectious diseases. This may lead to problems in fighting very common infectious diseases and become untreatable, and thus small surgical infections and minor infections can become severe and life-threatening [13, 14].

Viruses face a variety of challenges in the process of emergence of infectious diseases. Viruses require a host's cells to grow, replicate, and transmit. They cannot live and reproduce outside of the hosts, and this nature of replication is different from other pathogenic microorganisms like bacteria. This nature of dependence of viruses leads to frequent mutations to sustain the host's response to a virus, and this causes the frequent emergence of new strains. For example, influenza virus goes through several genetic mutations and leads to seasonal outbreaks and sometimes pandemics whenever a new strain comes. Likewise, SARS-CoV 2 virus showed several mutations in the RNA genome, and it leads to the emergence of new variants showing their new virulent effects [15].

The spread of infectious diseases is based on several interconnected factors including human nature, climate changes, and a variety of characteristics of pathogenic microorganisms. There are several human nature or behavior, including urbanization, global travel, and population density, causing the greater transmission of virus across the globe. Likewise, climate changes causes the transmission of vector-borne diseases like malaria and dengue. Thus, it is very essential to consider these factors to prevent and manage infectious diseases effectively [3].

After the huge development of research and science, infectious diseases are still a major challenge in human health. Complete surveillance on public health and research in the field of infectious diseases still need to be developed and conducted as there is the emergence of new pathogenic microorganisms in the world, emergence of frequent mutant viruses, and increased AMR. Considering all of the factors like how a new infectious disease emerge and spreads along with understanding the pathogenic microorganisms and through a comprehensive, multidisciplinary approach can make us understand, manage, and prevent the disease effectively, and thereby it can decrease the disease impact on public health globally.

1.2 Virology: Overview of Virus, Virus Structure, Classification, Replication, and Transmission

1.2.1 Overview of the Virus

Viruses are acellular pathogenic microorganisms that cause infectious diseases. Unlike other pathogenic agents such as bacteria, fungi, etc., viruses require a host cell to survive, replicate, and transmit disease, as they lack a proper cellular mechanism to perform metabolic function on their own to replicate and spread the diseases. It will proliferate by taking access of the host's cell through infection [16, 17].

Basically, every virus has a genetic material, i.e., either DNA or RNA, to replicate into viral particles inside the host's cell. Since viruses are acellular, the genetic material will be enveloped inside the protein case called capsid. In certain conditions, the virus additionally enveloped inside the lipid case which is generally derived from the host's cell. The envelop is usually arranged with glycoproteins that are essential for infectious process. Regardless of their simple nature, all viruses show extraordinary survival skills and adaptation, making them effective in infecting all kinds of hosts such as animals, plants, fungi, and bacteria [18–20].

The field of virology tells us that those virus particles are not simple in nature but designed in a complicated way to utilize the host's cell to replicate and transmit. Rapid changes or mutations of genetic material and its diversity pose a major challenge to public health, especially with newly emerging infectious diseases. Considering only basic factors such as basic structure and function of viruses makes it difficult to develop proper antiviral therapies, vaccines, and health strategies to fight against viral infections [18].

1.2.2 Virus Structure

The structure of a virus is intricately linked to its function and its ability to infect host organisms. At the core of every virus is its genome, which can consist of either DNA or RNA. This genetic material holds all the essential instructions needed for the replication and assembly of viral components. Depending on the virus, the genome can be either single-stranded or double-stranded and may exist in a linear or circular configuration. The variability in the viral genome is one of the factors that contribute to the diversity seen across different virus families. The viral genome is surrounded with a protective shell made of protein called capsid, which is composed of capsomere subunits. The capsid serves both to protect the genetic material from damage and to transport genetic materials inside the host cells. The arrangement of capsomere subunits is different, and this arrangement gives variety of the shapes and symmetries of viruses. The shape and symmetries of viruses include icosahedral, helical, and many other varieties observed in different virus families [17, 21].

Normally, some viruses contain another layer, which surrounds the capsid called the envelope. The basis of these envelopes is provided by the cell membrane of the host, when the virus is at the last stage of replication. The viral glycoprotein is useful to recognize and attach to exact receptors on the host cell surface. Generally, the envelop is embedded with viral glycoproteins. The host range and tissue tropism are decided by the viral glycoproteins. The viral glycoproteins of viruses control the process of infection to specific species and cell types. For example, the envelope in influenza virus uses one of its glycoproteins called hemagglutinin to bind to a receptor named sialic acid on the respiratory cells. The viral infection starts with the process of binding, which makes the viral envelope bind with the cell membrane of the host, and it facilitates the viral genome to get inside the host cell. The virus' shape and structure are not permanent. It can be changed through alterations in the mutations, which can finally influence

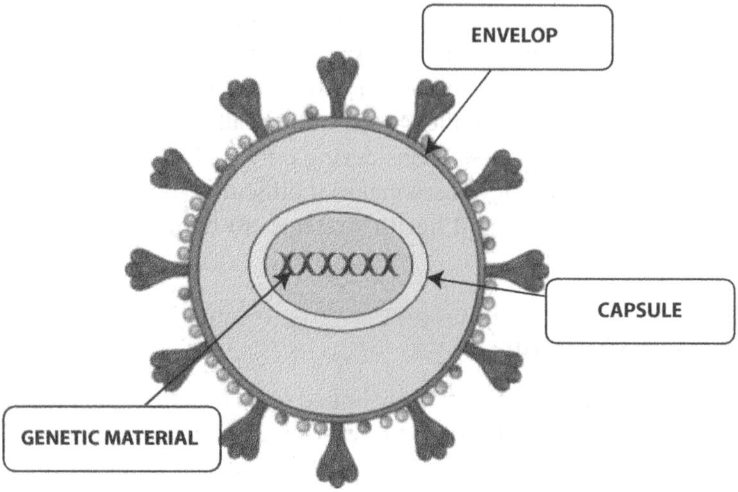

Figure 1.1 Structure of a virus.

the virulence effect and ability to enter the host cell. This continuous mutation is the major challenge that leads to viral evaluations and makes the requirement of effective vaccines and treatment [17, 18, 21].

Knowledge on the various structures of viruses is important to classify the viruses and also to develop proper therapeutic interventions. Commonly, therapeutic interventions like antivirals focus on components of the structure in viruses like capsid, an envelope that prevents the virus from infecting host cells or from replicating. Likewise, vaccines are used to increase or show an immunity action against the virus particles; thereby, it will prevent the infection by increasing the host's immunity. So, it is a very important aspect to study the structure of the virus which has direct connections with public health. Figure 1.1 depicted the structure of a virus, genetic material, capsule and envelop.

1.2.3 Classification of Viruses

Viruses are highly diverse infectious pathogens and acellular, and it must require the host cell to replicate. The structure has evolved over a period, classified into many varieties, and with that so much advancement happened to sustain their life. Commonly, viruses are classified based on morphology, chemical composition, and replication mode. This will help to understand the structure of the virus, genetic material, and mode of replication to generate new viruses within the host [17]. Table 1.1 presented the Baltimore viruses classification based on the genome type and replication strategy.

Table 1.1 Baltimore viruses classification [19, 22].

Group	Type of genome	Replication strategy	Example of viruses
I	Double-stranded DNA (dsDNA)	These viruses replicate their DNA within the host cell's nucleus. Transcription of mRNA occurs using the host's polymerase.	Herpesviruses, adenoviruses
II	Single-stranded DNA (ssDNA)	The ssDNA is converted into dsDNA by host enzymes before mRNA transcription and replication.	Parvoviruses
III	Double-stranded RNA (dsRNA)	These viruses carry an RNA-dependent RNA polymerase to transcribe their mRNA directly from their RNA genome.	Reoviruses
IV	Single-stranded RNA (ssRNA) positive-sense	Their RNA genome directly serves as mRNA, which is translated into viral proteins.	Polioviruses, coronaviruses
V	Single-stranded RNA (ssRNA) negative-sense	An RNA-dependent RNA polymerase found in the virus converts the RNA genome into a positive-sense mRNA.	Influenza viruses, rabies virus
VI	Single-stranded RNA (ssRNA) with reverse transcriptase	Through a process called reverse transcription, these viruses transform RNA into DNA, which is subsequently inserted inside the host genome to facilitate replication.	HIV (retroviruses)

(*Continued*)

Table 1.1 Baltimore viruses classification [19, 22]. (*Continued*)

Group	Type of genome	Replication strategy	Example of viruses
VII	Double-stranded DNA (dsDNA) with reverse transcriptase	They possess a DNA genome that replicates *via* an RNA intermediate, necessitating reverse transcription.	Hepatitis B virus

Morphological Classification

The morphology of a virus talks about the physical structure of the virus, which is very important to identify and classify the virus. Morphology is denoted by the size, shape, and symmetry of the virus. The viruses classified based on morphology are as follows:

Icosahedral Viruses

Structure: Icosahedral viruses have a geometric shape of 20 sides of capsid, similar to the geometry of equilateral triangle. These viruses are very stable to allocate more structural units and more effective in enclosing the viral genome [17, 19].

Examples:

Adenoviruses: It has an icosahedral symmetry and allows more stability and easy to infect using different environments. This virus will cause many infections such as respiratory infections, conjunctivitis, gastroenteritis, etc.
Polioviruses: These are icosahedral viruses that cause infections to the central nervous system, leading to poliomyelitis. The robust icosahedral structure of polioviruses helps them survive in harsh conditions, such as the acidic environment of the stomach.

Helical Viruses

Structure: Helical viruses have a rod-shaped or filamentous structure. The viral RNA or DNA is spirally coiled, with capsid proteins arranged around it in a helical manner, forming a cylindrical structure [17, 19].

Examples:

Tobacco mosaic virus (TMV): TMV was the first virus ever discovered and infects plants, particularly tobacco. Its helical structure allows it to efficiently penetrate plant cells and replicate.

Influenza virus: It is a significant human pathogen causing seasonal flu. The helical arrangement of the RNA genome within a lipid envelope aids in the virus' ability to adapt and evade the host immune system.

Complex Viruses

Structure: Complex viruses exhibit intricate structures that do not conform to simple helical or icosahedral patterns. They may possess multiple layers, such as a core containing the nucleic acid, surrounded by a lipid envelope, and additional proteins [17, 19].

Examples:

Poxviruses: These large viruses, including the smallpox virus, have a unique structure with multiple layers, including a complex outer envelope. They replicate in the cytoplasm, unlike most DNA viruses, which typically replicate in the nucleus.

Bacteriophages: These viruses infect bacteria and have a complex structure with an icosahedral head containing the genetic material and a tail structure used to inject this material into bacterial cells. The T4 bacteriophage, for instance, has a contractile tail that it uses to penetrate the bacterial cell wall [17, 19].

Chemical Composition Classification

Chemical composition is another crucial factor in viral classification, focusing on the type of nucleic acids, the nature of the capsid, and other structural components. The key elements include the following:

Nucleic acids: The genetic material that makes up viruses can be either single-stranded (ss) or double-stranded (ds) DNA or RNA. It is very core to a replication of virus and pathogenicity based on the type of nucleic acid.

Examples:

DNA viruses: These viruses contain the DNA as the genetic material, which may be single- or double-stranded DNA. For example:

Herpesviruses: It has an icosahedral shape capsid that encloses the double-stranded DNA and replication happens within the nucleus. These viruses cause infections such as oral herpes, infection to genitals, and genital herpes.

RNA viruses: These viruses contain the RNA as the genetic material with either single- or double-stranded RNA.

HIV (human immunodeficiency virus): It is a retrovirus having a single-stranded RNA that will convert into DNA within the host. This DNA later penetrates inside the host genome for control and replication and produces new viral particles.

Capsid Structure

The subunits of capsomeres combine to form the viral capsid. The stability of the virus and potential to infect the host depends on the capsomere arrangement that differs from one virus to another.

Examples:

Icosahedral capsids: Adenovirus is one of the examples of an icosahedral capsid, and capsomere is organized in symmetry to provide more durability and be effective in protecting the genetic material of the virus.

Helical capsids: Rabies contain the helical capsid and capsomeres that are arranged in a helical manner enclose the RNA genome. This will give more durability to allocate larger genetic material [17, 19].

Replication Mode Classification

Viral replication is probably the most complicated factor in virus classification. Various viruses have different strategies to overcome the host's cell to replicate and produce new viral particles. All of these strategies are based on the viral genome and virus carrier [17].

DNA Viruses

Replication process: Replication happens inside the nucleus of the host cell and where the virus uses DNA polymerase of the host to produce new viral particles.

Examples:

Herpesviruses: These viruses keep their DNA in the nucleus of the host and causes a latent form of infections, where the virus can activate later on to produce infectious particles.

Adenoviruses: These viruses cause infection to the respiratory system, eye, gastrointestinal system, etc. These viruses replicate inside the nucleus of the host.

RNA viruses: Commonly, RNA viruses replicate inside the cytoplasm of the host, and these viruses are categorized based on the RNA of the virus.

Positive-Sense RNA Viruses

The viral RNA genome will be converted into viral proteins by utilizing the host's ribosomes, and it allows very rapid viral replication.

Examples:
Coronavirus: The corona virus that causes COVID-19 contains the positive sense RNA genome. This viral genome rapidly translated to viral particles, making so fast the spread of the disease and its transmissibility.
Poliovirus: This viral genome contains the simple, single-stranded RNA genome acting as mRNA after entering the host cell and causing the infection that results in poliomyelitis.

Negative-Sense RNA Viruses
This virus is made of single-stranded RNA. The mRNA will be made by the virus genome called RNA-dependent RNA polymerase. The viral translation will happen only after the virus can be transcribed into positive-sense RNA.

Examples:
Influenza virus: This virus contains the segmented genome of RNA, which will facilitate more viruses to combine the genetic material of the virus within the host cell, and causes extra damage like pandemics.
Rabies virus: This bullet-shaped virus contains the negative-sense RNA genome that will be transcribed into positive-sense RNA before it becomes translated.

Retroviruses
Replication process: Retroviruses such as HIV have a single-stranded RNA genome and later it will convert into double-stranded DNA using enzyme reverse transcriptase. The integration of this DNA inside the host genome makes the virus stay longer in the host and it can be prevented from being detected by the host's immune system.

Examples:
HIV: This retrovirus has a single-stranded RNA genome that infects and causes AIDS (acquired immunodeficiency syndrome) in human beings. The viral genome is transcribed to RNA after inception into the host's DNA, generating the new virions.

Factors such as morphology, chemical composition, and replication mode helped to make the viral classification and provide a complete understanding of the huge variety of viruses. This classification may help the doctors and other healthcare providers in studying and understanding the complete viruses such as identifying, life cycles, and infection to the hosts and thereby developing the specific antiviral therapies. The advancement of molecular biology and genomics along with complete knowledge on viruses makes the subject more complex and sensitive and facilitates for a complete understanding of these viral pathogens [17, 19, 22].

1.2.4 Viral Replication

Viral replication is a complex process that primarily involves hijacking the host cells. Viral replication involves continuous steps that are different from each specific virus and these steps are useful for the multiplication of viruses.

Steps in Viral Replication

Attachment: It is the first step in viral replication, involving the attachment of the virus to the specific receptors on the surface of the host cell. The attachment of the virus to the host cell is highly specific and it infects only a specific type of cells. Recent reports say that viruses use not only specific receptors but more receptors for attachment and increase their host cells [23].

Penetration: After the attachment of the virus to the host cell, the virus will enter the host cell either through endocytosis in which the virus will be engulfed by the host cell membrane or through fusion with the host cell membrane. Recent studies say that viruses use different mechanisms to penetrate the host cell, providing a deeper understanding of the virus infection [23].

Uncoating: The viral outer covering layer capsid will be removed after entering the host cell and the viral genetic material will be released. This step is useful for the following steps in replication. Reports say that certain enzymes in viruses and some host factors help in uncoating the capsid and this provides the targets for antiviral treatment [23].

Genome replication: The viral DNA merges with the host cell machinery in the process of viral replication. Normally, DNA viruses reproduce their genome inside the nucleus and RNA viruses replicate within the cytoplasm of the host cell. The involvement of viral and host cell proteins is a complex process in viral genome replication and this has been the subject for the whole genome sequencing technology [23].

Transcription and translation: Primarily, the viral genome is transcribed to mRNA before being translated to viral protein particles. These viral proteins are useful in the replication, assembly, and release of viral particles.

Recent reports say that newly identified viral proteins and their functions provide a complete understanding of the underlying mechanism of viral replication [23].
Assembly: The newly formed viral components are merged together to become new virus particles. This procedure involves packing the genome to the capsid of the virus, and in some instances, it will become the envelope of the virus. Recent studies say that the invention of new antivirals is based on the intricate complex between host and viral proteins in viral replication [23].
Release: The virus particles formed through the process of assembly will be expelled from the host cell. This process can happen by either using lysis, in which the host cell splits or bursts and opens, or the budding process, where virus particles will be pumped out by tearing off from the cell membrane. Recent reports say that the process of viral particle release provides knowledge on spreading and causing the disease through the virus [23].

Understanding viral replication is easy after consideration of recent advancement in the field of virology. Studying the complete genetic material and the invention of new genes of viruses have happened through recent whole-genome sequencing technologies. Furthermore, recent advancements in structural biology made us understand the three-dimensional structure of viral proteins, allowing the understanding of viral proteins in good and potential harm of genes [22, 23].

1.2.5 Viral Transmission

Viral transmission is the process of the transferring or infecting a virus from one host to the other host. It is a major factor in the field of infectious disease epidemiology. This intricate procedure involves the interaction among virus, climate, and host.

Modes of Viral Transmission
There are several different modes of viral transmission, namely:

Respiratory transmission: It is very usual; most of the viruses use this way of transmission. Example: Some of the viruses like influenza and COVID-19 transmit through airborne droplets which can be from sneezing, coughing, or talking of an infected person.

Fecal–oral transmission: This transmission happens when a person eats food that got infected from the feces of infected person. Some of the viruses that use this way of transmission are hepatitis A, rotavirus, norovirus, etc.
Sexual transmission: The infection can be transmitted through sexual contact and some of the viruses use this mode of transmission such as HIV, herpes virus, hepatitis B, etc. These viruses will be found in bodily fluids such as semen, vaginal fluids, and anus.
Blood-borne transmission: Some of the viruses such as HIV, hepatitis B, and hepatitis C use blood-to-blood transfusion. This mode of transmission can happen through infected needles that were shared among people for blood transfusion or needles that are shared accidentally.
Vector-borne transmission: Some of the specific viruses such as Zika, dengue, and West Nile virus can transmit the infection through an infected insect bite, tick lice, mosquito, etc. Infected insects or mosquito can work as a vector in this way of transmission that will transmit from one host to another [24–27].

Factors Affecting Viral Transmission

The rate and effectiveness of viral transmission can be influenced by different factors such as the following:

Viral characteristics: The ability of the transmission of a disease can be affected through the virulence, infectivity, and stability of a virus. Some of the viruses which are more contagious, like measles, can be transmitted fast in a population. However, less contagious viruses need a very close contact for the transmission of diseases.
Host factors: The host sensitivity to infection can be different depending on factors such as age, immunity status, and genotype. Immuno-compromised people have less immunity and are more sensitive to viral infections.
Environmental factors: Factors such as air quality, temperature, and humidity can be predictors for the survival and transmission of viruses. Certain viruses are capable of surviving in very dry conditions; however, some viruses have vulnerability to very hot temperatures.
Social factors: High population density, cultural factors, and access to healthcare can be a key factor in viral transmission. Factor like highly populated areas allows the transmission of viruses. So, transmission can be controlled by proper public health measures.

Recent research reports provided important inputs on spreading of a virus. For instance, advance modeling techniques have been used to study the transmission of viruses among the public to find the important factors that cause outbreaks. Furthermore, recent studies found the importance of an infected person not showing symptoms, insisting the importance of taking public health measures such as maintaining social distance and using a facial mask in controlling the disease spread.

Knowledge on intrinsic factors that facilitate the viral disease spread is crucial in the development of effective strategies to prevent and control the outbreaks. By considering the role and connection among viruses, host, and the environment, physicians or other healthcare workers can find the potential factors and its sensitivity toward viral transmission and can invent the therapeutic or non-therapeutic intervention to prevent and control the chain of infection [28–30].

1.3 Viral Storms (Major Pandemics)

Viral storm is the fast and huge spread of viral disease among a larger population, causing the pandemics, and it has a major impact on the public health, social, and economic aspects of the public and countries. Various viral storms have had an impact on the development of human civilization throughout the history. These pandemics are typically marked by rapid transmission rates, considerable morbidity and mortality, and strain on healthcare systems [31, 32]. Table 1.2 presented the Major global pandemics and their impact.

1.3.1 The Spanish Flu (1918–1919)

Even after a century, the origins of the 1918 pandemic influenza virus remain elusive. The pandemic started in the last year of World War I,

Table 1.2 Major global pandemics and their impact [49–52].

Pandemic	Year	Origin	Regions affected	Estimated death	Virus type
1918 Spanish flu	1918 to 1919	USA	Worldwide	50–100 million	H1N1 influenza
HIV/AIDS	1981 to present	Africa	Worldwide	40 million people	HIV
2002 SARS	2002–2003	China	26 countries		SARS-CoV
2009 H1N1	2009–2010	Mexico	Worldwide	151,700–575,400	H1N1 influenza
COVID-19	2019–present	China	Worldwide	6.9 million+	SARS-CoV-2

spreading more rapidly than previous pandemics, which had primarily followed trade routes. The war's global scale facilitated the virus's spread through the mass movement of soldiers and civilians. Poor health and sanitation conditions in the trenches further exacerbated the spread of the disease. Public awareness of the pandemic's severity was limited, as many media outlets were restricted from reporting on the global health crisis and focused instead on morale-boosting topics. However, since Spain remained neutral during the war, its newspapers reported extensively on the virus's devastating effects within the country. This led to the widespread but inaccurate belief that the illness originated in Spain, giving rise to the misnomer "the Spanish flu" [33, 34].

The H1N1 influenza virus was the cause of the Spanish flu, one of the deadliest pandemics in recorded history. It caused an estimated 50 million fatalities and affected around one-third of the world's population. Three waves of the pandemic developed, the deadliest of which was the second. The virus spread swiftly due to troop movements during World War I and the absence of effective public health interventions [33–35].

1.3.2 HIV/AIDS Pandemic (1980s–Present)

The HIV/AIDS epidemic, which emerged in the 1980s, has been a devastating world health challenge. Initially concentrated in sub-Saharan Africa, the virus rapidly spread worldwide, disproportionately impacting marginalized populations. Associated with fear and misinformation, the initial years were spent by a lack of understanding about the disease. Fortunately, scientific advancements helped in the discovery of HIV as the causative agent, enabling the way for the diagnosis and prevention of the disease. The upcoming years improved the efficacy of the treatment with the introduction of antiretroviral therapy (ART), which helped in converting HIV into a tolerable chronic condition. However, there were still differences in the availability of ART between the rich and poor countries. The occurrence of newer infections is still a concern specially among the younger populations, and the epidemic is still a major worldwide issue for the health of people worldwide. The issues of drug-resistant HIV strains and their disproportionate impact on disadvantaged populations remain significant [36, 37].

To efficiently fight against HIV/AIDS, an appropriate approach is needed, involving prevention, treatment, care, and support services. This involves the promotion of sexual education, strategies for harm reduction, and increasing access to testing and treatment. Solving the underlying social problems such as poverty, inequality, and discrimination is also necessary. While progress has been seen, a world that is free from AIDS needs a thorough commitment and cooperation [36–40].

1.3.3 SARS (2002–2003)

The Severe Acute Respiratory Syndrome (SARS) outbreak was observed in the year between 2002 and 2003 and highlighted the world's potential interconnectedness for the rapid spread of the virus, alerting a serious global health emergency. SARS, being originated in Southern China, was a severe respiratory illness associated with high fever, cough, and shortness of breath. Over two dozen countries were affected by the rapid transmission of it through the air [41].

This outbreak was marked by a higher rate of uncertainty and fear. The lack of transparency and misinformation were the confounding factors that hindered the early attempts to contain the virus, thus the reason for it to become a global health alert. Healthcare personnel were the first ones to experience the impact during the patient's treatment. The consequences were so prominent that it led to economic crisis through travel restrictions and fear among the public, thus jeopardizing global trade and tourism [41, 42].

The SARS breakout caused a global emergence, which required international unity to help in combating the disease. The World Health Organization (WHO) was playing a major role in coordinating efforts, disseminating information, and offering guidance to countries. Thorough research into the virus helped in understanding its ability for transmission and pathogenesis, making a way for the development of effective prevention and control measures.

Finally, the SARS outbreak was controlled and eradicated through proper public healthcare measures, involving isolation, quarantine, and by tracing contacts. This event helped in critical learning opportunities for the global healthcare community, involving an essential role of detection and response and the support of the entire world in the prevention and management of infectious diseases. The lessons that were learned from SARS have played as a stepping stone in shaping global health policies and staying prepared for future outbreaks [41–43].

1.3.4 H1N1 Influenza (2009)

The H1N1 influenza, which is also known as the "swine flu", was a world health crisis that occurred in 2009. Unlike the previous influenza viruses, this new virus contained a unique genetic makeup composed of elements from human, swine, and avian influenza viruses. In June 2009, WHO classified HINI as a pandemic because it was spread rapidly and widely [44].

The pandemic had a great impact on young adults and children, age groups not typically affected by influenza-related illness. Even though the mortality rates were sometimes lower than during other pandemics of influenza, the virus still produced substantial morbidity in the form of respiratory complications, neurological conditions, and cardiovascular diseases. It was an economic burden as well due to disruption in healthcare systems, schools, and businesses around the world [44, 45].

The response of the international community to the 2009 H1N1 pandemic illustrated that cooperation among nations is indeed worth much in public health. Vaccine development and distribution happened quite quickly, which would otherwise have worsened the pandemic effects. However, weaknesses have also been pointed out, including vaccine shortages and uneven access to healthcare, by health systems in different countries. It has shed light on the necessity of potent surveillance systems, pandemic preparedness plans, and effective risk communication strategies. Though the pandemic was declared over, the 2009 H1N1 virus continues to circulate as a seasonal influenza strain, and this has put a highlight on the continued menace of influenza viruses and how vaccination plays a role [44–46].

1.3.5 COVID-19 Pandemic (2019–Present)

In late 2019, the COVID-19 pandemic which was caused by SARS-CoV-2 virus started spreading quickly and was the biggest worldwide emergency of the twenty-first century. Being originated in Wuhan, China, the virus started to quickly spread worldwide, affecting healthcare systems and economies. This virus can cause symptoms, from minor respiratory issues to fatal pneumonia, which had a disproportionate impact on elder people and those with pre-existing health complications [47].

The pandemic required unprecedented public health measures, including lockdowns, social distancing, and mandatory use of masks. Although these strategies were effective in slowing the virus' transmission, they also led to considerable economic and social repercussions. The development and swift distribution of COVID-19 vaccines marked a critical turning point in the pandemic, greatly decreasing hospitalization and death rates. Nonetheless, the appearance of new variants, including Delta and Omicron, continued to present challenges [48].

The COVID-19 pandemic has spurred advancements and innovation across several fields, such as virology, epidemiology, and vaccine development. It has underscored the crucial importance in global cooperation in tackling

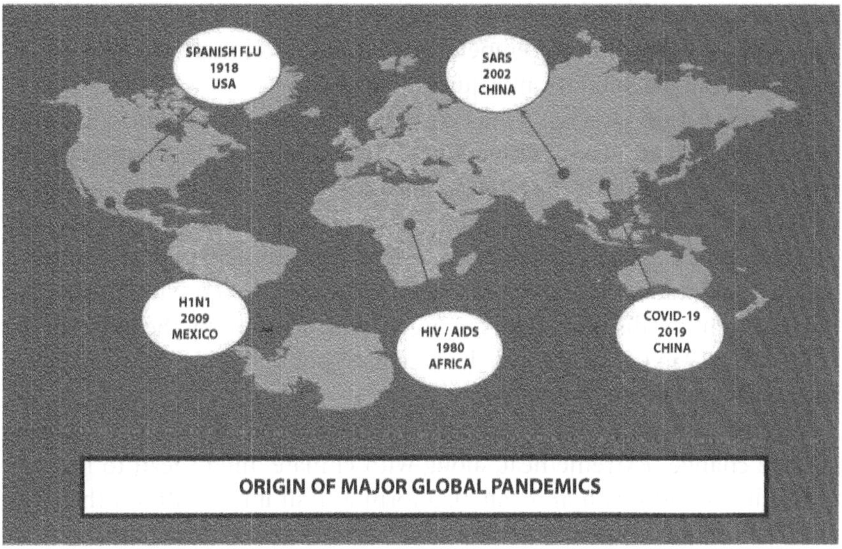

Figure 1.2 Origin of major pandemics.

public health problems. As many countries keep on addressing the complexities of the pandemic, sustained surveillance, research, and public health strategies are vital to managing its long-term effects [43, 47, 48]. Figure 1.2 illustrated the origin of major pandemics across the world.

1.4 Factors Contributing to Viral Storms

1.4.1 Environmental Factors, Climate Change, and Urbanization

The intricate relationship between the environment, climate change, and urbanization is increasingly recognized as a potent catalyst for the emergence and spread of infectious diseases. The term "viral storms," a metaphor for rapid and widespread outbreaks, has become synonymous with the challenges presented by these factors that are interconnected [53].

Climate change, marked by rise in temperatures, shuffling precipitation, and extreme climate changes, fosters favorable conditions for the proliferation of disease vectors. Mosquitoes, ticks, and other arthropods, which transmit pathogens, thrive in warmer, wet environments. The geographic range of these vectors is expanding, leading them to encounter new human populations and heightening the risk of disease transmission. An example for it is the Zika virus which was contained to tropical regions initially and then started spreading to temperate areas due to climate changes which made a suitable habitat for mosquitos [54, 55].

A major role in viral outbreaks is played by urbanization due to its dense population and rapid environmental modifications. The destruction of natural habitats through deforestation and interfering into the wildlife areas increases the rate of human–animal interactions, causing the emergence of zoonotic diseases. Furthermore, urban environments are deprived of adequate sanitation and waste management systems, supplying the perfect environment for disease vectors to breed. As evidenced by the early spread of COVID-19 in urban centers, the increased population in cities facilitates the rapid disease transmission [56, 57].

Additionally, urbanization's negative effects on health are worsened by climate change. Extreme heat, along with climate, might lead to heat-related illness and death and can also form a desirable condition that may lead to the transmission of pathogens. A by-product of urbanization is air pollution which is known to make the immune system weak and increase a person's susceptibility to infections [57].

To alleviate the risks shown by viral storms, a complex approach is necessary. To achieve this strengthening surveillance systems to identify emerging pathogens, development of effective vaccines and antiviral drugs, and improvement of public health infrastructure is necessary. In addition, to achieve long-term prevention, it is needed to tackle the underlying causes of climate change and promote sustainable urban development. By understanding the relation between environmental factors, climate change, and urbanization, we can stay prepared for and respond to any future outbreaks [56, 57].

It is very complex to investigate further in research to get the comprehensive understanding on the mechanism where these factors contribute to viral growth and transmission. This will facilitate to improve more specific interventions and a strong health system that can make us survive the difficulties caused by climate change and urbanization [53–59].

1.4.2 Social Factors: Global Travel, Population Density, and Healthcare Access

The significant spreading of an infectious disease is caused by the fast globalization of the world. Specific factors like rise in global travel, increased population density, poor healthcare access, and climate change are favorable to viral storms. Figure 1.3 summarized the factors contributing to viral storms.

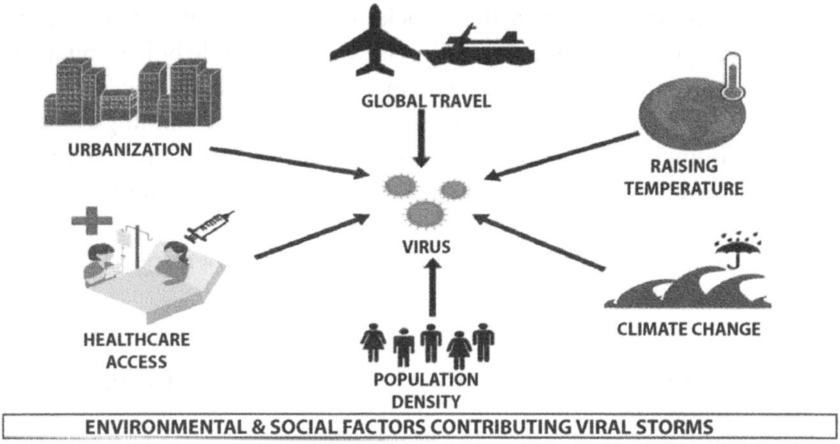

Figure 1.3 Factors contributing to viral storms.

Global travel has become an essential part of the modern society, which connects people from different countries within hours. Although this connection has given numerous benefits, it has also led to the rapid spread of the infectious disease. As evidenced by the rapid spread of SARS, H1N1, and COVID-19 pandemics, air travel has accelerated the global dissemination of pathogens. People affected by a virus are mostly likely to have traveled before exhibiting the symptoms, which may lead to widespread outbreaks in farther locations [7, 60].

Population density, which refers to the number of individuals living in a particular area, is also a critical factor that influences the transmission of the infectious disease. A crowded living place, which is usually seen in urban areas, makes a suitable environment for the faster spread of a pathogen. Closer distances between people have a higher incidence of person-to-person transmission. Upon that, highly dense areas often face challenges in imparting effective healthcare facilities, such as identification and tracking of contact source, which may further worsen the situation [7, 61].

Partiality in healthcare access and quality makes a population vulnerable to viral storms. In areas with very few healthcare resources, detection and appropriate response to outbreak might get delayed, which may cause the virus to spread undetected. Factors such as poverty, lack of education, and insufficient sanitation in the infrastructure worsen this issue. Additionally, differences in access to vaccines can lead to lack of protection against infectious diseases [7, 61].

To mitigate the impact of these social factors on viral storms, a comprehensive approach is necessary. Crucial actions include bolstering international surveillance networks, enhancing public health facilities, and advocating for fair access to healthcare. Furthermore, funding research and development for medicines, vaccines, and diagnostics is essential to stopping such outbreaks in the future. By addressing the underlying social determinants of health and strengthening international cooperation, we can build a more resilient global health system capable of responding effectively to viral threats [7, 60–62].

1.4.3 Biological Factors: Viral Mutations and Host Immune Response

Viruses are often susceptible to mutations that can lead to the spread of newer strains with different characteristics, such as increased ability in transmission, pathogenicity, and also resistance to all currently available treatments. A suitable example for this is the influenza virus, due to its antigenic shifts leading to epidemics and pandemics. Another factor responsible is the host immune response. A good immune system can clear a virus, while a weakened immune system can also exacerbate the disease and even lead to death. Some viruses such as HIV have mutated and evolved mechanisms to devour the host immune system, making it challenging to control.

The viral outbreaks are majorly linked to the biological factors, usually viral mutations and host immune response. These components interact with each other in a complicated manner, which causes a prominent trajectory for an epidemic [63, 64].

The mutation in viruses is the backbone of viral evolution. It plays a prominent role in paving a way for viral storms. Viruses are often characterized by sudden replication rates, and error-prone replication mechanisms are prone to accumulate genetic alterations. These mutations may result in a change of the virus' structure, infectivity, and other capabilities. For example, if we take SARS-CoV-2 variants with increased ability of transmission, it has played a critical role in mutations in the virus, thereby leading to pandemics [64].

The immune response of the host, a network of cellular and humoral defensive mechanisms, is the primary line of defense against foreign bodies. A strong immune response can easily manage a viral outbreak by nullifying the virus and thereby preventing it from spreading. However, certain factors such as age, health condition, and immune suppression are responsible for compromising a host's quality to show proper immune response, causing susceptibility to the disease and worsening the viral growth. Some viruses such as influenza are known to cause antigenic drift, changes in viral genome, which can evade the already existing immunity [65, 68].

The interaction between the host immune system and the mutations in virus can have various outcomes. In certain cases, a virus can also become less virulent, as a deadly virus can easily deplete the host population, reducing its options to transmission. Similarly, an increase in the ability of a virus's transmission without increase in its virulence can be subjected to more widespread outbreaks. Additionally, the mutations that act on immune mechanisms of virus can diminish the activity of vaccines, which may prolong the pandemic.

It is important to understand the biological factors that are responsible for viral storms to develop and the effective prevention and strategies to control it. Various systems for the surveillance to monitor viral evolutions, development of vaccines that may target the viral components, and immune response enhancement research are the crucial areas which can be focused. By unraveling the relationship between the viruses and their hosts, scientists can help to contain future outbreaks [63–69].

1.5 Norovirus

1.5.1 Introduction to Norovirus: Overview of the Virus and Its Significance

Noroviruses are a subfamily of non-enveloped, single-stranded RNA viruses which belong to the Calciviridae family; they also exhibit genetic diversity. These are known for their high infectivity and rapid spread, being a leading cause of gastroenteritis worldwide. They have been classified into 10 genogroups (G), having genogroups I, II, and IV being the most prominent for human diseases [70].

Being first identified in 1968 at Norwalk, Ohio causing an outbreak of gastroenteritis, it was named "Norwalk virus". This name was later changed to norovirus as researchers identified related viruses and recognized their broader classification within the Caliciviridae family [70].

Norovirus is often mistaken for the "stomach flu," although it is unrelated to influenza. It spreads primarily through the fecal–oral route, which can involve contaminated food, water, or surfaces. The virus is notably resilient and can persist on surfaces and in the environment despite standard cleaning practices [71].

The virus is renowned for causing outbreaks in settings where people are in close quarters, such as cruise ships, healthcare facilities, schools, and nursing homes. The rapid transmission results in explosive outbreaks characterized by acute onset of vomiting, diarrhea, abdominal cramps, and nausea. These symptoms, while generally self-limiting, can lead to severe dehydration, particularly in vulnerable populations like young children, the elderly, and immunocompromised individuals. Efforts to combat norovirus focus on improving sanitation and hygiene practices, particularly in food handling and preparation environments. There is currently no specific antiviral treatment for norovirus infections; management primarily involves supportive care to prevent dehydration and alleviate symptoms [70–72].

1.5.2 Epidemiology: Global Impact and Spread of Norovirus

Norovirus is a major global pathogen responsible for a high burden of gastroenteritis. Worldwide, it causes an estimated 685 million cases of gastroenteritis each year, making it a leading cause of foodborne illness. In developed nations, norovirus is particularly prevalent, with around 19 to 21 million cases annually in the United States alone. The prevalence rate in these regions is approximately 5.4 cases per 1,000 people. Despite the typically self-limiting nature of norovirus infections, the virus is responsible for approximately 200,000 deaths globally each year, predominantly affecting the elderly and young children. Norovirus outbreaks are known to differ in duration, ranging from a few days to several weeks. These usually occur in a seasonal pattern, showing higher incidence rates during the winter in temperate climate. Various factors might affect the duration and impact of outbreaks such as effectiveness of public healthcare and the scale of outbreak [72].

Recently in India, the focus has been increased on the epidemiological profile of norovirus. Norovirus is a major cause of gastroenteritis in children under 5 years old, accounting for roughly 10% of all cases. Norovirus outbreaks occur seasonally, especially during the monsoon season, increasing along with the global patterns of the incidence. Although the exact mortality details for norovirus in India are less documented, it anyways has helped substantially to gastroenteritis-related hospitalizations and death, especially in vulnerable population with lesser access to healthcare. Outbreaks in India typically last from a few days to several weeks, with higher frequency and duration observed in densely populated areas and during periods of inadequate sanitation and hygiene practices. Overall, norovirus remains a significant public health challenge both globally and within India, underscoring the need for continued vigilance, improved sanitation, and effective public health interventions to manage and mitigate its impact [72–74].

1.5.3 Clinical Presentation: Symptoms, Severity, and Complications

Following an incubation period of 12 to 48 h, the typical symptoms of norovirus infection include a sudden onset of vomiting, abdominal cramps, and watery diarrhea. Additionally, constitutional symptoms such as low-grade fever, generalized muscle aches, malaise, headache, and chills often accompany gastroenteritis. Vomiting and diarrhea typically occur together, though one may be present without the other. Most individuals experience a short, self-limiting illness, with symptoms generally resolving within 2 to 3 days, and recovery usually occurs without medical treatment [72, 73].

The clinical spectrum of illness from norovirus infection varies widely; up to one-third of those infected may show no symptoms at all. Conversely, those most at risk for severe outcomes include individuals with underlying medical conditions, young children, the elderly, and those with weakened immune systems. These vulnerable groups are more susceptible to severe symptoms and complications, such as severe dehydration, which may necessitate hospitalization. Although complications are infrequent, they can include electrolyte imbalances, secondary infections like acute renal failure requiring hemodialysis, cardiac issues such as arrhythmias, acute graft organ rejection in transplant recipients, and, in rare cases, death [72, 73].

In healthy adults, the complications are usually less frequent, but if they occur, they can be severe. For example, post-infection inflammatory bowel syndrome cases have been documented, with symptoms staying for up to 3 months after the infection. This condition has been seen more frequently in people who were already infected compared to people who were not. US military recruits who had contracted gastroenteritis during norovirus outbreaks had seen long-term health issues such as dyspepsia, constipation, and GERD. In spite of the fact that neurological symptoms are rare, there have been cases documented. For instance, certain symptoms such as severe headaches, neck stiffness, photophobia, and altered levels of consciousness along with gastrointestinal distress had been recorded by British military personnel. In one severe case, a patient developed disseminated intravascular coagulation, a serious disorder affecting blood clotting, and required ventilator support for respiratory complications [72, 73].

In addition to these instances, other rare complications have been observed in otherwise healthy individuals. These include necrotizing enterocolitis, a serious gastrointestinal condition, seizures, and post-infectious arthritis, particularly in children. Isolated cases among adults have also reported severe conditions such as ischemic colitis, transient hepatocellular injury, and haemolytic uremic syndrome, a serious condition affecting the blood and kidneys [72, 73].

1.5.4 Transmission and Prevention: Modes of Transmission and Strategies to Prevent Outbreaks

The genetic diversity of noroviruses has significant implications for public health. Different genotypes are linked to various transmission modes and, potentially, the severity of disease outcomes. Genogroup I viruses are commonly linked to foodborne and waterborne outbreaks, with the newly emerged GI.6 virus being particularly linked to foodborne illness. In contrast, GII.4 viruses are closely related the spreading from person-to-person and are frequently observed in healthcare settings [70, 71].

Norovirus is extremely contagious and can cause infection with just a small number of viral particles. It can be transmitted through contaminated food and water, direct contact with an infected person, or by contact with area contaminated with the virus [70].

Given the diverse methods by which norovirus is transmitted and the absence of a single, definitive solution to halt its spread, a comprehensive array of strategies is necessary for preventing and managing norovirus outbreaks. Effective prevention hinges on rigorous adherence to hygiene protocols. It is crucial to actively promote hand hygiene practices in healthcare individuals, patients, and visitors where areas contaminated with norovirus gastroenteritis. During such outbreaks, it is crucial to wash hands completely with soap and wash under the running water for at least 20 seconds after caring for suspected patients or with confirmed infections. Proper handling and preparation of food, along with diligent disinfection of contaminated surfaces, are critical for controlling the spread of the virus. Research indicates that proper hand hygiene is a valuable measure in reducing the risk of norovirus. However, despite its widespread use, the effectiveness of hand sanitizers with alcohol based against norovirus remains uncertain. Consequently, during outbreaks, hand sanitizers should be used in conjunction with traditional hand-washing. The risk of transmission is further increased by the aero droplets of noroviruses infected individuals and close contact with them. Therefore, isolating infected patients and using appropriate PPE, such as gloves and masks, most importantly when clearing up the vomit, are crucial for reducing further transmission within healthcare settings. Furthermore, during outbreaks, PPE should be disposable and designed for one use only [70, 71, 73].

Proper sanitation is essential following incidents of vomiting or diarrhea, particularly in the context of viral infections such as norovirus. The affected area should be immediately cleaned and disinfected to prevent further transmission. The process begins with the individual donning rubber or disposable gloves to protect against contamination. The area should be thoroughly wiped with paper towels, which must then be discarded in a sealed plastic trash bag. For disinfection, bleach solution of chlorine with 1,000 to 5,000 ppm concentration that equivalent to 5 to 25 table spoons of household bleach with 5% to 8% per gallon of water—or an EPA-registered disinfectant that has been proven against norovirus should be utilized. The disinfectant must remain on the contaminated surface for at least five minutes to ensure maximum efficacy. After disinfection, the area must be washed again by soap and hot water to remove any residual chemicals. Furthermore, all contaminated laundry should be promptly washed, trash properly disposed of, and hands thoroughly washed to minimize the risk of infection [71, 73].

Measures like social distancing, including isolating or grouping symptomatic patients, have proven effective in curbing norovirus transmission during significant outbreaks. However, the optimal duration for isolation and the application of contact precautions remains unclear. Even after symptoms subside, patients may continue to excrete norovirus in their stool, leading to recommendations that contact with patients must be minimized when the illness is at acute phase and for 24 to 72 hours after recovery when viral shedding is still high. This is crucial for long duration care facilities to interrupt the transmission cycle, reduce secondary spread, and shorten the outbreak duration. Most guidelines suggest grouping patients by symptomatic, exposed but asymptomatic, and unexposed asymptomatic statuses, with designated healthcare staff assigned to care for infected individuals. To reduce the risk of transmission from incubating or asymptomatic cases, such patients should not be moved to unaffected areas, usually within 48 h of exposure [70, 71, 74–77].

1.6 Summary of Key Points

This chapter offers a comprehensive examination of infectious diseases, with a specific focus on viral infections and their impact on global health.

It begins by defining infectious diseases and emphasizing the crucial role that viruses play in shaping public health outcomes worldwide. The chapter is giving the full detailed explanation on structure of the virus, classification, mode of replication, and transmission types and giving the clear view on mechanisms on how viruses cause and spread diseases.

The viruses causing viral storms—this concept has been comprehensively discussed and introduced. The huge damage or disruption caused due to the pandemics has been discussed, highlighting historical examples with consequences that can happen when viral spread went globally. Different factors contributing to the cause and spread of viral storms have been explored and discussed in this chapter. Factors such as environmental changes like climate change and deforestation are found to be important key factors to increase viral outbreaks. Social factors such as world travel, increasing population density, and differences in the availability of healthcare facilities are key role points in allowing the fast spread of viruses. Furthermore, biological factors like mutations in the viral genome that increases the virulence effect and the immune response of host are discussed.

The infectious disease norovirus has been discussed to signify how the virus can rapidly transmit and spread and causes considerable burden on public health. This discussion serves to demonstrate the real-life example from the concept provided in the book chapter, highlighting the challenges created by viral infections. In conclusion, this chapter provides understanding of pandemics and the factors responsible for these outbreaks and has highlighted the need for advancement in research, strict vigilance, and mitigation strategies to address these persistent threats to global health.

1.7 Conclusion

To conclude, this book chapter has explained the intricate relationship among factors such as environmental, social, and biological that contribute to the transmission and spreading of viral storms. The explanation of pandemics happened in the past, and it has been evident in norovirus outbreaks how all of these factors cumulatively lead to the huge spread of infections. In the future, the research can be focused on understanding the factors like environment change, rise in global travel, and viral mutations that facilitate the dynamics of a pandemic. Considering all of these and understanding them comprehensively can be useful for the development of effective strategies for early detection, prevention, and fast response to future emerging viral threats. Furthermore, strengthening collaboration across the world and investing in strong, adaptable public health systems are crucial steps in reducing the impact of future viral pandemics. Implementing strict vigilant measures, updating and increasing infrastructure in healthcare, and promoting transparent healthcare information will cumulatively strengthen the ability to predict and control the outbreaks of infectious diseases, thereby protecting and safeguarding global health.

AI disclosure statement: The author acknowledges the use of AI-based tools for minor language editing or grammatical correction. However, all intellectual contributions, analyses, and conclusions presented in this chapter are entirely the author's own.

References

1. Huremović, D., Brief History of Pandemics (Pandemics Throughout History), in: *Psychiatry of Pandemics*, vol. 16, pp. 7 35, 2019 May, doi: 10.1007/978-3-030-15346-5_2.
2. Cole, L. and Kramer, P.R., Bacteria, Virus, Fungi, and Infectious Diseases, in: *Human Physiology, Biochemistry and Basic Medicine*, pp. 193–6, 2016.
3. Morens, D.M., Folkers, G.K., Fauci, A.S., The challenge of emerging and re-emerging infectious diseases. *Nature*, 430, 6996, 242–9, 2004 Jul 8, doi: 10.1038/nature02759.
4. Eccles, R., Understanding the symptoms of the common cold and influenza. *Lancet Infect. Dis.*, 5, 11, 718–25, 2005 Nov, doi: 10.1016/S1473-3099(05)70270-X.
5. Fajgenbaum, D.C. and June, C.H., Cytokine storm. *N. Engl. J. Med.*, 383, 23, 2255–73, 2020 Dec 3.
6. Mehta, P., McAuley, D.F., Brown, M., Sanchez, E., Tattersall, R.S., Manson, J.J., HLH Across Speciality Collaboration, UK. COVID-19: consider cytokine storm syndromes and immunosuppression. *Lancet*, 395, 10229, 1033–1034, 2020 Mar 28, doi: 10.1016/S0140-6736(20)30628-0.
7. Baker, R.E., Mahmud, A.S., Miller, I.F., Rajeev, M., Rasambainarivo, F., Rice, B.L., Takahashi, S., Tatem, A.J., Wagner, C.E., Wang, L.F., Wesolowski, A., Metcalf, C.J.E., Infectious disease in an era of global change. *Nat. Rev. Microbiol.*, 20, 4, 193–205, 2022 Apr, doi: 10.1038/s41579-021-00639-z.
8. Bloom, D.E., Black, S., Rappuoli, R., Emerging infectious diseases: A proactive approach. *Proc. Natl. Acad. Sci. U.S.A.*, 114, 16, 4055–4059, 2017 Apr 18, doi: 10.1073/pnas.1701410114.

9. Fauci, A.S. and Morens, D.M., The perpetual challenge of infectious diseases. *N. Engl. J. Med.*, 366, 5, 454–61, 2012 Feb 2.
10. Desjardins, J., World's rising life expectancy: How medicine has cheated death, World Economic Forum, 2020, May 11, Retrieved from https://www.weforum.org/agenda/2020/05/worlds-rise-life-expectancy-medicine-health/.
11. Wu, F., Zhao, S., Yu, B., Chen, Y.M., Wang, W., Song, Z.G., *et al.*, A new coronavirus associated with human respiratory disease in China. *Nature*, 579, 7798, 265–269, 2020 Mar, doi: 10.1038/s41586-020-2008-3.
12. Karesh, W.B., Dobson, A., Lloyd-Smith, J.O., Lubroth, J., Dixon, M.A., Bennett, M., *et al.*, Ecology of zoonoses: natural and unnatural histories. *Lancet*, 380, 9857, 1936–45, 2012 Dec 1, doi: 10.1016/S0140-6736(12)61678-X.
13. World Health Organization, Antimicrobial resistance: global report on surveillance [Internet], 2014, Available from: https://www.who.int/publications/i/item/9789241564748.
14. Ventola, C.L., The antibiotic resistance crisis: part 1: causes and threats. *P T*, 40, 4, 277–83, 2015 Apr.
15. Cohen, F.S., How Viruses Invade Cells. *Biophys. J.*, 110, 5, 1028–32, 2016 Mar 8, doi: 10.1016/j.bpj.2016.02.006.
16. Skvortsov, T., Maillard, J.Y., Viruses and other acellular infectious agents: Characteristics and control, in: Gilmore BF, Denyer SP, editors. *Hugo and Russell's Pharmaceutical Microbiology*, 9th ed. Chichester, United Kingdom, John Wiley & Sons Ltd., 2023 Jan 1.
17. Gelderblom, H.R., Chapter 41 Structure and Classification of Viruses, in: *Medical Microbiology*, 4th edition, S. Baron (Ed.), University of Texas Medical Branch at Galveston, Galveston (TX), 1996, Available from: https://www.ncbi.nlm.nih.gov/books/NBK8174/.
18. Kausar, S., Said Khan, F., Ishaq Mujeeb Ur Rehman, M., Akram, M., Riaz, M., Rasool, G., Hamid Khan, A., Saleem, I., Shamim, S., Malik, A., A review: Mechanism of action of antiviral drugs. *Int. J. Immunopathol. Pharmacol.*, 35, 20587384211002621, 2021 Jan-Dec.
19. Louten, J., Virus Structure and Classification, in: *Essential Human Virology*, pp. 19–29, 2016, doi: 10.1016/B978-0-12-800947-5.00002-8.
20. Mite, S.M., (Acari Acarina) vectors involved in transmission of plant viruses, in: *Applied Plant Virology*, pp. 257–273, Amsterdam, Netherlands, Academic Press, 2020.
21. Chan, C.M., Chu, H., Zhang, A.J., Leung, L.H., Sze, K.H., Kao, R.Y., *et al.*, Hemagglutinin of influenza A virus binds specifically to cell surface nucleolin and plays a role in virus internalization. *Virology*, 494, 78–88, 2016 Jul 1.
22. Koonin, E.V., Krupovic, M., Agol, V.I., The Baltimore Classification of Viruses 50 Years Later: How Does It Stand in the Light of Virus Evolution? *Microbiol. Mol. Biol. Rev.*, 85, 3, e0005321, 2021 Aug 18, doi: 10.1128/MMBR.00053-21.

23. Louten, J., Virus Replication, in: *Essential Human Virology*, pp. 49–70, 2016, doi: 10.1016/B978-0-12-800947-5.00004-1.
24. Burrell, C.J., Howard, C.R., Murphy, F.A., Virus Replication, in: *Fenner and White's Medical Virology*, pp. 39–55, 2017, doi: 10.1016/B978-0-12-375156-0.00004-7.
25. Limeres Posse, J., Diz Dios, P., Scully, C., Viral Diseases Transmissible by Kissing, in: *Saliva Protection and Transmissible Diseases*, pp. 53–92, 2017, doi: 10.1016/B978-0-12-813681-2.00004-4.
26. Tauro, L.B., Bandeira, A.C., Ribeiro, G.S., Reis, M.G., Pizarro, C.P., Araujo, K.A., Castro, A.P., Queiroz, J.M., Simon, S.S., Menezes, G., Campos, G.S., Sardi, S., Potential use of saliva samples to diagnose Zika virus infection. *J. Med. Virol.*, 89, 1, 1–2, 2017 Jan, doi: 10.1002/jmv.24696.
27. Piazza, M., Chirianni, A., Picciotto, L., Cataldo, P.T., Borgia, G., Orlando, R., Blood in saliva of patients with acquired immunodeficiency syndrome: possible implication in sexual transmission of the disease. *J. Med. Virol.*, 42, 1, 38–41, 1994 Jan, doi: 10.1002/jmv.1890420108.
28. Wang, W.K., Chen, S.Y., Liu, I.J., Chen, Y.C., Chen, H.L., Yang, C.F., *et al.*, Detection of SARS-associated coronavirus in throat wash and saliva in early diagnosis. *Emerg. Infect. Dis.*, 10, 7, 1213–9, 2004 Jul, doi: 10.3201/eid1007.031113.
29. Pica, N. and Bouvier, N.M., Environmental factors affecting the transmission of respiratory viruses. *Curr. Opin. Virol.*, 2, 1, 90–5, 2012 Feb 1.
30. Meyerowitz, E.A., Richterman, A., Gandhi, R.T., Sax, P.E., Transmission of SARS-CoV-2: a review of viral, host, and environmental factors. *Ann. Internal Med.*, 174, 1, 69–79, 2021 Jan.
31. Morens, D.M. and Fauci, A.S., Emerging Pandemic Diseases: How We Got to COVID-19. *Cell*, 182, 5, 1077–1092, 2020 Sep 3, doi: 10.1016/j.cell.2020.08.021.
32. Ferguson, N., Cummings, D., Cauchemez, S., *et al.*, Strategies for containing an emerging influenza pandemic in Southeast Asia. *Nature*, 437, 209–214, 2005, https://doi.org/10.1038/nature04017.
33. Vázquez-Espinosa, E., Laganà, C., Vázquez, F., The Spanish flu and the fiction literature. *Rev. Esp. Quimioter.*, 33, 5, 296–312, 2020 Oct, doi: 10.37201/req/049.2020.
34. Akin, L. and Gözel, M.G., Understanding dynamics of pandemics. *Turk. J. Med. Sci.*, 50, SI-1, 515–519, 2020 Apr 21, doi: 10.3906/sag-2004-133.
35. Simonetti, O., Martini, M., Armocida, E., COVID-19 and Spanish flu-18: review of medical and social parallelisms between two global pandemics. *J. Prev. Med. Hyg.*, 62, 3, E613–E620, 2021 Sep 15, doi: 10.15167/2421-4248/jpmh2021.62.3.2124.

36. De Cock, K.M., Jaffe, H.W., Curran, J.W., Reflections on 40 Years of AIDS. *Emerg. Infect. Dis.*, 27, 6, 1553–1560, 2021 Jun, doi: 10.3201/eid2706.210284.
37. Centers for Disease Control and Prevention (CDC), HIV: About HIV. [Internet], Centers for Disease Control and Prevention, Atlanta, 2023 [cited 2024 Aug 29], Available from: https://www.cdc.gov/hiv/about/index.html.
38. Bhagavathula, A.S., Clark, C.C.T., Sharma, R., Chhabra, M., Vidyasagar, K., Chattu, V.K., Knowledge and attitude towards HIV/AIDS in India: A systematic review and meta-analysis of 47 studies from 2010-2020. *Health Promot. Perspect.*, 11, 2, 148–160, 2021 May 19, doi: 10.34172/hpp.2021.19.
39. Bandera, A., Gori, A., Clerici, M., Sironi, M., Phylogenies in ART: HIV reservoirs, HIV latency and drug resistance. *Curr. Opin. Pharmacol.*, 48, 24–32, 2019 Oct, doi: 10.1016/j.coph.2019.03.003.
40. Monteiro, S.S., Villela, W.V., Soares, P.S., The interaction between axes of inequality in studies on discrimination, stigma and HIV/AIDS: Contributions to the recent international literature. *Glob. Public Health*, 8, 5, 519–33, 2013, doi: 10.1080/17441692.2013.779738.
41. World Health Organization (WHO), Severe Acute Respiratory Syndrome (SARS). [Internet], World Health Organization, Geneva, 2024 [cited 2024 Aug 29], Available from: https://www.who.int/health-topics/severe-acute-respiratory-syndrome#tab=tab_1.
42. Cherry, J.D., The chronology of the 2002–2003 SARS mini pandemic. *Paediatr. Respir. Rev.*, 5, 4, 262–9, 2004.
43. Yang, Y., Peng, F., Wang, R., Guan, K., Jiang, T., Xu, G., *et al.*, The deadly coronaviruses: The 2003 SARS pandemic and the 2020 novel coronavirus epidemic in China. *J. Autoimmunity*, 109, 102434, 2020 May 1.
44. Jilani, T.N., Jamil, R.T., Nguyen, A.D., Siddiqui, A.H., H1N1 Influenza, in: *StatPearls* [Internet], StatPearls Publishing, Treasure Island (FL), 2024 Mar 4, 2024 Jan–. PMID: 30020613.
45. Smith, G., Vijaykrishna, D., Bahl, J., *et al.*, Origins and evolutionary genomics of the 2009 swine-origin H1N1 influenza A epidemic. *Nature*, 459, 1122–5, 2009, doi: 10.1038/nature08182.
46. Gatherer, D., The 2009 H1N1 influenza outbreak in its historical context. *J. Clin. Virol.*, 45, 3, 174–8, 2009.
47. Hu, B., Guo, H., Zhou, P., Shi, Z.L., Characteristics of SARS-CoV-2 and COVID-19. *Nat. Rev. Microbiol.*, 19, 3, 141–154, 2021 Mar, doi: 10.1038/s41579-020-00459-7. Epub 2020 Oct 6. Erratum in: Nat Rev Microbiol. 2022 May;20(5):315. doi: 10.1038/s41579-022-00711-2.
48. Umakanthan, S., Sahu, P., Ranade, A.V., Bukelo, M.M., Rao, J.S., Abrahao-Machado, L.F., Dahal, S., Kumar, H., Kv, D., Origin, transmission, diagnosis and management of coronavirus disease 2019 (COVID-19). *Postgrad. Med. J.*, 96, 1142, 753–758, 2020 Dec, doi: 10.1136/postgradmedj-2020-138234.

49. Taubenberger, J.K. and Morens, D.M., Influenza: the Mother of All Pandemics. *Emerg. Infect. Dis.*, 12, 1, 15–22, 19182006, https://doi.org/10.3201/eid1201.050979.
50. Joint United Nations Programme on HIV/AIDS (UNAIDS), UNAIDS homepage. [Internet], UNAIDS, Geneva, [cited 2024 Aug 29]. Available from: https://www.unaids.org/en.
51. Dawood, F.S., Iuliano, A.D., Reed, C., Meltzer, M.I., Shay, D.K., Cheng, P.Y., *et al.*, Estimated global mortality associated with the first 12 months of 2009 pandemic influenza A H1N1 virus circulation: a modelling study. *Lancet Infect. Dis.*, 12, 9, 687–95, 2012 Sep, doi: 10.1016/S1473-3099(12)70121-4.
52. World Health Organization (WHO), COVID-19: WHO's response. [Internet], World Health Organization, Geneva, 2024 [cited 2024 Aug 29], Available from: https://www.who.int/emergencies/diseases/novel-coronavirus-2019.
53. Carlson, C.J., *et al.*, Climate change will drive novel cross-species viral transmission, Preprint at bioRxiv, 2020, https://doi.org/10.1101/2020.01.24.918755.
54. Patz, J.A., Epstein, P.R., Burke, T.A., Balbus, J.M., Global climate change and emerging infectious diseases. *JAMA*, 275, 217–223, 1996.
55. Climate change could increase the geographic extent of Hendra virus spillover risk. *Ecohealth*, 15, 509–525, 2018.
56. Yuen, K.Y., Fraser, N.S., Henning, J., Halpin, K., Gibson, J.S., Betzien, L., Stewart, A.J., Hendra virus: Epidemiology dynamics in relation to climate change, diagnostic tests and control measures. *One Health*, 12, 100207, 2021 Jun, doi: 10.1016/j.onehlt.2020.100207.
57. Dengue, urbanization and globalization: the unholy trinity of the 21st century. *Trop. Med. Health*, 39, S3–S11, 2011.
58. Li, Y., *et al.*, Urbanization increases Aedes albopictus larval habitats and accelerates mosquito development and survivorship. *PLoS Negl.Trop. Dis.*, 8, e3301, 2014.
59. Rose, N.H., *et al.*, Climate and Urbanization drive mosquito preference for humans. *Curr. Biol.*, 30, 3570–3579.e6, 2020.
60. Crépey, P. and Barthélemy, M., Detecting robust patterns in the spread of epidemics: a case study of influenza in the United States and France. *Am. J. Epidemiol.*, 166, 1244–1251, 2007.
61. Brownstein, J.S., Wolfe, C.J., Mandl, K.D., Empirical evidence for the effect of airline travel on inter-regional influenza spread in the United States. *PLoS Med.*, 3, e401, 2006.
62. Tayoun, A.A., *et al.*, Multiple early introductions of SARS-CoV-2 into a global travel hub in the Middle East. *Sci. Rep.*, 10, 17720, 2020.
63. Deng, X., *et al.*, Genomic surveillance reveals multiple introductions of SARS-CoV-2 into northern California. *Science*, 369, 582–587, 2020.
64. Miller, I.F. and Metcalf, C.J.E., Evolving resistance to pathogens. *Science*, 363, 1277–1278, 2019.

65. Park, M., Loverdo, C., Schreiber, S.J., Lloyd-Smith, J.O., Multiple scales of selection influence the evolutionary emergence of novel pathogens. *Philos. Trans. R. Soc. Lond. B Biol. Sci.*, 368, 20120333, 2013.
66. Pitzer, V.E., *et al.*, High turnover drives prolonged persistence of influenza in managed pig herds. *J. R. Soc. Interface*, 13, 20160138, 2016.
67. Kemp, S.A., *et al.*, SARS-CoV-2 evolution during treatment of chronic infection. *Nature*, 592, 277–282, 2021.
68. Zhan, J., Thrall, P.H., Papaïx, J., Xie, L., Burdon, J.J., Playing on a pathogen's weakness: using evolution to guide sustainable plant disease control strategies. *Annu. Rev. Phytopathol.*, 53, 19–43, 2015.
69. Olival, K.J., *et al.*, Possibility for reverse zoonotic transmission of SARS-CoV-2 to free-ranging wildlife: a case study of bats. *PLoS Pathog.*, 16, e1008758, 2020.
70. Centers for Disease Control and Prevention (CDC), Norovirus: About. [Internet], Centers for Disease Control and Prevention, Atlanta, 2024 [cited 2024 Aug 29], Available from: https://www.cdc.gov/norovirus/about/index.html.
71. Centers for Disease Control and Prevention (CDC), Norovirus: Prevention. [Internet], Centers for Disease Control and Prevention, Atlanta, 2024 [cited 2024 Aug 29], Available from: https://www.cdc.gov/norovirus/prevention/index.html.
72. Cubitt, W.D., McSwiggan, D.A., Moore, W., Winter vomiting disease caused by calicivirus. *J. Clin. Pathol.*, 32, 8, 786–93, 1979 Aug, doi: 10.1136/jcp.32.8.786.
73. Lopman, B.A., Reacher, M.H., Vipond, I.B., Sarangi, J., Brown, D.W., Clinical manifestation of norovirus gastroenteritis in health care settings. *Clin. Infect. Dis.*, 39, 3, 318–24, 2004 Aug 1, doi: 10.1086/421948.
74. Sandmann, F.G., Shallcross, L., Adams, N., Allen, D.J., Coen, P.G., Jeanes, *et al.*, Estimating the Hospital Burden of Norovirus-Associated Gastroenteritis in England and Its Opportunity Costs for Nonadmitted Patients. *Clin. Infect. Dis.*, 67, 5, 693–700, 2018 Aug 16, doi: 10.1093/cid/ciy167.
75. UK Health Security Agency, National norovirus and rotavirus bulletin: Management information. [Internet], UK Health Security Agency, London, 2024 [cited 2024 Aug 29], Available from: https://www.gov.uk/government/-statistical-data-sets/national-norovirus-and-rotavirus-bulletin-management-information–2.

76. Winder, N., Gohar, S., Muthana, M., Norovirus: An Overview of Virology and Preventative Measures. *Viruses*, 14, 12, 2811, 2022 Dec 16, doi: 10.3390/v14122811.
77. Yunus, M.A. and Yunus, M.A., Molecular Mechanisms for Norovirus Genome Replication. *Norovirus*, 23, 278–91, 2021 Mar 23.

2

Norovirus: Its Role in Gastroenteritis Outbreaks

Ranjana Rohilla[1]* and Shriyansh Srivastava[2]

[1]Department of Microbiology, AII India Institute of Medical Sciences, Rishikesh, India
[2]Department of Pharmacology, Delhi Pharmaceutical Sciences and Research University, New Delhi, India

Abstract

Noroviruses are a heterogeneous group of genetically related viruses causing acute gastroenteritis (AGE). Cases in extremes of ages and immunocompromised people suffer from severe infection. An individual experiences an average of three to eight norovirus infection episodes in his/her lifetime, the first episode of which will occur before or at 5 years of age. As per the Centre for Disease Control and Prevention (CDC), every year noroviruses account for about 600 million total cases of AGE worldwide, with a 0.025% mortality rate, mostly in developing countries. This acute infection is estimated to cost more than $60 billion globally due to healthcare expenditure and lost quality of life. Overall, norovirus is one of the stellar causes of AGE outbreaks, which are usually food- or water-borne and associated with seasonal variations in different latitudes and longitudes. Outbreaks are associated with poorer prognosis in clusters of immunocompromised individuals. Norovirus outbreaks commonly culminate in healthcare facilities, schools, daycare centers, etc. Outbreaks occur quickly as the infection spreads from person to person directly, through food or water, and *via* touching environmental surfaces. Mass public awareness is also increasing because of frequent reporting of outbreaks. Using reverse transcriptase polymerase chain reaction (RT-PCR) and gene sequencing, a particular strain of the virus can be traced as source of an outbreak in a particular area. Immunity is strain-specific or genotype-specific and there is either little or no immune protection across genotypes. As the virus is highly infectious in nature, vaccination is the only suitable strategy to control outbreaks. Recovery is almost eventless in majority of outbreaks, but immunocompromised

**Corresponding author*: ranjana86choudhary@gmail.com; ORCID: 0000-0002-4851-6933

Sachin Kumar, Shriyansh Srivastava, Rishabha Malviya and Ranjit Sah (eds.) Emerging Infectious Disease: Unravelling Norovirus and Pandemic Dynamics, (39–60) © 2026 Scrivener Publishing LLC

individuals and outbreaks in healthcare facilities have been observed to be followed by increased morbidity and mortality. Appropriate healthcare interventions like improvements in hygienic water supply, sanitation, and use of oral rehydration solution (ORS) are mandatory to mitigate such outbreaks in early stages.

Keywords: Norovirus, non-bacterial AGE, outbreaks, diarrhea, viral diarrhea

2.1 Infectious Etiology of Gastroenteritis

Gastroenteritis is defined as the inflammation of the mucosal lining of the stomach and small and large intestines and is a common infectious disease syndrome followed by symptoms like nausea, vomiting, diarrhea, abdominal aches, and cramps [1]. Acute gastroenteritis (AGE) lasts for less than 2 weeks and is self-limiting in most cases; persistent gastroenteritis lasts for 2–4 weeks, whereas chronic cases can last for more than 1 month.

Table 2.1 Etiological agents of acute gastroenteritis.

Viruses (>70%)	**Bacteria (up to 20%)**	**Parasites (10%–15%)**	**References**
Norovirus	*Shigella*	*Giardia*	[1–6]
Rotavirus	*Salmonella enterica*	*Entamoeba*	
Adenovirus type 40 and 41	*Campylobacter*	*Cryptosporidium*	
Astrovirus	*Escherichia coli* ➢ Shiga-toxin producing *Escherichia coli* (STEC) ➢ Enteropathogenic *Escherichia coli* (EPEC) ➢ Enterotoxigenic *Escherichia coli* (ETEC) ➢ Enteroaggregative *Escherichia coli* (EAEC)	*Cystoisospora*	
Picornavirus		*Cyclospora*	
		Microsporidium	
	Vibrio		
	Clostridioides difficile		
	Yersinia		

The etiology agents of acute gastroenteritis are varied and categorized under viruses, bacteria, and certain parasites, with most being viruses (Table 2.1).

As per the Centers for Disease Control and Prevention (CDC), Atlanta, USA, AGE is persistently a gigantic global public health challenge in all parts of the world particularly in the pediatric age group. According to the World Health Organization (WHO), there are more than 1.5 billion cases of childhood diarrhea associated with morbidity, with approximately five lakh deaths every year. Asia alone has an estimated mortality rate of 100 per lakh population due to AGE [7]. Viruses are the leading cause of such episodes, leading to death of nearly two lakh children annually worldwide. Enteropathogenic viral agents such as Rotavirus, Norovirus, and Adenovirus account for the maximum number of pediatric AGE cases [8].

2.2 Introduction to Norovirus Gastroenteritis

The family *Calciviridae* has 11 genera of viruses; the genera *Norovirus* and *Sapovirus* cause AGE in all age groups. As per the updated classification in 2019, noroviruses are classified into 10 established (GI–GX) and two temporary genogroups, which are further divided into 49 genotypes [9]. Noroviruses, like other Calciviridae members, contain RNA as genetic material. This ubiquitous virus accounts for a majority, i.e., >90%, of viral AGE outbreaks and 50% of all AGE cases globally [7, 10]. Kapikian and others discovered this agent of diarrhea outbreaks for the first time using an immune electron microscope in the year 1972 from a group of students and it was subsequently named as "Norwalk-like virus". Transmission was also established *via* contaminated food, water, and surfaces and person to person. Additional risk factors for increased morbidity and mortality are compromised quality of drinking water supply, inadequate health infrastructure, and prolonged breast feeding that is exceeding 6 months [11, 12].

In developed as well as developing countries, norovirus is being implicated as a common cause of pediatric AGE requiring medical attention. Outbreaks are common in all ages, but incidence is highest in children. A report on the global burden of norovirus states that each individual does acquire three to eight events of norovirus AGE in a lifetime, with one occurring before 5 years of age [12, 13]. Norovirus outbreaks have been reported in many different settings like healthcare facilities, restaurants, catering events, day care centers, schools or residential homes, cruise ships, etc. AGE outbreak is said to occur when there are two or more than two similar cases of gastroenteritis after exposure from a common source, which is confirmed by lab testing to be caused by norovirus. Outbreaks can also result from sharing food or utensils. It is the top cause of outbreaks resulting from contaminated food in the USA; it can also spread by touching ready-to-eat food before serving or handling raw food without precautions [14]. In healthcare facilities, the outbreaks can result from infected patients, visitors, attendants, or the food served. Outbreaks are challenging to manage as it is easy for the virus to spread. Not only outbreaks but norovirus also accounts for most sporadic cases of AGE [14].

2.3 History of Norovirus and Outbreaks

Norovirus was first established as a source of AGE outbreak after such an outbreak took place in an elementary school in the USA in 1968. Table 2.2 describes the details of discovery of norovirus, its linkage with AGE outbreaks, and the laboratory tests used so far for its diagnosis in different scenarios.

Table 2.2 History of norovirus discovery and its linkage with outbreaks.

Year	Event description	References
1968	Outbreak of AGE among school staff and students in Norwalk, Connecticut, USA, in Brosnon Elementary school. 50% of all attending schools including teachers, attendants, staff, and one-third of family members of patients reported symptoms. No growth was obtained in stool culture.	[15]
1970	Fecal filtrates from four different AGE outbreaks were tested on a group of volunteers. As mentioned earlier, one of these outbreaks took place in a school, with all falling sick of an AGE with a secondary attack rate of 30%. The disease was also called "winter vomiting disease" and more cases were reported in winter months. A preserved sample from one secondary case was prepared as 2% filtrate and administered to volunteers when two-thirds developed a similar GI illness. Although no success could be achieved in isolation of the etiological agent, it was hypothesized that this winter vomiting bug has a size of <50 nm depending upon the size of filers used to prepare filtrates.	[16]

(*Continued*)

Table 2.2 History of norovirus discovery and its linkage with outbreaks. (*Continued*)

Year	Event description	References
1972	The search for the etiological agent for nonbacterial AGE began in 1960 and intensified until 1970s. Albert Z. Kapikian in National Institute of Health (NIH) extended immune electron microscopic studies using the Norwalk agent present in the stool filtrates and convalescent sera from a volunteer as source of antibodies. After incubating stool filtrate with serum sample, aggregates of non-enveloped antibody-coated 27-nm virus-like particles were visualized on an immune electron microscope. Therefore, in a series of experimental tests, i.e., artificially and naturally, infected individuals demonstrated serological evidence of infection which led to the conclusion that the etiological agent was Norwalk-like virus causing AGE, thus fulfilling Koch's postulates were related to infectious etiology.	[16]
1978	A.Z. Kapikian *et al.* developed an immune adherence hemagglutination test for producing antibodies to the Norwalk agent of the AGE outbreak using viral antigens purified from the stool of experimentally infected volunteers. Antibodies to Norwalk agent was found in 50% of the population.	[17]
1981	It was established that the Norwalk virus belongs to the Calciviridae family and has a single structural protein with a mass of 59 kDa.	[18]
1982	Kaplan *et al.* established four criteria to identify an outbreak of Norwalk virus AGE. 1. Vomiting in >50 % cases 2. Mean incubation period of 24–48 h 3. Mean duration of illness: 1–3 days 4. No bacterial pathogen isolated in the stool culture testing Kaplan criteria was evaluated as highly specific when assessed in 2000 by CDC, Atlanta, USA, among food-related AGE outbreaks. In 1982, radioimmunoassay was used to identify the Norwalk virus in outbreaks of AGE.	[18–20]

(*Continued*)

Table 2.2 History of norovirus discovery and its linkage with outbreaks. (*Continued*)

Year	Event description	References
1986	Enzyme immunoassay was used to investigate Norwalk outbreaks.	[18]
1992	Real-time polymerase chain reaction (RTPCR) was developed by X Jiang *et al.* to detect Norwalk virus in human stool samples. It detected virus in stool sample diluted 10(-4). They used cetylmethyl ammonium bromide to remove inhibitors of RTPCR from stool extracts.	[21]
1998	An analysis of 500 patients' samples from 90 outbreaks showed that the Norwalk-like virus was proven to cause >90% of foodborne and epidemic AGE in the USA.	[18]
2000	In an AGE outbreak among students at Texas University, 125 fell sick; for the first time, RT-PCR and sequence analysis confirmed the presence of the Norwalk virus from a food item contaminated by a food handler.	[22]

2.4 Epidemiology and Common Settings of Norovirus Outbreaks

Data analysis of many outbreaks has revealed that norovirus of genogroup II is the most common AGE-causing strain globally [23]. The new emerging variants are found to be more and more associated with outbreaks. The variants are classically named after geographic locations where they were isolated for the first time and the year, e.g., GII.4 Sydney 2012 [24]. GII.4 viruses are agents behind most outbreaks or sporadic cases in the last 15 years. GII.2, GII.3, and GII.6 have become more common recently [25, 26]. The virus is contagious; norovirus AGE outbreaks have high attack rates, resulting in substantial economic and healthcare burdens. Outbreaks are also reported in developed and industrialized countries. Healthcare costs and disability-adjusted life years (DALYs) lost for the care of ill persons account for 60 billion dollars globally every year. A model estimated the annual global expenditure to be 4.2 billion dollars in direct healthcare systems dealing with norovirus infection. Children <5 years of age account for 33% of all norovirus-associated deaths (NAD) and 56% of DALY globally, whereas the elderly >70 years account for 40% of NADs globally and 11% of DALYs. There was a noticeable reduction in AGE incidence during the COVID era, but now the AGE incidence is equivalent to the pre-COVID pandemic level [25]. In a study from China on the epidemiology of 419 norovirus outbreaks from the year 2000–2018, a majority, i.e., more than 60%, were reported from schools and kindergarten day cares. Person-to-person transmission was observed in 40% of the outbreaks, food-borne in 23%, and water-borne in 22%. GII was the overall most reported genogroup [27]. The virus is easily transmitted by consuming fecally contaminated water; the infective dose is also low, i.e., up to 18 viruses are sufficient to initiate an infection. Moreover, 5 billion infectious particles are present in 1 g of feces during peak shedding, contaminating the environment, surfaces, food, and water. Asymptomatic people are also the main drivers of outbreaks as they shed virions without having symptoms. Therefore, outbreaks are non-amenable to control [28]. The standard settings for the outbreaks are shown in Figure 2.1. Healthcare facilities are high-risk centers for outbreaks, leading to increased morbidity and high secondary attack rates. Severe cases are expected in extremes of ages and immunocompromised individuals. Norovirus attack rates can be up to 40%, resulting in 4% hospitalization and 2% mortality rates [25, 29–31].

Temperate zones experience more outbreaks which occur throughout the year, yet cases peak during winters, i.e., October to February, in the northern part of the hemisphere. In tropical countries, norovirus follow

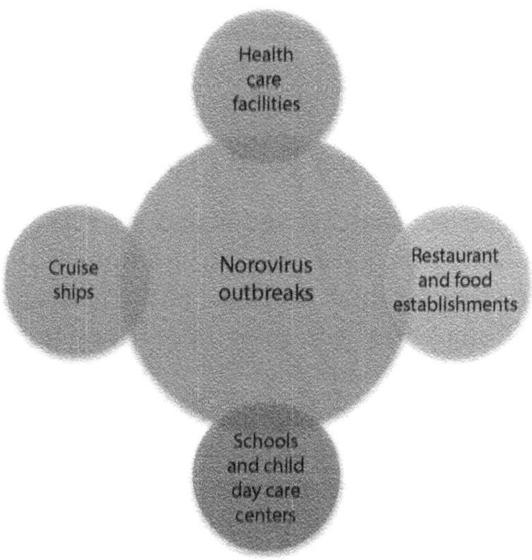

Figure 2.1 Standard settings of norovirus acute gastroenteritis outbreaks.

less seasonality, though peaks can be observed in colder months or monsoons; cases are reported throughout the year [25, 32]. As per CDC, more than 2,000 AGE outbreaks are reported yearly in the USA alone. Every year, there are nearly 1,000 deaths, more than one lakh hospitalizations, >4 lakh emergency department visits, and 22 lakh outpatient visits due to NAD [29].

2.4.1 Healthcare Facilities

Norovirus outbreaks typically occur in semi-closed areas, which are easy settings for rapid transmission, like wards or multi-occupied rooms. Shared washrooms, hand washing basins or kitchens between wards, faulty ward closures, and lack of proper ventilation are other risk factors for spread [33]. Long-term care facilities for chronic elderly patients are centers for enhanced transmission of norovirus. A study on a large data set of 107 long-term care facilities from five states of the USA from 2015 to 2019 revealed some interesting facts. Exposure to vomitus was a significant risk factor in comparison to exposure to feces. Residents are generally admitted with some underlying health condition, therefore making residents prone to severe illness and death [24, 25, 34]. Facility residents usually have higher secondary attack rates than staff [35–37]. Staff members are also impacted in hospital and long-term care facilities, which can lead to workforce shortages and temporary closure of wards to control outbreaks. Some studies indicate that healthcare professionals (HCPs) can be infected and that the number can also exceed the number of sick patients. Also, a lack of cooperation and support between clinicians and infection control teams may lead to increased cases among HCPs in a hospital [38].

2.4.2 Restaurants and Other Food Establishments

Nearly 10,000 food establishments and restaurant-related outbreaks have been reported to CDC from 1998 to 2004, and norovirus has been implicated in more than half of them. Because of the ease of spread *via* food handlers' hands, if norovirus gets entry into a food preparation facility, it is difficult to control the outbreak. The virus is viable on surfaces for more than 10 days and is also not destroyed by most commercially available disinfectants or surface cleansing agents. Once an outbreak happens, there is a staff shortage, temporary closure of establishments, and related expenditures [39]. In most outbreaks, the culprit is the mishandling of ready-to-eat food, such as sandwiches, salads, baked potatoes, cake icing, and salad dressing. Shellfish and oysters have also been linked to norovirus outbreaks. Norovirus seawater contamination results from dumping sewage and improper wastewater treatment [40–42]. Similarly, contaminated and untreated groundwater in food preparation can also lead to outbreaks [43–45]. There are records of ample food-related outbreaks of norovirus available from the CDC. On November 26, 2022, an outbreak was reported in a food establishment in Illinois, USA, where more than 300 cases fell sick and dined at restaurants. Probable cases and controls (who dined but did not fall sick) were interviewed, the suspected food was salad, and contamination was found to occur during food preparation for the salad portion. A food handler was found to be preparing without gloves; he had on vomited the previous day and was declared the source [46]. Frozen food products like raspberries and bilberries contaminated by polluted water or picked by symptomatic patients can also become a source of outbreaks [47–50]. The virus can enter food at the production point and during food preparation. In Japan, seven food-borne outbreaks of norovirus in different rural regions were linked to a commercially sold shredded red algae edible product that got contaminated during processing [51]. Also, re-outbreaks are reported from some facilities within a short duration, depending on whether the reservoir of the virus stays in a facility or the virus is reintroduced. For example, two successive outbreaks of norovirus AGE due to the GII.6 variant were reported from a holiday camp house in Catalonia, Spain, within a week. The attack rate in both often was 90% and 70%, respectively. The first was due to a green salad, and the second was a cheese omelet [52]. Similarly, in October 2017, two outbreaks occurred in a food establishment within 1 week; the investigation disclosed confirmed cases among food handlers who prepared food between and during the two outbreaks [53]. After analyses of 120 more AGE outbreaks linked to restaurants in the USA, an association was drawn between restaurant policies and the size and duration of outbreaks. In restaurants having small outbreaks, it was observed that food safety training was given to managers, fewer managers were handling the workflow, and these establishments were serving food with less preparation time. Safety guidelines, such as the FOOD CODE, have been issued by the Food and Drug Administration (FDA), USA, and can help prevent food-borne illness in food establishments. Appropriate hand washing, training on food safety norms, food storage temperature, and cleaning of slicers or chop boards are a few steps mandated by the organization. The virus has a short incubation period. Therefore, all outbreaks point to the source, and if an ill worker or reservoir is taken out, the outbreak typically dies [54].

2.4.3 School and Child Day Care

With clustering of many children at one place, with many shared surfaces, enteric pathogens spread at a very rapid rate in schools and child day care centers. As per the National Outbreak Reporting System (NORS), USA, from 2009 to 2019, in >2,000 schools, nearly 1,900 child day care centers reported AGE outbreaks. Almost all had spread person-to-person and norovirus was the predominant pathogen followed by *Shigella*. To prevent the spread of enteric pathogens, the American Academy of Pediatrics recommends the exclusion of symptomatic wards as well as staff. AGE outbreaks can disrupt school operations and lead to the closure of schools and day cares, resulting in direct or indirect financial burden. After the introduction of rotavirus vaccination in 2006 in the USA, norovirus has become a dominant pathogen in non-bacterial AGE outbreaks [55, 56].

2.4.4 Cruise Ships

As outbreaks commonly occur in crowded settings with a clustering of cases and the environment is semi-closed, cruise ships are ideal epicenters for outbreaks of enteric pathogens. CDC has its vessel sanitation program (VSP) to monitor maritime illnesses or AGE outbreaks in cruise ships [57]. The latest cruise ship norovirus outbreak can be traced to https://www.cruisemapper.com/, which gives information on COVID-19, norovirus, and viral epidemics on different vessels. In 2020, five reports of 876 patients and crew members infected on board were submitted to CDC. Because of the high frequency of norovirus outbreaks on cruise ships, it is also called the "cruise ship virus", although reported outbreaks on cruise ships are 1% of all the outbreaks. Close living quarters, shared kitchens and dining tables, intermingling passengers, and repeated outbreaks are reported from infected crew or contaminated environments [58, 59].

2.5 Clinical Features

Symptoms develop 12–48 h after exposure to norovirus. Common symptoms are diarrhea, nausea, vomiting, and abdominal cramps. Less common symptoms are low-grade fever, headache, and body aches. Symptoms typically stay for 1–3 days. Many remain asymptomatic and do not manifest symptoms but may pass the infection to others. In some cases, diarrhea might be a predominant symptom; in others, it is nausea/vomiting. If there is gross blood in stool, it is not consistent with a diagnosis of norovirus infection.

Depending on episodes of vomiting and diarrhea, the patient may suffer from dehydration, resulting in signs and symptoms such as tachycardia, decreased skin turgor on pinch test, orthostatic hypotension, dry mucosa, and dry mouth. Very rarely, some can present with neurological symptoms like seizure or encephalopathy [7]. While taking a history from cases, the patient's living conditions, source of drinking water, recent food intake history, and history of contact with people showing similar symptoms should be sought. Kaplan criteria is an essential criterion for diagnosing cases of AGE. If all Kaplan criteria are satisfied, there is a high probability of a norovirus outbreak. In the case of unavailable laboratory testing, Kaplan criteria are still considered very useful. Some complications associated with norovirus are electrolyte imbalance like hypokalemia, hyponatremia, metabolic alkalosis, and renal failure [7].

2.6 Gastroenteritis in Immunocompromised Individuals

Immune responses play a part in reducing the severity of infection, period of viral shedding, and viral load. Humoral immunity protects from norovirus-associated illness. An increasing number of transplants and anti-cancer therapeutics have increased the immunocompromised cohort. These patients are always at risk of acquiring infection in hospitals and the community. In primary immunodeficiency syndromes like common variable immune deficiency (CVID) and severe combined immunodeficiency (SCID) and in secondary immunodeficiency following solid organ transplant (SOT) or hematopoietic stem cell transplant, norovirus infection can cause prolonged diarrhea, graft rejection, and even death [60]. There is prolonged viral shedding in such patients, even increasing the risk of outbreaks manifolds. In a study conducted in a cohort of pediatric oncology patients, the median duration of viral shedding was 20 days. In contrast, it was even longer in a cohort of SOT-associated chronic norovirus infection, i.e., 218 days. Immunocompromising conditions might be followed by complications like high-frequency diarrhea episodes, weight loss and malnutrition, and pathological changes such as altered intestinal mucosal barrier, blunting of villus, and cell infiltration in lamina propria. Outbreaks have been reported in cancer and transplant units [60]. Also, HIV patients are prone to getting a chronic infection, leading to a high risk of epidemics. These patients are exposed to the virus *via* family members, HCPs, and contaminated food or water and surfaces.

In a study of 2 years in duration among renal transplant patients, 17% were having chronic norovirus infection and all of those infected had diarrhea on and off. The infection lasts for weeks to years. The vomitus and feces of these hosts might have a high viral load. Management of chronic norovirus AGE in these patients with impaired immune system is even more challenging and complicated with high morbidity rates and compromised quality of life. Even the seasonal variation, as observed in general population, is not there in immunocompromised hosts [61]. Chronic infections in immunocompromised hosts, in whom prolonged viral shedding of the virus is associated, have high chances of genetic variations in viral genome and evolution of new strains.

2.7 Gastroenteritis in Immunocompetent Individuals

Norovirus outbreak in immunocompetent individuals is a self-limiting condition with symptoms subsiding within 48 h, the symptomatic patients

Table 2.3 Comparison of norovirus infection in immunocompetent and immunocompromised host.

Characteristics	Immunocompetent	Immunocompromised	References
Viral shedding duration and levels	20–40 days; 108–109 genome copies/ gram of stool	Weeks to months; 105–108 genome copies/ gram of stool	[61, 62]
Seasonality	Peak in winters	Year-round prevalence	[61]
Tendency to form new strains	Very rare, stable variants	Due to prolonged shedding and other factors, high tendency of the virus to acquire mutations and form diverse viral populations or new strains	[61]
Complications	None	High-frequency diarrhea episodes, weight loss, electrolyte imbalance and malnutrition	[61, 62]
Prognosis	Usually good prognosis, self-limiting infection	Depends on the immune status; associated with chronicity	[61]

are isolated, and the outbreaks slowly die on its own. In contrast, it is associated with prolonged illness and complications if the host is immunocompromised and the outbreaks persist longer. The different characteristics of norovirus illness are described in Table 2.3.

2.8 Economic Burden of Infection

A report on estimated worldwide economic burden of norovirus published by Sarah M. Bartsch in 2016 quotes that the total annual global expenditure related to norovirus-related illness is 4.2 billion dollars in direct healthcare and 60.3 billion dollars is the societal cost. Illness-related societal cost in age group less than 5 years is $39.8 billion, whereas it is 20 billion dollars for other age groups [63]. This cost is even higher than other diarrheal agents like rotavirus which also gravely impacts the pediatric age group [64]. No such report data is available after 2016. As per a report on the clinical and economic burden of norovirus AGE outbreaks in USA published in 2020, every year approximately $173.5 million is spent in the management of such outbreaks, which comprised of direct medical expenditure and productivity losses. Productivity losses account for >90% of the total cost. Half of the economic burden is caused by adults >45 years of age, more than half is caused by the winter months, and >905 outbreaks are due to person-to-person transmission [63].

2.9 Preparedness for Outbreaks

Norovirus is a highly contagious virus. Outbreak is suspected when two or more people are sick within 24–48 h of each other with the same exposure source (like a school, day care, and long-term care facility), with abrupt onset of suggestive symptoms. An outbreak is said to end when it is 5 days since the last case recovered and there are no more cases [65]. If the disease is a notifiable disease in a region, cases should be reported to the health department as per regulations. CDC has the National Outbreak Reporting System (NORS) for all gastrointestinal pathogens and CaliciNet for norovirus outbreaks which should be notified for the outbreak. The critical points for initiating a response to a norovirus outbreak in a facility are shown in Figure 2.2 [30, 66, 67].

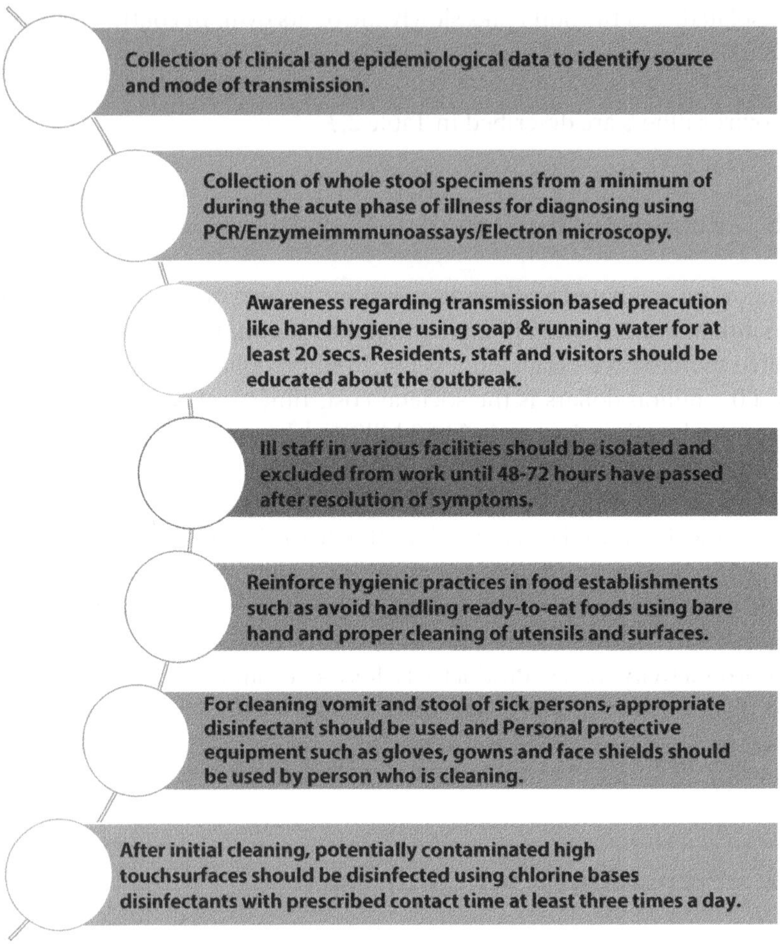

Figure 2.2 Key points for initiating a norovirus outbreak response [30, 68–70].

2.10 Conclusion

Norovirus, being highly contagious, remains the prevalent cause of AGE outbreaks worldwide in community as well as healthcare settings and is a huge economic burden on any country. People sick with this virus can shed billions of viral particles in stool, and it takes only a few particles to make others sick, resulting in an outbreak in a short period. Additionally, the emergence of new strains puts forth a challenge for health departments. Outbreaks are common in semi-closed settings such as healthcare facilities, schools, day care centers, cruise ships, etc. There is a need for robust surveillance networks so that the epidemiology of norovirus outbreaks can be studied in a particular geographical location in a specific time frame. Also, laboratories play a significant role in diagnosing cases and the management of outbreaks. Large-scale application of different diagnostic assays helps to delineate the exact disease burden of this viral infection and ongoing research will add to learning of the existing epidemiology of outbreaks and recent evolving trends. Using RT-PCR and gene sequencing, diverse strains have been classified into various clusters. Using such methods, a particular strain of the virus can be traced as source of an outbreak in a specific area. The critical steps in controlling outbreaks are cleaning and disinfection of the area, frequent hand hygiene practices as prescribed, segregation of ill persons from others until the symptoms resolve, and limitation of movement of cases. As the virus is highly infectious, developing and deploying vaccines is the core strategy to control such outbreaks, which has a substantial economic impact. Thus, outbreaks can be contained by initiating an early response as per "an agreed upon outbreak management plan" in place. Establishment of an outbreak control team, including members of an infection prevention and control (IPC) team, is essential.

AI disclosure statement: The author acknowledges the use of AI-based tools for minor language editing or grammatical correction. However, all intellectual contributions, analyses, and conclusions presented in this chapter are entirely the author's own.

References

1. Graves, N.S., Acute gastroenteritis. *Prim Care*, 40, 3, 727–41, 2013 Sep, doi: 10.1016/j.pop.2013.05.006. Epub 2013 Jul 19. PMID: 23958366; PMCID: PMC7119329.
2. Shane, A.L., Mody, R.K., Crump, J.A., Tarr, P.I., Steiner, T.S., Kotloff, K., *et al.*, Infectious Diseases Society of America Clinical Practice Guidelines for the Diagnosis and Management of Infectious Diarrhea. *Clin. Infect. Dis.*, 65, 12, e45–e80, 20172017 Nov 29, doi: 10.1093/cid/cix669. PMID: 29053792; PMCID: PMC5850553.
3. Chen, S.M., Ni, Y.H., Chen, H.L., Chang, M.H., Microbial etiology of acute gastroenteritis in hospitalised children in Taiwan. *J. Formos. Med. Assoc.*, 105, 12, 964–70, 2006 Dec, doi: 10.1016/S0929-6646(09)60280-1. PMID: 17185238.
4. Burkhart, D.M., Management of acute gastroenteritis in children. *Am. Fam. Physician*, 60, 9, 2555–63, 2565-6, 1999 Dec, Erratum in: Am Fam Physician 2000 May 1;61(9):2614. PMID: 10605991.

5. Nicholson, M.R., Van Horn, G.T., Tang, Y.W., Vinjé, J., Payne, D.C., Edwards, K.M., Chappell, J.D., Using Multiplex Molecular Testing to Determine the Etiology of Acute Gastroenteritis in Children. *J. Pediatr.*, 176, 50–56.e2, 2016 Sep, Doi: 10.1016/j.jpeds.2016.05.068. Epub 2016 Jun 18. PMID: 27329497; PMCID: PMC5215462.
6. Fleckenstein, J.M., Matthew Kuhlmann, F., Sheikh, A., Acute Bacterial Gastroenteritis. *Gastroenterol. Clin. North Am.*, 50, 2, 283–304, 2021 Jun, doi: 10.1016/j.gtc.2021.02.002. Epub 2021 Apr 23. PMID: 34024442; PMCID: PMC8793005.
7. Capece, G. and Gignac, E., Norovirus, in: *StatPearls* [Internet], StatPearls Publishing, Treasure Island (FL), 2024 Jan, [Updated 2023 Aug 14]. Available from: https://www.ncbi.nlm.nih.gov/books/NBK513265/.
8. Hassan, E. and Baldridge, M.T., Norovirus encounters in the gut: multifaceted interactions and disease outcomes. *Mucosal Immunol.*, 12, 6, 1259–1267, 2019 Nov, doi: 10.1038/s41385-019-0199-4. Epub 2019 Sep 9. PMID: 31501514; PMCID: PMC7479810.
9. Omatola, C.A., Ogunsakin, R.E., Onoja, A.B., Okolo, M.O., Abraham-Oyiguh, J., Mofolorunsho, K.C., *et al.*, Enteropathogenic viruses associated with acute gastroenteritis among African children under 5 years of age: A systematic review and meta-analysis. *J. Infect.*, 88, 6, 106169, 2024 Jun, doi: 10.1016/j.jinf.2024.106169. Epub 2024 Apr 30. PMID: 38697269.
10. Calcivirustypingtool, Available from: https://calcivirustypingtool.cdc.gov.info.html. Accessed online in 01.08.2024s.
11. Gaythorpe, K.A.M., Trotter, C.L., Lopman, B., Steele, M., Conlan, A.J.K., Norovirus transmission dynamics: a modelling review. *Epidemiol. Infect.*, 146, 2, 147–158, 2018 Jan, doi: 10.1017/S0950268817002692. Epub 2017 Dec 22. PMID: 29268812; PMCID: PMC5851036.
12. Patel, M.M., Hall, A.J., Vinjé, J., Parashar, U.D., Noroviruses: a comprehensive review. *J. Clin. Virol.*, 44, 1, 1–8, 2009 Jan, doi: 10.1016/j.jcv.2008.10.009. Epub 2008 Dec 11. PMID: 19084472.
13. Zhang, X., Chen, C., Du, Y., Yan, D., Jiang, D., Liu, X., *et al.*, Global Burden and Trends of Norovirus-Associated Diseases From 1990 to 2019: An Observational Trend Study. *Front. Public Health*, 10, 905172, 2022 Jun 17, doi: 10.3389/fpubh.2022.905172. PMID: 35784210; PMCID: PMC9247406.
14. Norovirus outbreaks, Available from: www.cdc. gov/Norovirus/outbreak-basics/index.html. Accessed online 0n 02.08.2024.
15. Tizard, I.R. and Musser, J.M.B., Panic and neglect—2000–2018, in: *Great American Diseases*, pp. 319–44, 2022, doi: 10.1016/B978-0-323-98925-1.00004-0. Epub 2022 Jan 21. PMCID: PMC8866134.
16. Kapikian, A.Z., The discovery of the 27-nm Norwalk virus: an historic perspective. *J. Infect. Dis.*, 181 Suppl 2, Suppl 2, S295–302, 2000 May, doi: 10.1086/315584. PMID: 10804141; PMCID: PMC7110248.
17. Kapikian, A.Z., Greenberg, H.B., Cline, W.L., Kalica, A.R., Wyatt, R.G., James, H.D., Jr, *et al.*, Prevalence of antibody to the Norwalk agent by a newly

developed immune adherence hemagglutination assay. *J. Med. Virol.*, 2, 4, 281–94, 1978, doi: 10.1002/jmv.1890020402. PMID: 215713.

18. Kaplan, J.E., Gary, G.W., Baron, R.C., Singh, N., Schonberger, L.B., Feldman, R., Greenberg, H.B.. Epidemiology of Norwalk gastroenteritis and the role of Norwalk virus in outbreaks of acute nonbacterial gastroenteritis. *Ann. Intern. Med.*, 96, 756–61, 1982, doi: 10.7326/0003-4819-96-6-756.
19. Herrmann, J.E., Nowak, N.A., Blacklow, N.R., Detection of Norwalk virus in stools by enzyme immunoassay. *J. Med. Virol.*, 17, 2, 127–33, Oct. 1985, doi: 10.1002/jmv.1890170205.
20. Turcios, R.M., Widdowson, M.A., Sulka, A.C., Mead, P.S., Glass, R.I., Reevaluation of epidemiological criteria for identifying outbreaks of acute gastroenteritis due to Norovirus: United States, 1998-2000. *Clin. Infect. Dis.*, 42, 7, 964–9, 2006 Apr 1, doi: 10.1086/500940. Epub 2006 Feb 27. PMID: 16511760.
21. Jiang, X., Wang, J., Graham, D.Y., Estes, M.K., Detection of Norwalk virus in stool by polymerase chain reaction. *J. Clin. Microbiol.*, 30, 10, 2529–34, 1992 Oct, doi: 10.1128/jcm.30.10.2529-2534.1992. PMID: 1383265; PMCID: PMC270473.
22. Daniels, N.A., Bergmire-Sweat, D.A., Schwab, K.J., Kate, A., *et al.*, A Foodborne Outbreak of Gastroenteritis Associated with Norwalk-like Viruses: First Molecular Traceback to Deli Sandwiches Contaminated during Preparation. *J. Infect. Dis.*, 181, 4, 1467–1470, April 2000, https://doi.org/10.1086/315365.
23. Glass, R.I., Parashar, U.D., Estes, M.K., Norovirus gastroenteritis. *N. Engl. J. Med.*, 361, 18, 1776–85, 2009 Oct 29, doi: 10.1056/NEJMra0804575. PMID: 19864676; PMCID: PMC3880795.
24. Chadwick, P.R., Trainor, E., Marsden, G.L., Mills, S., Chadwick, C., O'Brien, S.J., Evans, C.M., Mullender, C., Strazds, P., Turner, S., Weston, V., Toleman, M.S., de Barros, C., Kontkowski, G., Bak, A., Guidelines for the management of Norovirus outbreaks in acute and community health and social care settings. *J. Hosp. Infect.*, 136, 127–191, 2023 Jun, doi: 10.1016/j.jhin.2023.01.017. Epub 2023 Feb 15. PMID: 36796728.
25. Carlson, K.B., Dilley, A., O'Grady, T., Johnson, J.A., Lopman, B., Viscidi, E., A narrative review of Norovirus epidemiology, biology, and challenges to vaccine development. *npj Vaccines*, 9, 1, 94, 2024 May 29, doi: 10.1038/s41541-024-00884-2. PMID: 38811605; PMCID: PMC11137017.
26. Khan, M.K. and Alam, M.M., Norovirus Gastroenteritis Outbreaks, Genomic Diversity and Evolution: An Overview. *Mymensingh Med. J.*, 30, 3, 863–873, 2021 Jul. PMID: 34226482.
27. Yu, F., Jiang, B., Guo, X., Hou, L., Tian, Y., Zhang, J., Li, Q., Jia, L., Yang, P., Wang, Q., Pang, X., Gao, Z., Norovirus outbreaks in China, 2000-2018: A systematic review. *Rev. Med. Virol.*, 32, 6, e2382, 2022 Nov, doi: 10.1002/rmv.2382. Epub 2022 Aug 10. PMID: 35946340.
28. Wang, J., Gao, Z., Yang, Z.R., Liu, K., Zhang, H., Global prevalence of asymptomatic norovirus infection in outbreaks: a systematic review and

meta-analysis. *BMC Infect. Dis.*, 23, 1, 595, 2023 Sep 12, doi: 10.1186/s12879-023-08519-y. PMID: 37700223; PMCID: PMC10496210.

29. Norovirus, Available from: https://www.cdc.gov/Norovirus/index.html. Accessed online 25.08.24.
30. Updated Norovirus Outbreak Management and Disease Prevention Guidelines, Available from: https://www.cdc.gov/mmwr/pdf/rr/rr6003.pdf. Accessed online on 25.08.24.
31. Yasmin, F., Ali, S.H., Ullah, I., Norovirus outbreak amid COVID-19 in the United Kingdom; priorities for achieving control. *J. Med. Virol.*, 94, 3, 1232–1235, 2022 Mar, doi: 10.1002/jmv.27426. Epub 2021 Nov 9. PMID: 34713915; PMCID: PMC8662166.
32. Parrón, I., Barrabeig, I., Soldevila, N., Bartolomé, R., Guix, S., Rius, C., Cornejo-Sánchez, T., Izquierdo, C., Domínguez, A., Working Group for the Study of Outbreaks of Acute Gastroenteritis in Catalonia. Outbreaks of Gastroenteritis Due to Norovirus in Schools and Summer Camps in Catalonia, 2017-2019. *Microbiol. Spectr.*, 10, 3, e0011922, 2022 Jun 29, doi: 10.1128/spectrum.00119-22. Epub 2022 May 11. PMID: 35543555; PMCID: PMC9241749.
33. Fraenkel, C.J., Inghammar, M., Söderlund-Strand, A., Johansson, P.J.H., Böttiger, B., Risk factors for hospital norovirus outbreaks: impact of vomiting, genotype, and multi-occupancy rooms. *J. Hosp. Infect.*, 98, 4, 398–403, 2018 Apr, doi: 10.1016/j.jhin.2018.01.011. Epub 2018 Jan 31. PMID: 29355578.
34. Adams, C., Young, D., Gastañaduy, P.A., Paul, P., Marsh, Z., Hall, A.J., Lopman, B.A., Quantifying the roles of vomiting, diarrhea, and residents vs. staff in norovirus transmission in U.S. nursing home outbreaks. *PLoS Comput. Biol.*, 16, 3, e1007271, 2020 Mar 25, doi: 10.1371/journal.pcbi.1007271. PMID: 32210423; PMCID: PMC7135310.
35. Calderwood, L.E., Wikswo, M.E., Mattison, C.P., Kambhampati, A.K., Balachandran, N., Vinjé, J., Barclay, L., Hall, A.J., Parashar, U., Mirza, S.A., Norovirus Outbreaks in Long-term Care Facilities in the United States, 2009-2018: A Decade of Surveillance. *Clin. Infect. Dis.*, 74, 1, 113–119, 2022 Jan 7, doi: 10.1093/cid/ciab808. PMID: 34523674; PMCID: PMC8978331.
36. Alsved, M., Fraenkel, C.J., Bohgard, M., Widell, A., Söderlund-Strand, A., Lanbeck, P., Holmdahl, T., Isaxon, C., Gudmundsson, A., Medstrand, P., Böttiger, B., Löndahl, J., Sources of Airborne Norovirus in Hospital Outbreaks. *Clin. Infect. Dis.*, 70, 10, 2023–2028, 2020 May 6, doi: 10.1093/cid/ciz584. PMID: 31257413; PMCID: PMC7201413.
37. Sakon, N., Sadamasu, K., Shinkai, T., Hamajima, Y., Yoshitomi, H., Matsushima, Y., Takada, R., Terasoma, F., Nakamura, A., Komano, J., Nagasawa, K., Shimizu, H., Katayama, K., Kimura, H., Foodborne Outbreaks Caused by Human Norovirus GII.P17-GII.17-Contaminated Nori, Japan, 2017. *Emerg. Infect. Dis.*, 24, 5, 920–923, 2018 May, doi: 10.3201/eid2405.171733. PMID: 29664371; PMCID: PMC5938760.
38. Hofmann, F.M., Olawumi, E., Michaelis, M., *et al.*, Significance of Norovirus in occupational health: a review of published Norovirus outbreaks in Central

and Northern Europe. *Int. Arch. Occup. Environ. Health*, 93, 911–923, 2020, https://doi.org/10.1007/s00420-020-01543-4.

39. Symptoms of Norovirus and the role of Restaurants, Available from: https://www.foodpoisonjournal.com/food-poisoning-information/symptoms-of-Norovirus-and-the-role-of-restaurants/ Accessed online on 25.08.24.
40. Silverberg, R., Jones, M.K., Schneider, R.G., Sreedharan, K.R., Schneider Preventing Foodborne Illness: Norovirus, Available from: https://www.nifa.usda.gov/sites/default/files/resource/Preventing-Foodborne-Illness-Norovirus.pdf. Accessed online on 25.08.24.
41. Hardstaff, J.L., Clough, H.E., Lutje, V., McIntyre, K.M., Harris, J.P., Garner, P., O'Brien, S.J., Foodborne and Food-Handler Norovirus Outbreaks: A Systematic Review. *Foodborne Pathog. Dis.*, 15, 10, 589–597, 2018 Oct, doi: 10.1089/fpd.2018.2452. Epub 2018 Aug 15. PMID: 30109958; PMCID: PMC6201779.
42. Kittigul, L., Pombubpa, K., Rupprom, K., Thasiri, J., Detection of Norovirus Recombinant GII.2[P16] Strains in Oysters in Thailand. *Food Environ. Virol.*, 14, 1, 59–68, 2022 Mar, doi: 10.1007/s12560-022-09508-1. Epub 2022 Jan 24. PMID: 35075605.
43. Kim, J.G., Kim, J.S., Kim, J.G., Characteristics of Norovirus Food Poisoning Outbreaks in Korea in the 2000s. *J. Food Prot.*, 84, 3, 472–480, 2021 Mar 1, doi: 10.4315/JFP-20-093. PMID: 33108448.
44. Kauppinen, A., Pitkänen, T., Miettinen, I.T., Persistent Norovirus Contamination of Groundwater Supplies in Two Waterborne Outbreaks. *Food Environ. Virol.*, 10, 1, 39–50, 2018 Mar, doi: 10.1007/s12560-017-9320-6. Epub 2017 Oct 11. PMID: 29022247.
45. Meghnath, K., Hasselback, P., McCormick, R., Prystajecky, N., Taylor, M., McIntyre, L., Man, S., Whitfield, Y., Warshawsky, B., McKinley, M., Bitzikos, O., Hexemer, A., Galanis, E., Outbreak Team, Outbreaks of Norovirus and Acute Gastroenteritis Associated with British Columbia Oysters, 2016-2017. *Food Environ. Virol.*, 11, 2, 138–148, 2019 Jun, doi: 10.1007/s12560-019-09374-4. Epub 2019 Mar 21. PMID: 30900141.
46. Hanley, M.N., Altman, S.M., Phillips, A., Notes from the Field: Outbreak of Norovirus Linked to a Food Establishment - Illinois, November 2022. *MMWR Morb. Mortal Wkly. Rep.*, 72, 33, 897–898, 2023 Aug 18, doi: 10.15585/mmwr.mm7233a4. PMID: 37590287; PMCID: PMC10441824.
47. Torén, K., Schiöler, L., Nenonen, N.P., Hannoun, C., Roth, A., Andersson, L.M., Westin, J., Bergström, T., Risk factors for norovirus infection in healthcare workers during nosocomial outbreaks: a cross-sectional study. *Antimicrob. Resist. Infect. Control*, 10, 1, 107, 2021 Jul 22, doi: 10.1186/s13756-021-00979-8. PMID: 34294149; PMCID: PMC8299649.
48. Nasheri, N., Vester, A., Petronella, N., Foodborne viral outbreaks associated with frozen produce. *Epidemiol. Infect.*, 147, e291, 2019 Oct 18, doi: 10.1017/S0950268819001791. PMID: 31625499; PMCID: PMC6813648.

49. Raymond, P., Paul, S., Perron, A., Bellehumeur, C., Larocque, E., Charest, H., Detection and Sequencing of Multiple Human Norovirus Genotypes from Imported Frozen Raspberries Linked to Outbreaks in the Province of Quebec, Canada, in 2017. *Food Environ. Virol.*, 14, 1, 40–58, 2022 Mar, doi: 10.1007/s12560-021-09507-8. Epub 2022 Jan 23. PMID: 35066807; PMCID: PMC8881426.
50. Summa, M., Tuutti, E., Al-Hello, H., Huttunen, L.M., Rimhanen-Finne, R., Norovirus GII.17 Caused Five Outbreaks Linked to Frozen Domestic Bilberries in Finland, 2019. *Food Environ. Virol.*, 16, 2, 180–187, 2024 Jun, doi: 10.1007/s12560-024-09587-2. Epub 2024 Mar 11. PMID: 38466479.
51. O'Brien, S.J., Sanderson, R.A., Rushton, S.P., Control of norovirus infection. *Curr. Opin. Gastroenterol.*, 35, 1, 14–19, 2019 Jan, doi: 10.1097/MOG.0000000000000491. PMID: 30346315.
52. Alsedà, M., Godoy, P., Bach, P., Soldevila, N., Cornejo, T., Corominas, L., Grau, M., Domínguez, A., Working Group for the Study of Outbreaks of Acute Gastroenteritis in Catalonia. Two successive outbreaks of acute gastroenteritis due to Norovirus GII.6 in a holiday camp house. *Sci. Rep.*, 13, 1, 15558, 2023 Sep 20, doi: 10.1038/s41598-023-42622-z. PMID: 37730810; PMCID: PMC10511710.
53. Monini, M., Ostanello, F., Vignolo, E., Pagani, E., Gamper, S., Spertini, S., Masi, E., Rabini, M., Stenico, A., Poznanski, E., Di Bartolo, I., Occurrence of two Norovirus outbreaks in the same cafeteria in one week. *New Microbiol.*, 42, 3, 156–160, 2019 Jul. Epub 2019 Jul 15. PMID: 31305934.
54. Hoover, E.R., Hedeen, N., Freeland, A., Kambhampati, A., Dewey-Mattia, D., Scott, K.W., Hall, A., Brown, L., Restaurant Policies and Practices Related to Norovirus Outbreak Size and Duration. *J. Food Prot.*, 83, 9, 1607–1618, 2020 Sep 1, doi: 10.4315/JFP-20-102. PMID: 32421792; PMCID: PMC8140480.
55. Mattison, C.P., Calderwood, L.E., Marsh, Z.A., Wikswo, M.E., Balachandran, N., Kambhampati, A.K., Gleason, M.E., Lawinger, H., Mirza, S.A., Childcare and School Acute Gastroenteritis Outbreaks: 2009-2020. *Pediatrics*, 150, 5, e2021056002, 2022 Nov 1, doi: 10.1542/peds.2021-056002. PMID: 36278284; PMCID: PMC10061552.
56. Havumaki, J., Eisenberg, J.N.S., Mattison, C.P., Lopman, B.A., Ortega-Sanchez, I.R., Hall, A.J., Hutton, D.W., Eisenberg, M.C., Immunologic and Epidemiologic Drivers of Norovirus Transmission in Daycare and School Outbreaks. *Epidemiology*, 32, 3, 351–359, 2021 May 1, doi: 10.1097/EDE.0000000000001322. PMID: 33652444.
57. Jenkins, K.A., Vaughan, G.H., Jr, Rodriguez, L.O., Freeland, A., Acute Gastroenteritis on Cruise Ships - Maritime Illness Database and Reporting System, United States, 2006-2019. *MMWR Surveill. Summ.*, 70, 6, 1–19, 2021 Sep 24, doi: 10.15585/mmwr.ss7006a1. PMID: 34555008; PMCID: PMC8480991.
58. Cruise Ship Norovirus Outbreaks, Available from: https://www.cruisemapper.com/wiki/1056-cruise-ship-norovirus-outbreaks2020 Accessed online on 08.08.24.

59. Cruisemapper, Available from: www.cruisemapper.com/ Accessed online on 20.08.24.
60. Chong, P.P. and Atmar, R.L., Norovirus in health care and implications for the immunocompromised host. *Curr. Opin. Infect. Dis.*, 32, 4, 348–355, 2019 Aug, doi: 10.1097/QCO.0000000000000557. PMID: 31107251; PMCID: PMC8191235.
61. Bok, K. and Green, K.Y., Norovirus gastroenteritis in immunocompromised patients. *N. Engl. J. Med.*, 367, 22, 2126–32, 2012 Nov 29, doi: 10.1056/NEJMra1207742. PMID: 23190223; PMCID: PMC4944753.
62. Green, K.Y., Norovirus infection in immunocompromised hosts. *Clin. Microbiol. Infect.*, 20, 8, 717–23, 2014 Aug, doi: 10.1111/1469-0691.12761. PMID: 25040790; PMCID: PMC11036326.
63. Bartsch, S.M., Lopman, B.A., Ozawa, S., Hall, A.J., Lee, B.Y., Global Economic Burden of Norovirus Gastroenteritis. *PLoS One*, 11, 4, e0151219, 2016, https://doi.org/10.1371/journal.pone.0151219.
64. The high cost of Norovirus worldwide, Available from: https://publichealth.jhu.edu/2016/the-high-cost-of-Norovirus-worldwide. Accessed online on 20.08.24.
65. Norovirus and managing outbreaks, Available from: https://www.healthvermont.gov/disease-control/foodborne-diarrheal-illnesses/Norovirus-and-managing-outbreaks. Accessed online on 20.08.24.
66. Burke, R.M., Shah, M.P., Wikswo, M.E., Barclay, L., Kambhampati, A., Marsh, Z., Cannon, J.L., Parashar, U.D., Vinjé, J., Hall, A.J., The Norovirus Epidemiologic Triad: Predictors of Severe Outcomes in US Norovirus Outbreaks, 2009-2016. *J. Infect. Dis.*, 219, 9, 1364–1372, 2019 Apr 16, doi: 10.1093/infdis/jiy569. PMID: 30445538; PMCID: PMC6939307.
67. Park, G.W., Williamson, K.J., DeBess, E., Cieslak, P.R., Gregoricus, N., De Nardo, E., Fricker, C., Costantini, V., Vinjé, J., High Hand Contamination Rates During Norovirus Outbreaks in Long-Term Care Facilities. *Infect. Control Hosp. Epidemiol.*, 39, 2, 219–221, 2018 Feb, doi: 10.1017/ice.2017.274. Epub 2018 Jan 14. PMID: 29331156; PMCID: PMC9007178.
68. Wainaina, E., Otieno, C.A., Kamau, J., *et al.*, Norovirus infections and knowledge, attitudes and practices in food safety among food handlers in an informal urban settlement, Kenya 2017. *BMC Public Health*, 20, 474, 2020, https://doi.org/10.1186/s12889-020-8401-x.
69. Foodborne disease outbreaks: Guidelines for investigation and control, World Health Organization, Available from: https://iris.who.int/bitstream/handle/10665/43771/9789241547222_eng.pdf. Accessed online on 7.10.24.
70. Summary of recommendations III. Guideline for the Prevention and Control of Norovirus Gastroenteritis Outbreaks in Healthcare Settings, 2011, Available from: https://www.cdc.gov/infection-control/hcp/norovirus-guidelines/summary-recommendations.html. Accessed on 5.10.24.

3

The Genetics of Norovirus: Understanding Its Genome and Genogroups

Naga Rani Kagithala[1], Rahul Sharma[1], Sasikanth Kothamasu[2], Shriyansh Srivastava[3] and Ismail A. Ibrahim[4*]

[1]*Department of Pharmacy, School of Medical and Allied Sciences, Galgotias University, Greater Noida, India*
[2]*Department of Pharmaceutics, SIMS College of Pharmacy, Magaldasnagar, Guntur, India*
[3]*Department of Pharmacology, Delhi Pharmaceutical Sciences and Research University, New Delhi, India*
[4]*Faculty of Health Sciences, Fenerbahçe University, Istanbul, Türkiye*

Abstract

Norovirus is one of the main consequences of the global spread of acute gastroenteritis. From a genomic standpoint, this chapter offers valuable insights into the genetics of noroviruses, including gene diversity and classification. A single positive-strand ribonucleic acid, norovirus is split within three ORFs through a sequential manner. ORF1 makes non-structural proteins (nsps), which are needed for the virus to replicate. ORF2 and ORF3 make the structure of the proteins in the capsid that are VP1 and VP2, respectively. Genetic variability, exhibited by high rates of mutation with recombination events that cause the division of the virus into genotypes and genogroups, is one of the defining features of norovirus. Gene types I (GI) and II (GII) are the main ones that infect humans. This essay also talks about how molecular methods can be used to investigate norovirus genetics. Whole-genome sequencing (WGS) and next-generation sequencing (NGS) are two of the best ways to find out how viruses have changed over time and how they spread. The genetic diversity of norovirus is a significant issue that hinders the development of vaccines and leads to resistance to synthetic antivirals. The constant emergence of new strains and genogroups, as a result of new genomic research, has further expanded knowledge on the virus' evolutionary process. This article discusses the relevance of genetic studies to

**Corresponding author*: ismailaiaibrahim@gmail.com

Sachin Kumar, Shriyansh Srivastava, Rishabha Malviya and Ranjit Sah (eds.) Emerging Infectious Disease: Unravelling Norovirus and Pandemic Dynamics, (61–98) © 2026 Scrivener Publishing LLC

address norovirus' health problem and explains the move toward genetic studies for control and prevention strategies.

***Keywords*:** Norovirus genome, genetic diversity, genogroups, RNA virus, next-generation sequencing (NGS), vaccine development

3.1 Introduction

Acute gastroenteritis is primarily caused by human noroviruses (HuNoVs) and can occur in epidemic and occasional forms globally. Among the Caliciviridae family, they belong to the genus Norovirus. Viruses are microscopic organisms sometimes referred to as "transportation shells" because they are able to multiply and spread disease because they carry genetic material specific to their host species. Over time, many viruses have developed increasingly complex strategies for surviving both inside and the outside of their host system, setting up a significant burden regarding our healthcare system [1, 2]. The Norwalk viruses, also known as the norovirus (NoV), is one virus that is particularly good at this. For almost 50 years, human populations have been infected with this endemic virus. Norovirus (NoV) is the most prevalent enteric non-enveloped virus that makes people of all ages get gastroenteritis. About 200,000 people die from it yearly, and small towns with care and educational facilities are often the sites of outbreaks. It is mostly in emergency and long-term care facilities where NoV outbreaks happen, where 40% of outbreaks occur [3, 4]. In 1972, a foodborne illness outbreak in Norwalk, Ohio, led to the first microscopic identification of norovirus. After being dubbed the Norwalk virus at first, it was expanded to include additional isolates and named Norwalk-like viruses (now known as noroviruses). It has not been grown yet. It belongs to the Caliciviridae family. Another name for it is SRSV, which stands for "small round structured viruses." According to estimates, noroviruses cause 64,000 hospitalized episodes of diarrhea, 200,000 children under years old die every year in poor countries, and 900,000 children in developed countries go to the clinic [5]. States that took part in NoroSTAT reported 1,459 cases of norovirus outbreaks between August 1, 2023 and July 16, 2024. In the same time frame of the previous seasonal year, these states reported 1,225 cases of norovirus outbreaks [6]. Some pathogens like bacteria, parasites, and viruses may cause diarrhea. In developing nations, bacterial agents are comparatively more significant, while in developed nations, viral agents are more significant. These agents are important because they affect the population's sanitation and hygiene standards. At the moment, the majority of pediatric gastroenteritis are thought to be brought on by viruses belonging to four distinct families: Adenoviridae (adenovirus), Astroviridae (astrovirus),

Caliciviridae (norovirus and sapovirus), and Reoviridae (rotavirus) [7]. Studying norovirus genetics is important. An understanding of norovirus genetics is necessary to comprehend the evolution, transmission, and implications for public health. For their high mutation rates and capacity for recombination, noroviruses can produce novel strains and genetic diversity [8]. These viruses are now common causes of gastroenteritis-related hospital admissions worldwide, especially in the elderly, young, and immunocompromised people [9]. Next-generation sequencing has revealed that transmission events and chronic infections in immunocompromised hosts contribute to viral diversity, while acute infections maintain relatively stable populations [10]. Intergenic capsid-RNA-dependent RNA polymerase gene recombination plays a vital role in norovirus evolution, particularly in genotype GII.3. This recombination allows for higher evolutionary rates, enhanced genetic diversity and adaptation. Understanding how GII.3 noroviruses evolved is crucial to understanding their role in sporadic gastroenteritis in children. Vaccine design benefits from understanding norovirus evolution since vaccines target high-risk groups, including children, pertains to a pediatric genotype [11]. In order to create effective preventative and intervention measures, such as vaccinations and antiviral medications, it is imperative to comprehend these genetic mechanisms. The remarkable genetic diversity and swift evolution of norovirus impede attempts to contain outbreaks and create efficacious vaccinations. Through the analysis of its genetic composition, scientists can monitor the evolution of viral strains, comprehend their strategies for evading immune responses, and detect novel variants that could present fresh risks to public health [8]. As the chapter's scope, a thorough examination of the genetic composition and categorization of noroviruses will be covered in this chapter, with particular attention paid to the genome structure, classification, evolution, and implications for public health of the different identified genogroups. The chapter would begin with an introduction to the genomic structure of noroviruses, detailing their RNA genome and the organization of ORFs. The capsid protein (VP1) and RNA-dependent RNA polymerase (RdRp) are essential for viral categorization and function, and it would explain their importance. The division of noroviruses into different genogroups (GI to GVII) and genotypes according to genetic diversity would be a major area of focus. This chapter will address recent revisions to the schemes of classification, emphasizing the addition of new genogroups and the standards by which they were categorized. RdRp and VP1 areas have nucleotide and amino acid diversity. In order to wrap up the chapter, we will discuss some of the current issues facing norovirus genetics research as well as some new technologies that are helping us learn more about these dangerous infections.

3.2 Norovirus Genome

3.2.1 Structure and Organizational Forms of the Norovirus Genome

The family of viruses called Caliciviridae includes noroviruses, which are viruses possessing a single strand of RNA and a positive sense that does not have a casing [12]. The norovirus is a diminutive virion with a density that is buoyant ranging from 1.33 to 1.41 g/cm^3 in calcium chloride and a diameter of 27 to 32 nm [13]. The genome of this positive-sense, single-stranded, non-enveloped RNA virus measures 7.4–7.7 kb in length [14, 15]. Polyadenylation occurs at the 3′ end of RNA. The 5′ terminus of each calicivirus genome begins with GU nucleotides. Capsid gene (VP1) starts subgenomic RNA and has a 5′ end sequence that is internally repeated within the genome and varies in length by genus. This area may be in the RNA-dependent RNA polymerase promoter [16].

3.2.2 RNA Genome Characteristics

Human noroviruses belong to the Caliciviridae family. Linear RNA of approximately 7,500 nucleotides is a positive-sense, single-stranded genome of NOV. In 1990, the initial prototype norovirus (GI.1), which was the first complete genome of the Norwalk virus, was sequenced [15]. The genome is made up of three overlapping open reading frames (ORFs). The ORF1 gene encodes a polyprotein that includes the structural protein VPg and various nonstructural proteins. These proteins comprise the 3D-like RNA-dependent RNA polymerase as well as the 3C-like protease. The projected nucleotide total of this polyprotein is 5,100. The ORF2 gene, approximately 1,600 nucleotides in length, encodes the viral protein 1 (VP1), which is the primary capsid protein. The S, P1, and P2 subdomains represent additional subdivisions. The genome contains three overlapping open reading frames: ORF1, ORF2, and ORF3. Nonstructural proteins are encoded by the 5,100-base ORF1 gene and consist of RNA-dependent RNA polymerase, which exhibits a 3D structure, the structural protein VPg, and the 3C-like protease. The ORF2, consisting of about 1,600 bases, encodes viral protein 1 (VP1), the primary capsid protein involved. The S subdomain, P1 subdomain, and P2 subdomain represent additional subdivisions [15]. The minor capsid protein VP2, roughly 720 bases in length, encoded by ORF3, may aid in the stabilization of the capsid virion [17]. The 5′ terminus of the genome is covalently attached to VPg, while the 3′ terminus features a polyadenylated tail. The entire genome can be transfected to generate complete virus particles, indicating that the norovirus RNA genome is intrinsically infectious [3].

3.2.3 Open Reading Frames (ORFs)

There are three conserved ORFs in the norovirus genome; however, the murine norovirus (MNV) contains a fourth alternative ORF. The only known member of the Caliciviridae family with a fourth ORF is the human sapovirus. The fourth ORF is exclusive to the GV MNV cluster and has not been detected in any other norovirus [18]. Sections of the viral RNA genome known as open reading frames (ORFs) have the ability to be translated into proteins. There is a limited amount of overlap between the three (ORFs) that make up the Norovirus, which are located at the 5' and 3' junctions. Apart from the murine norovirus (MNV-1), which possesses a fourth open reading frame (ORFs), the majority of noroviruses have a genome that is made up of three ORFs that overlap with one another [12, 19]. The norovirus genome consists of three ORFs: ORF1, ORF2, and ORF3. The initial and concluding open ORFs are located in reading frame 2 of the complementary DNA (cDNA), whereas ORF 2 is situated in reading frame 3. It is anticipated that the main ORF 1 will code for the nonstructural proteins. Sequence analysis has revealed similarities with the picornavirus' 3D RNA-dependent RNA polymerase, 2C helicase, and 3C protease [12]. The second open reading frame, commonly referred to as ORF 2, encodes the protein of the capsid. When insect cells produce baculovirus recombinant particles, the Capsid protein prompts the self-assembly of empty recombinant virus-like particles (rVLPs) [20, 21]. At the 3′ end of the genome lies the third open reading frame (ORF 3), which encodes a protein comprised of 212 amino acids whose function is not well understood. The No V ORF 3 protein is expected to have a molecular weight of 22,479 Da. The key protein ORF 3 suggests a potential function in nucleic acid binding, owing to its expected isoelectric point of 10.99 [22].

3.2.4 Non-Structural Proteins (ORF1)

Upstream of the RdRp, the virus-encoded 3CLpro cleaves ORF1's 200 kDa polyprotein. Fast and co-translational proteolytic processing produces six proteins that are not structured. ORF1, the initial two-thirds of the genome, encodes seven non-structural proteins that are essential for genome replication and are longer than five kilobases. NTPase (NS3), RNA-dependent RNA polymerase (NS7), Some of the parts are the viral genome-linked protein (NS5), the 3C-like protease (NS6), the N-terminal protein with an unknown function (NS1-2), and p22, which also has an unknown function (NS4) [23].

3.2.5 Principal Capsid Protein (VP1): ORF2

The main capsid protein, VP1, is produced by norovirus ORF2. The three domains of VP1's capsid protein are the N-terminal, P, and S domains. A pliable joint separates the S and P regions. A key factor in the formation of the icosahedral capsid structure is the well-preserved S domain. It is essential that this framework be assembled. The P domain consists of two subdomains: P1, characterized by low conservation, and P2, recognized for its highly variable nucleotide sequence. Due to the presence of binding sites for cellular and immune recognition in the P2 subdomain of the norovirus genome, this part of the capsid structure is the most exposed to light [24–27].

3.2.6 ORF3: VP2 or Minor Capsid Protein

The tiny, structurally important auxiliary protein VP2 is encoded by ORF3 and is present in the Norovirus genome. VP2 exhibits a length between 208 and 268 amino acids, characterized by significant sequence variability among different strains. VP2 has two functions: stabilizing VP1 inside the viral structure and augmenting VP1 expression. Furthermore, VP2 shields VP1 from the harmful effects of proteolytic degradation and disintegration processes [27]. The role of VP2 in viral replication remains unclear; however, its pronounced basicity may facilitate interactions with RNA, thereby assisting in the packaging of the viral genome. Furthermore, it has been documented that the synthesis of infectious viral particles entails the VP2 protein [28, 29].

3.3 Genetic Diversity in Norovirus

3.3.1 Genogroups and Genotypes

Acute gastroenteritis in both humans and animals is caused by RNA viruses known as noroviruses. These viruses display distinct genetic configurations. The variation of capsid proteins (VP1) forms the basis for the later classification into genotypes and genogroups (GI-GX). The Norovirus Classification For the genotyping of GI and GII noroviruses, the NCWG proposed a widely recognized naming and typing system grounded in the phylogenetic classification of the complete VP1 amino acid sequence [30]. Noroviruses were classified into six genogroups (GI to GVI) following an analysis of the sequence data that was accessible then. A seventh genogroup (GVII) has also been suggested [2]. By further subdividing these genogroups, more than forty genotypes were produced [31]. Human beings are susceptible to infection from the viruses known as GI, GII, and GIV. On the other hand, the genotypes GII.11, GII.18, and GII.19 that are carried by GII viruses are only found in fecal specimens from pigs, but the genotype GIV.2 is only found in carnivorous animals like cats and dogs [2, 32, 33]. Owing to the absence of naming criteria for GII.4 variants, they will be classified through phylogenetic clustering, and new variants will only be acknowledged once they become epidemic in at least two different geographic regions. In the last 20 years, eight GII.4 variants have circulated, with each variant supplanting its predecessor as the dominant strain. The most recent GII.4 variation as of 2012 is GII.4 Sydney [34].

3.3.2 Mechanisms of Genetic Variation

Noroviruses are highly diverse RNA viruses recognized for inducing gastroenteritis in humans. The two principal mechanisms of genetic variation in noroviruses are point mutations and recombination. Numerous investigations have demonstrated that viral biology can be altered by single-point mutations, impacting biochemical aspects that influence replication and enhancing transmissibility [35]. Some genotypes, such GII.2 and GII.3, have different capsid/polymerase combinations. However, even small genetic modifications can make norovirus stronger and more likely to spread. The fact that certain GII.4 and GII.17 variants are more common and spread more widely around the world than other variants with similar RNA-dependent RNA polymerase (RdRp) types that are only found in certain areas is an important factor in the evolution and spread of norovirus [36].

Recombination is a crucial process that boosts the genetic diversity of noroviruses. In this process, which often occurs at the junction of the ORF1 and ORF2 regions, genetic material from different viral strain exchanges. By this interchange, viruses can become chimeric, with potentially beneficial characteristics that increase their ability to survive and spread. In contrast, recombinant strains of non-GII.4 noroviruses are highly prevalent and circulate alongside other variants, adding to their genetic diversity. Homologous recombination and mutation are the two methods of variation used by NoVs. It has been proposed that both processes promote pandemic GII.4 lineage [37, 38]. People with weakened immune systems and recurrent infections may harbor new variants of the virus, even though acute infections share the same viral population. There are four main factors that can impact how quickly GII evolves. Replication kinetics, host receptor recognition, sequence space, and herd immunity duration are the four strains. Together these mechanisms increase the norovirus' ability to persist and cause recurrent epidemics in humans [10]. Although it has been evidenced for murine noroviruses, the mechanism by which human noroviruses internalize themselves after attaching to HBGAs is unknown and may depend on dynamin and cholesterol. The virus replicates in the cytoplasm in close proximity to membrane complexes derived from the host (see Figure 3.1). It is still mostly unknown how human noroviruses assemble and release themselves.

The norovirus genome encodes a three-ORF polyprotein. This polyprotein contains six proteins that are non-structural and VP1 and VP2. An RNA strand called (+) RNA, which stands for positive-sense, is one of the parts that help make the capsid that protects the DNA. The capsid sticks to the cell surface in the first and second phases using VP1 and HBGAs. During stages three and four, receptor binding, entrance, and uncoating are all processes that take place. The host cell's cytoplasm contains (+) RNA that has been translated and transcribed. For the purpose of promoting transcription in step 5 and recruiting host translation components, the non-structural viral protein known as VPg attaches to the 5′ end of the genome. After translation, the ORF1 polyprotein is cleaved, producing several proteins such p48 (NS1/2 or N-terminal), NTPase (NS3 or 2C-like), p22 (NS4 or 3A-like), VPg, Pro, and RNA-dependent RNA polymerase. The virus protease Pro (NS6 or 3C-like) helps with this process. The process of genome replication, which happens in steps 8 and 9, makes a pattern for making new genomic and subgenomic (+) RNAs. This template is used to transform (+) RNA into (-) RNAs. In order to produce VP1 and VP2, subgenomic (+) RNAs are utilized, which only contain ORF2 and ORF3 in their sequences. In the 10th step of the virion building process, genomic and maybe subgenomic (+) RNAs are wrapped up inside new virions. In the 11th step, virions are expelled from the infected host cell using a mechanism that has not yet been described.

3.3.3 Mutation Rates and Hotspots

Most cases of acute diarrhea around the world are caused by noroviruses. Their RNA genomes exhibit diversity, encompassing two principal genogroups (GI and GII) that have a minimum of 28 genotypes linked to human disease [39]. Estimates of the norovirus mutation frequency range from 3×10−33×10−3 to 4.64×10−34.64×10−3 substitutions per site per

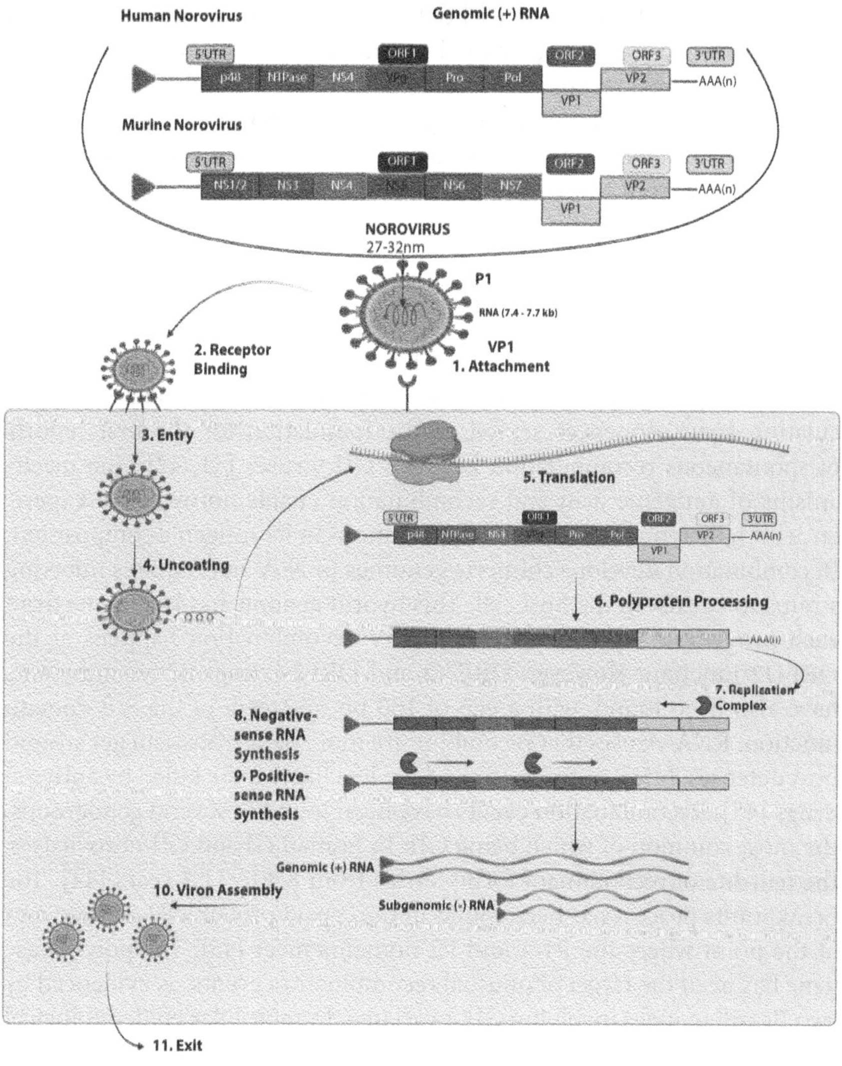

Figure 3.1 Human norovirus composition and life cycle.

year, indicating a notably high frequency. Because noroviruses mutate as quickly as other RNA viruses do, they can quickly adapt to changes in host immunity and environmental factors, among other selective pressures [8, 31]. Comparing RNA viruses to most DNA viruses, their mutation rates can be up to 1,000 times higher, ranging from 10–3 to 10–4 nucleotide (nt) substitutions/year [39]. Because of its high contagiousness, norovirus spreads quickly in crowded, enclosed spaces. Hospitals, assisted living facilities, daycare facilities, and schools are frequently the sites of norovirus outbreaks because of how quickly the virus can spread among young children. The limited area and communal amenities on cruise ships facilitate the virus' easy spread among passengers. High turnover and shared dining areas are two factors that can lead to outbreaks in hotels and resorts. Keep your distance from those who are infected, wash your hands frequently, and sanitize surfaces [40].

3.3.4 Recombination Events

One of the factors contributing to the growing diversity of noroviruses is the widespread recombination found in them. The examination of circulating virus sequences served as the foundation for the first reports of spontaneous recombination between GII viruses [41, 42]. The mechanisms of antigenic drift and recombination enable norovirus to experience both incremental and sudden alterations in its antigenic composition. Recombination develops chimeric genomes in NoV but requires infecting strains to coexist in the host cell. This hybrid genome has fragments from each progenitor. Most of the time, NoV recombination happens at the ORF1/2 junction. However, ORF2/3 and ORF2-intragenic recombinants have also been found, with a gap of 150 bp upstream of the S/P domain junction. RNA viruses that recombine are more dangerous, can get around host defenses, infect more hosts, and are less likely to be killed by antiviral drugs [43]. Recombination events have been found in several genogroups, the most common of which being GII. In human GI and GII noroviruses, the half-life of recombinant forms varies from 8.4 to 11.3 years [44]. The breakpoints process has been found in the capsid protein's P2 domain and at the point where the P1-1 and P2 domains meet [45]. The polymerase gene has been the target of unusual recombination events, as evidenced by two Brazilian cases involving GII.4 variants. In molecular epidemiological studies, these recombination events may result in increased virulence and genotyping errors [46].

3.3.5 Evolutionary Dynamics of Norovirus

For more than 30 years, noroviruses, especially strain GII.4, have been the main cause of gastroenteritis. Noroviruses, emerging RNA viruses, possess a diverse genetic structure and persistently evolve in humans due to a combination of point mutations and genomic recombination [47]. Virus strains GII.4 and GII.3 change $4.16 \times 10-3$ and $4.3 \times 10-3$ nucleotides per site per year, respectively, which means that they evolve very quickly [48, 49].

New GIIs emerge due to the ongoing accumulation of mutations in the VP1 capsid protein. Every 2 to 8 years, four distinct strains of norovirus emerge. Non-GII can nonetheless precipitate substantial epidemics. Despite limited genetic diversity, four norovirus genotypes have been predominantly identified [50]. According to a Bayesian analysis, the average evolutionary rate for the VP1 gene is 1.21 × 10–2 nucleotide substitutions/site/year, indicating that norovirus populations are growing rapidly. The RNA virus is rapidly evolving and comprises 10 distinct genogroups (GI–GX). These genogroups are subdivided into 49 genotypes and 60 unique P-types, determined by changes in the amino acid gene sequence of the main viral protein shell (VP1) and the nucleotide sequence of the (RdRp) [30].

3.4 Norovirus Genogroups

3.4.1 Classification and Characteristics of Norovirus Genogroups

The Caliciviridae family encompasses Norovirus, previously referred to as Norwalk-like viruses [51]. Norovirus belongs to the Caliciviridae family of viruses, which encompasses those previously referred to as Norwalk-like viruses. Within the family of viruses known as the Caliciviridae, the Norovirus, Sapovirus, Lagovirus, and Vesivirus are all members [14]. Lagovirus and Vesivirus genera include rabbit hemorrhagic sickness virus and feline calicivirus. Bovine enteric calicivirus genera have been added to the Caliciviridae family: Naboviral Becoviruses [52–54]. People and other animals can only get sick from the groups Norovirus and Sapovirus. The full nucleotide sequence of the capsid gene VP1 was used to divide norovirus into five genogroups: GI, GII, GIII, GIV, and GV. Acute gastroenteritis in mammals is induced by genetically diverse RNA noroviruses. The amino acid makeup of the VP1 gene (GI–GX) tells us the genotypes and genogroups. Sequence data resulted in the categorization of noroviruses into six genogroups (GI to GVI) [2, 34], proposed seventh genogroup (GVII). More than 40 genotypes were created from these genogroups [31]. GI, GII, and GIV virus strains infect human hosts, but GII has three genotypes (GII.11, GII.18, and GII.19) found in swine feces, and GIV has a genotype (GIV.2) found only in carnivores [2, 33, 55]. Human infections are linked to genotypes I, II, and IV, with GII.4 being the most common [56]. The latest classification has 10 genogroups, 49 genotypes, and 60 P-types based on polymerase diversity [30]. Cattle and sheep infected with genogroup III (GIII) have two P-types and four genotypes (GIII.1–4) [57].

3.4.2 Classification of VP1 Sequences

Employing 2Hsd criteria, the comprehensive VP1 amino acid sequences (totaling 305 sequences) are categorized into established and potential norovirus genotypes. The four genogroups that were recently confirmed are GVII (canine), GVIII (human), GIX (human) (reclassified from GII.15), and GX (chiropteran). GNA1 (*Phocoena phocoena*) and GNA2 (*Zalophus californianus*) are the two experimental genogroups. For the GNA1 and GNA2 viruses, which belong to a new genogroup, the 2×sd criterion does not apply because it needs two genes per cluster to figure out the standard deviation for distances within the cluster. The genes GII.23–27, GII.NA1, GII.NA2, GIIV.NA1, a possible new GIV genotype, GII.24–GII.25, GII.26, and GII.27 were all confirmed to be real [30].

3.4.3 RdRp Designations

P-groups and P-types are the names for genotypes and genotypes that are classed as RdRp regions. VP1 genogroups and genotypes were set apart from P-groups and P-types by adding a capital P to the Roman genogroups (GI.P, GII.P, GIII.P) or genotypes (GI.P1, GII.P). P-type numbers were allocated to all RdRp nucleotide sequences possessing a minimum of two sequences from distinct geographic areas, irrespective of their VP1 genotype, to eradicate "orphan" P-types from the norovirus categorization system. "NA" was designated to P-types (GI.PNA1 and GII. NA2) with a numeral following NA when just a single RdRp sequence was accessible [30]. Table 3.1 shows the classification of norovirus.

Norovirus P-groups, P-types, genotypes, variants, and genogroups: The term "non-assigned" refers to sequences that are either unique or consist of several non-identical sequences from the same geographic location that together suggest possible genogroups, genotypes, P-groups, and P-types.

3.4.4 Genogroup I (GI)

Norovirus humans get gastroenteritis: Genetic diversity: GI has nine genotypes, with GI.1 being the most notable, originally known as the Norwalk virus [58]. This genotype group includes GI.2–GI.9. Epidemiology: Though rarer than GII strains, GI strains cause outbreaks in nursing homes and cruise ships. Some GI strains can cause severe gastroenteritis, especially in vulnerable populations. Infections cause vomiting, diarrhea, and abdominal pain, like other norovirus strains [59].

Table 3.1 Norovirus classification.

VP1 typing			RdRp typing		
Genogroups	Confirmed	GI, GII, GII…etc.	P-groups	Confirmed	GI.P, GII.P, GIII.P… etc.
	Tentative	GNA1, GNA2… etc.		Tentative	GNA1.P, GNA2.P…etc.
Genotypes	Confirmed	GI.1, GII.1…etc.	P-types	Confirmed	GI.P1, GII.P1… etc.
	Tentative	GII.NA1, GII. NA2…etc.		Tentative	GII.PNA1, GII. PNA2… etc.
Variants	GII.4 variants	GII4 New Orleans, GII.4 Sydney… etc.			

doi: 10.1016/B978-0-12-800947-5.00002-8

3.4.5 Genogroup II (GII)

Norovirus GII comprises human and swine strains. Dominance: The most common genogroup, GII, is responsible for most norovirus cases around the world. There are 27 identified genotypes in it, the most prevalent of which is GII.4, which is linked to widespread epidemics. Genetic variability: GII is more genetically diverse than GI, which helps to create new variants that can evade immunity that already exists [30]. There are now new forms of the GII.4 genotype because of many mutations and recombination events. GII infections can cause vomiting, diarrhea, and stomach pain, among other similar symptoms. Genetic variations can affect the frequency of particular symptoms, with some exhibiting distinct symptom profiles [58].

3.4.6 Genogroup III (GIII)

GIII contains only bovine strains [54]. Genetic diversity: Three genotypes are known to exist in genogroup III. Animal noroviruses, especially those that infect cattle and other mammals, are the main source of these genotypes. Epidemiology: Although GIII strains are not as common in human infections as GI and GII, they have been found in human cases, suggesting a possibility of zoonotic transmission. There are gaps in our knowledge regarding the epidemiological significance of GIII in human gastroenteritis because it has been less studied than GI and GII. Transmission: Although humans have occasionally become infected due to exposure to tainted food or water sources, especially in agricultural environments, genome group III noroviruses are mainly found in animals. The symptoms of GIII norovirus infections can resemble those of other strains of the virus and may include nausea, vomiting, diarrhea, and abdominal pain. The overall effect on human health, however, has not received as much documentation. Recombination is a mechanism by which GIII strains, like those of other norovirus genogroups, may give rise to new variants. This variation in genetic makeup can make monitoring and management more difficult [60].

3.4.7 Epidemiology of Norovirus Genogroups

Noroviruses can infect dogs, pigs, people, mice, bats, and sea lions due to their genetic diversity [36]. Most people who get norovirus illness are young children. Few industrialized countries have population-based national estimates of norovirus disease incidence, which vary from 3.8% to 10.4% per year. Regional study findings have proven to be trustworthy. Over the course of an 80-year lifetime, the average person will contract three to eight norovirus infections. Young children are most commonly affected by norovirus gastroenteritis. Population-based national estimates of norovirus disease incidence, which range from 3.8% to 10.4% annually, are available for only a small number of developed nations. Results from regional studies have been reliable. Table 3.2 shows the norovirus numbers by country. An individual will typically have three to eight norovirus infections during their 80 years of life [61]. Human norovirus AGE cases are primarily associated with a particular GII norovirus genotype, GII.4. It has been at least 15 years since GII.4 viruses were the main cause of outbreaks and rare cases [62, 63]. Norovirus diversity was not well understood prior to 2000, when improved surveillance efforts were made; however, in the 1980s and 1990s, GII was the cause of widespread outbreaks. There were four variations, in agreement with current findings [64, 65]. New variations are given names after they spread to two or more geographically distinct areas. Since the middle of the 1990s, six new GII.4 mutations have been linked to large-scale breakouts: Den Haag in 2006, New Orleans in 2009, Sydney in 2012, Grimsby in 1996, Farmington Hills in 2002, and Hunter in 2004. Even though GII.4 is the most common genotype, the number of illnesses varies depending on the population, the surveillance method, the location, the year the data was collected, and environmental factors. In recent years, genotype GI.3 has been the most prevalent in a variety of studies, including 58% in US outbreaks from 2013 to 2016, 65% in global laboratory-based surveillance from 2005 to 2016, 67% in observational studies of child cases unrelated to outbreaks from 2004 to 2012, and 41% in a group of babies who were followed prospectively for the first 2 years of

Table 3.2 A summary of norovirus numbers by country.

Country	State/region	% rate of disease	Notes on outbreaks	References
United States	Various states	18% of gastroenteritis cases globally	2,500 annual outbreaks; the highest between November–April	[80, 81]
United Kingdom	England	Increase post-COVID, double pre-pandemic levels	High peaks of GII.17 strain; hospital outbreaks reported weekly	UKHSA Norovirus Reports [82]
India	Kerala	Estimated 1.4%–44.4% of gastro cases	Notable outbreaks in Kerala are linked to food and water contamination	[83]
Canada	National	Not explicitly stated; foodborne outbreaks	Outbreaks reported through national surveillance	PHAC Norovirus Surveillance [84]
Australia	National	20% of acute gastroenteritis outbreaks	Increased reports in colder months; GII.4 most common	Australian Department of Health [83]

life. GII.2, GII.3, GII.6, and, right now and mostly in Asia, GII.17 are the next most common genotypes found in recent outbreak monitoring from high- and middle-income countries. The annual percentages of disease attributed to these genotypes vary, but they are often lower than those of GII.4 [66].

3.5 Molecular Techniques for Norovirus Genetic Analysis

3.5.1 Sequencing Technologies

Advancements in sequencing technologies have greatly enhanced the analysis of norovirus genomes. Full-genome sequencing of a variety of norovirus genotypes from stool samples has shown high sensitivity and coverage when paired with Illumina sequencing and SureSelect target enrichment [67]. Clinical samples can be sequenced for norovirus in two ways: sequencing suitable PCR fragments [68–70] and whole RNA sequencing directly [71–73]. The first method generates a pristine viral template, hence improving sequencing quality, although it necessitates many PCR amplifications; to get to the target norovirus DNA, the second method needs very deep sequencing. This article describes the implementation of a third technology known as sure select target enrichment (Agilent), which has been useful for making full disease genomes from patient samples for DNA and RNA viruses as well as bacteria that are hard to grow in the lab [74–76]. Specially constructed 120-mer RNA traps that match all public norovirus sequences are employed to extract norovirus genomes from stool RNA extracts. Various baits are employed to target specific loci within the genome. This method resolves challenges associated with nontarget sequencing in RNA-Seq and the design of PCR primers. Systematically sequencing clinical and shellfish samples collected during norovirus outbreaks linked to eating shellfish is done to find the cause. Because Sanger sequencing does not work very well, PCR isolates have to be cloned to find genetic differences within just one specimen. The examination of bacteriomes using 16S rRNA amplicons in high-throughput sequencing (HTS) demonstrates that HTS is an economical method for collecting nucleotide diversity and generating extensive data. In contrast to Sanger sequencing, high-throughput sequencing (HTS) does not necessitate the cloning of isolates from environmental samples to differentiate numerous sequences per amplicon. Various genotypes in oyster samples can be detected at low quantities *via* an improved amplicon-based high-throughput sequencing method that employs LunaScript reverse transcriptase and AmpliTaq Gold DNA polymerase [77]. Nanopore's long-read sequencing technology has recovered near-complete norovirus genomes from single reads, improving speed and read length [78]. Whole-genome deep sequencing of tropical city norovirus transmission and evolution revealed new evolutionary pressure in the VP1, p48, and p22 protein-coding domains [79]. These advances improve norovirus genomics and epidemiology research.

3.5.2 Next-Generation Sequencing (NGS)

NGS can quickly and cheaply sequence a lot of genetic material at the same time, which makes it useful for studying the genomic variety of viruses [85].

More and more, NGS technologies are being used to find new and known viruses in environmental or clinical data. Next-generation sequencing (NGS) made virus discovery possible without culture [86]. NGS is a powerful tool for studying norovirus (NoV) diversity and evolution. NGS and bacterial rRNA depletion can identify multiple genotypes and recombinants in outbreak samples with near-full-genome coverage of NoV [87, 88]. This method discovered greater intra-host genetic diversity in immunocompromised chronic NoV patients than acute cases, indicating viral evolution reservoirs [88]. NGS has been used to study NoV diversity in shellfish, revealing a wide range of genotypes and emphasizing their role in NoV epidemiology [89]. NGS has limitations in clinical settings due to its speed and the requirement for precise diagnostic algorithms, despite its potential to improve outbreak investigations and surveillance. NGS facilitates comparison with the broader virome and microbiome, potentially affecting norovirus expression and transmission, along with viral diversity profiling across various sample types. There are two common ways to do NoV NGS: neutral, randomly primed sequencing and primed amplicon sequencing. "Agnostic" refers to methods that do not depend on facts about virus sequences. Forward and reverse primers are put in a specific way across the viral genome to allow reverse transcription and PCR amplification. To get short read coverage, overlapped amplicons from across the viral genome are collected and sequenced. On the plus side, sequencing resources can be directed on the target virus, which is usually NoV, if the primers are well made. Another problem with the method is that it can only find target viruses and hide co-infecting viruses that the primers cannot find. NGS possesses significant clinical and public health implications, including infection monitoring and source identification. NGS may assist in resolution; nonetheless, acquiring a dependable sequence will be the principal hurdle in error detection. NGS can assist in diagnosis and elucidate the complexity of environmental infections. Until certain substantial hurdles are resolved, NGS is improbable to swiftly supplant accurate diagnostic methods with established positive and negative predictive values. Table 3.3 shows that technologies currently in use for diagnosing norovirus. Cost may also be a constraint. The cost of a sample processed for NGS includes amplification, library preparation, and sequencing [90]. However, NGS may help us understand NoV transmission, evolution, and strain emergence.

Table 3.3 The technologies currently in use for diagnosing norovirus.

Technology	Principle	Time	Applications	Sensitivity	Cost	References
RT-qPCR (real-time quantitative PCR)	Amplifies viral RNA for quantification using reverse transcription	2–4 h	Clinical diagnostics, environmental testing	~10 viral RNA copies/reaction	High (~$10,000 equipment)	[107]
ELISA (enzyme-linked immunosorbent assay)	Detects viral antigens using antibodies and colorimetric signals	1 to 2 h	Food, water, clinical samples	103–105 virus particles/mL	Moderate (~$500–$1,500 equipment)	[108]
LAMP (loop-mediated isothermal amplification)	Amplifies viral RNA under isothermal conditions	1 h	Point-of-care diagnostics, food testing	Comparable to RT-qPCR (~10 viral RNA copies/reaction)	Low (~$500 equipment)	[109]
CRISPR-based detection	Cas enzymes target and cleave viral RNA, with a fluorescent signal	1 to 2 h	Clinical diagnostics, future applications	~10 copies/μL	Moderate to high (~$1,000–$2,000 equipment)	[110]

(Continued)

Table 3.3 The technologies currently in use for diagnosing norovirus. (*Continued*)

Technology	Principle	Time	Applications	Sensitivity	Cost	References
Immunochromatographic assay (lateral flow test)	Detects viral antigens on a test strip using labeled antibodies	15–30 min	Point-of-care, food safety, environmental	105–106 virus particles/mL	Low (~$2–$10 per test)	[110]
Next-generation sequencing (NGS)	Sequencing of viral genomes in clinical samples	Several days	Research, epidemiology, outbreak analysis	Ultra-sensitive (~1–10 viral particles)	Very high (~$1,000 per run)	[111]
Biosensors (e.g., SPR, electrochemical sensors)	Detect viral particles or RNA through surface interactions	Minutes to hours	Food safety, environmental monitoring	102–103 virus particles/mL	Moderate (~$1,000–$10,000 equipment)	[112]
Cell culture-based assays	Growing viruses in cell cultures and detecting cytopathic effects	Several days	Research, virology labs	Depends on viral replication efficiency	High (~$1,000–$5,000)	[113]

3.5.3 Whole-Genome Sequencing (WGS)

Comprehensive genome sequencing (WGS) is an effective method for characterizing, detecting, and monitoring noroviruses. WGS enhances strain genetic differentiation over Sanger sequencing, improving the accuracy of epidemiological studies and outbreak tracking. By examining the complete genome, whole-genome sequencing (WGS) of NoV can enhance the surveillance and investigation of outbreaks. By finding SNVs, WGS is a good way to tell the difference between viral types that are genetically similar [138]. The use of whole-genome sequencing is growing in popularity as a means of streamlining the surveillance and investigation of bacterial outbreaks [91]. Target enrichment with RNA baits is one of the many WGS techniques that have been developed [67] as direct feces sequencing. Capsid and polymerase gene PCR genotyping costs £50 less than whole-genome sequencing with target enrichment. Stool RNA or mRNA is sequenced by RNA-Seq. RNA-Seq is impartial because it does not need PCR primers [92]. These techniques can produce whole norovirus genomes from clinical specimens with excellent sensitivity, even for low viral loads. WGS identified positive selection zones to study norovirus evolution in urban tropical contexts. In order to improve public health responses to this significant pathogen, WGS is promising for norovirus genetic diversity and evolution, surveillance, and outbreak investigations [94]. WGS is a promising tool for improving public health responses to norovirus infections because of its ability to provide data for surveillance, genetic diversity and evolution, and outbreak investigations.

3.5.4 Phylogenetic Analysis

Tracking transmission and comprehending genetic diversity of norovirus (NoV) require phylogenetic analysis. Common NoV strains and their evolutionary patterns have been found through regional studies. From 2014 to 2016, the predominant genotypes in Seoul were GII.4, GII.17, and GII.3, with new strains appearing annually [93]. A total of 49 genotypes, some with multiple variants, and 10 genogroups (GI–GX) were found through phylogenetic analysis of VP1 sequences [94]. NoV prevalence was 9% in Japanese oysters meant for raw consumption, and both genogroups I and II were found [95]. The phylogenetic analysis shows that the Sverdlovsk sequences had the smallest genetic distance, supporting their clustering. The genetic diversity and evolutionary processes of HuNoV will be monitored through surveillance and phylogenetic analysis of circulating norovirus infections. This will improve anti-epidemic measures and reduce infectious disease burden. A standardized HuNoV genotyping method and phylogenetic analysis will assess viral genetic diversity and evolutionary divergence in Sverdlovsk's endemic districts to regulate infection [96]. It was found in sewage and river water in Argentina that GII.4 was the most common genotype. The strains found there were more closely linked to new global variants than to local sequences [97]. The studies mentioned above highlight how important it is to conduct continuous surveillance and phylogenetic analysis in order to track the genetic diversity of NoV and predict potential outbreaks.

3.5.5 Genotyping and Molecular Epidemiology

Norovirus genotyping is crucial for outbreak surveillance and investigation. A number of different ways have been created, including the two-region strategy, which targets the regions of the polymerase and the capsid. Sanger sequencing is used after conventional RT-PCR to genotype norovirus strains for routine surveillance. Since the middle of the 1990s, finding and typing norovirus has been mostly about a small area in the middle of the polymerase gene [65, 98], A region at ORF1's 3′ end is where genotyping first emerged [99] to a specific area at the 5′-end of ORF2, which has become the typing assay used by numerous laboratories worldwide in the past decade [100]. Two ORF2-based protocols were compared in a multicenter study, and region C was suggested for routine genotyping and region D for the identification of new GII. four variations [101]. A pilot study in the Czech Republic revealed the prevalence of GII.4 genotype, particularly in children aged 0–4 years [102]. In order to make genotyping more accurate, a one-step RT-PCR method for typing both GI and GII noroviruses was created and confirmed. Using this technique, 90% of 2,663 samples from different countries were successfully genotyped [103]. This assay detected most known human genotypes and P types, with a detection limit of 5–50 viral RNA copies per reaction. Standardized genotyping protocols enable rapid data exchange *via* surveillance networks that support vaccine development and outbreak management. Studies on the molecular epidemiology of norovirus have revealed a significant impact of the virus on global gastroenteritis outbreaks. Norovirus has a low contagious dose, sheds for a long time, is very stable, and changes often [104]. Genogroup II, particularly GII4, has been the predominant strain in many outbreaks [105]. New variants and recombinants can appear over time, causing the molecular epidemiology pattern to fluctuate quickly [104]. Norovirus outbreaks are common in closed settings like hospitals and care center [105]. In addition to food and water, the virus may also be transferred through the air. By using molecular epidemiological techniques, it has been shown that norovirus can be detected in both host and environmental samples. These studies give us important information for making vaccines and figuring out what new types will appear [106].

3.6 Implications of Norovirus Genetic Variability

3.6.1 Impact on Vaccine Development

The norovirus causes most cases of acute gastroenteritis worldwide. It is very hard to make a vaccine against it because it has a lot of different genetic forms and is grown in uncertain ways. The creation of a broadly successful norovirus vaccine is still hard because noroviruses have a lot of different genetic and antigenic variants that can be found at the same time. It is impossible to exaggerate the significance of norovirus vaccines. The vaccinations used now mostly use virus-like particles (VLPs), which are based on the VP1 protein and have shown promise in clinical studies [114, 115]. There are currently no approved pharmacological treatments for norovirus infections, nor are there licensed vaccines to prevent the disease, which is due in part to our limited understanding of human norovirus biology. Scientists have not been able to look into how human norovirus gets into cells, which is a key step in the disease's development because they do not have any working molecular clones or the right cell culture platforms for *in vitro* replication [85, 116]. Three norovirus vaccines are in development. P particles, recombinant adenoviruses, and nonreplicating virus-like particles (VLPs) are possible vaccination candidates [117]. In spite of the fact that certain human norovirus vaccines were withdrawn from circulation during the preclinical development phase, a significant number of other vaccine candidates advanced to various stages of clinical testing. Additionally, more and more people are calling for the creation of human norovirus vaccines that would be given along with other immunogens during vaccination. A special three-in-one vaccine was created to protect against rotavirus, norovirus, and VP6 viruses, which cause most serious cases of gastroenteritis in children [118].

3.6.1.1 Monovalent Vaccines

Most people who got a GI.1 VLP vaccine that was mixed with monophosphoryl lipid A had norovirus-specific antibodies in their blood after being challenged through the nose. A comparison of the vaccination and control groups showed a significant drop in the number of people getting norovirus infections and acute gastroenteritis (infection rate: 61% in vaccinated people vs. 82% in placebo recipients; disease rate: 37% in vaccinated people vs. 69% in placebo recipients) [119].

3.6.1.2 Bivalent Vaccines

Current research focuses on bivalent vaccinations with human norovirus GI and GII VLPs. To find out if the bivalent norovirus vaccine is safe, if it works, and if it can cause antigen interaction, an animal model received the vaccine intraperitoneally in the form of a locally gelling dry powder. This work sparked new research ideas. The formulation is both safe and immunogenic, as evidenced by dose-dependent increases in systemic and mucosal immune responses to each of the VLPs. Furthermore, the GI and GII VLP-based vaccine was demonstrated to have a boosting effect after a second dose without immunological interference [119]. The main challenges are finding reliable immune correlates of protection, managing strain mismatches, and developing vaccines that are effective for all age groups. Even with these problems, there is still hope for making a norovirus vaccine because several options are currently being tested in people. The potential dual market in developed and developing countries may help with global distribution and commercial development [120].

3.6.2 Antiviral Drug Resistance

Drug resistance against norovirus: The norovirus is a major global cause of acute gastroenteritis, and it is known to cause long-lasting illnesses in people whose immune systems are weak. The development of possible treatments for norovirus is raising concerns about antiviral drug resistance. Rupintrivir, a protease inhibitor, has shown efficacy against norovirus replication, but resistant mutants with amino acid substitutions in the viral protease NS6 can rapidly develop [121]. There is a severe lack of antivirals, and research into human norovirus replication and pathogenesis is impeded by the inadequate cell culture system. Effective antiviral therapies and preventative vaccinations are still unavailable, which is acknowledged as a public health concern given the substantial global health burden. Since HuNoVs are hard to culture in the lab, the similarly related group V murine norovirus (MNV) is employed as a model virus for analyzing them in organisms and in the lab. This slows down the process of making effective treatments [122]. Moreover, the only treatment available for immunocompromised patients with persistent infection is the compassionate off-label use of repurposed antivirals, which has limited efficacy. This underscores the pressing need for effective and targeted antivirals against HRV. Additionally, off-label antivirals such as favipiravir, nitazoxanide, ribavirin, and intravenous immunoglobulin have been used in an effort to treat the infection, but the results have been inconsistent and unpredictable, underscoring the urgent need for targeted antivirals [123]. The calcineurin inhibitors tacrolimus and cyclosporine moderately inhibited human norovirus (HuNV) replication, while mycophenolic acid (MPA), an inhibitor of isosine-5′-monophosphate dehydrogenase (IMPDH), potently and quickly inhibited norovirus replication without causing resistance. The study also examined the effects of various immunosuppressants and ribavirin on norovirus and investigated their mechanisms of action. When combined with immunosuppressants, the competitive IMPDH inhibitor ribavirin also effectively inhibited the norovirus and had an additive effect [124]. An additional investigation revealed that the protease inhibitor rupintrivir exhibits activity against various genogroups of noroviruses, implying that it may be a viable therapeutic intervention. However, nitazoxanide, which had been proposed as a possible antiviral treatment for norovirus, was not successful in treating chronic norovirus gastroenteritis in an immunocompromised patient, according to a case study [125]. A new group of norovirus inhibitors that target viral RNA replication has shown a high level of tolerance, with only minor mutations being seen after long-term exposure [126]. To stop the spread of the murine norovirus, the nucleoside polymerase inhibitor 2′-C-methylcytidine (2CMC) worked well. It did this without creating drug-resistant forms [127]. Current antiviral research faces challenges due to the lack of efficient human norovirus cultivation systems, relying on replicon-bearing cells and murine norovirus models [128]. These findings show that more research need to be done on the effectiveness and mechanism of action of possible antiviral treatments in order to solve the problem of norovirus infections becoming resistant to antiviral drugs. Persistent efforts are necessary to develop efficient antivirals with low resistance potential since norovirus infections represent major global health and economic burdens.

3.6.3 Public Health Challenges

Acute gastroenteritis is primarily caused by norovirus, which presents serious public health issues. Immunocompromised people, the elderly, and young children are among the vulnerable groups that are disproportionately affected by norovirus. There are currently no FDA-approved vaccinations, and supportive care is the only available form of treatment. It is challenging to contain outbreaks due to the virus' genetic diversity, quick evolution, and ease of transmission [66]. Public health challenges in norovirus: The enormity of the financial and social costs that the norovirus can impose on the National Health Service is intolerable [129]. With the virus' genetic diversity and the lack of known immune indicators of protection, vaccine development is hampered. Nonetheless, multivalent vaccination strategies appear promising for producing antibodies that are broadly neutralizing [114]. Ongoing research focuses on understanding viral evolution, improving detection methods, and developing effective vaccines to address this significant public health concern [31]. Raising public awareness of the norovirus and how it spreads, taking the appropriate safety measures, and filing more case reports are critical. Over the past ten years, the cruise industry has encountered numerous difficulties in controlling norovirus outbreaks, underscoring the necessity for risk and crisis management expertise in the hospitality sector [130].

3.7 Recent Advances in Norovirus Genetics

3.7.1 New Discoveries in Norovirus Genomic Research

Significant new understandings of viral diversity and evolution have been made possible by recent developments in norovirus genomic research. Similar evolutionary rates have been found across genomic regions and genotypes by extensive genome-wide analyses, with emerging viruses evolving from transiently circulating intermediate [58]. The utilization of complete genomic sequences in sequence analysis has become progressively beneficial for norovirus research [131]. A whole genome study found genomic data for 24 genotypes, the oldest from 12. We examined 281 ancient norovirus types from 10 countries and four continents during the 1970s. An in-depth study helps us comprehend norovirus evolution and spread. In genotypes with many variants, selection can modify the VP1 protein, new viruses can develop from intermediate strains that are momentarily in circulation, and genomic area recombination is limited [132]. New residues on the viral capsid protein (VP1) that are important for the emergence and antigenic topology of GII have been discovered by researchers. Four variations may help identify new variants with pandemic potential early on [133]. It has been found that noroviruses can infect both immune system cells and enterocytes, with bile acids making the infection more likely. New *in vitro* culture methods for human noroviruses have been created [134]. Furthermore, it has been established that the bacteria in the gut microbiome may influence the risk of contracting norovirus infection. These discoveries advance our knowledge of the biology and evolution of noroviruses, which may aid in the development of vaccines and treatment plans [134].

3.7.2 Emerging Genogroups and Strains

Latest progress in studying noroviruses has helped us learn more about this group of viruses. The method for grouping things has been updated to include 10 GI–GX genogroups, with 49 genotypes. More possible groupings are still being considered while more sequences are generated [30]. Point mutations in ORF1 and ORF2 genes and chimeric viral recombination generate norovirus diversification. GII.4 noroviruses, which are the most common type, form when amino acid changes build up in VP1. Non-GII.4 types, on the other hand, depend more on recombination [36]. Recent discoveries have enhanced our understanding of virus–host interactions, including cell tropism, attachment factors, and host immunity. These advances, along with improved *in vitro* cultivation techniques using human intestinal enteroids, are guiding vaccine development efforts [135].

3.7.3 Future Directions in Norovirus Genetics

New developments in norovirus research have centered on creating instruments to examine this difficult infection. Reverse genetics systems have become effective tools for producing contagious viral particles and learning about viral mechanisms [8, 136]. These systems offer promising avenues for studying norovirus pathogenesis and developing therapeutics and vaccines [136]. More infections in people with weak immune systems and the globalization of the food industry are becoming bigger problems, but the fast genetic and antigenic evolution of noroviruses makes it harder to make medicines [31]. Molecular processes, replication, pathophysiology, and host immunity of noroviruses have been studied despite these challenges. Non-replicating virus-like particle vaccinations show potential in clinical trials [137]. Ongoing research aims to improve detection methods, trace viral evolution, and explore infection mechanisms to combat this significant cause of gastroenteritis worldwide [8].

3.8 Conclusion

Norovirus continues to represent a significant danger to worldwide public health due to its strong transmissibility, genetic variation, and swift evolution. Understanding the genetic structure of norovirus is crucial for developing effective strategies to reduce its widespread effects. The proteins, both structural and non-structural, that are essential for the norovirus' pathogenicity and life cycle are encoded within its genome, which consists of several (ORFs). Thanks to the genetic variation resulting from mutation and recombination, multiple norovirus genogroups and genotypes have emerged; genogroups I and II (GI and GII) are the most common in human infections. Recent advancements in molecular techniques, particularly in technology sequencing like WGS and NGS, have greatly enhanced our ability to study norovirus genomes. These instruments have illuminated the virus' evolutionary dynamics and enhanced molecular epidemiology and genotyping through facilitating thorough phylogenetic analyses. The genetic diversity of norovirus makes vaccine development and combating drug resistance more challenging. It also complicates the management of outbreaks. Ongoing research is uncovering new details about the genomics of noroviruses, yet there remains optimism for future findings. Continuous oversight and adjustment of public health approaches are essential due to the discovery of new strains and genogroups. As insights into norovirus genetics grow, advancements in research will be essential in steering the creation of effective vaccines, antiviral treatments, and public health measures.

Abbreviations

NoV	norovirus
GI	genogroup I
GII	genogroup II
RdRp	RNA-dependent RNA polymerase
ORF	open reading frame
VP1	viral protein 1
VP2	viral protein 2
NGS	next-generation sequencing
WGS	whole-genome sequencing
NS	non-structural protein
HBGA	Histo-blood group antigen

AI disclosure statement: The author acknowledges the use of AI-based tools for minor language editing or grammatical correction. However, all intellectual contributions, analyses, and conclusions presented in this chapter are entirely the author's own.

References

1. Dolin, R., Blacklow, N.R., DuPont, H., Buscho, R.F., Wyatt, R.G., Kasel, J.A., *et al.*, Biological properties of Norwalk agent of acute infectious nonbacterial gastroenteritis. *Proc. Soc. Exp. Biol. Med.*, 140, 2, 578–83, 1972.
2. Vinjé, J., Advances in laboratory methods for detection and typing of norovirus. *J. Clin. Microbiol.*, 53, 2, 373–81, 2015.
3. Chan, M.C., Kwan, H.S., Chan, P.K., *Structure and genotypes of Noroviruses*, pp. 51–63, Elsevier, The Norovirus, 2017.
4. Winder, N., Gohar, S., Muthana, M., Norovirus: an overview of virology and preventative measures. *Viruses*, 14, 12, 2811, 2022.
5. Kwan, H.S., Chan, P.K., Chan, M.C., Overview of norovirus as a foodborne pathogen, in: *The Norovirus*, pp. 21–30, Elsevier, 2017, https://doi.org/10.1016/B978-0-12-804177-2.00002-6.
6. About the current norovirus season.
7. Morillo, S.G., Tlmenetsky MdCST. Norovirus: an overview. *Rev. Assoc. Med. Bras. (English Edition)*, 57, 4, 453–8, 2011.
8. Lin, S.-C., Bai, G.-H., Lin, P.-C., Chen, C.-Y., Hsu, Y.-H., Lee, Y.-C., *et al.*, Molecular and Genetics-Based Systems for Tracing the Evolution and Exploring the Mechanisms of Human Norovirus Infections. *Int. J. Mol. Sci.*, 24, 10, 9093, 2023.
9. Koopmans, M., Progress in understanding norovirus epidemiology. *Curr. Opin. Infect. Dis.*, 21, 5, 544–52, 2008.
10. Bull, R.A., Eden, J.-S., Luciani, F., McElroy, K., Rawlinson, W.D., White, P.A., Contribution of intra-and interhost dynamics to norovirus evolution. *J. Virol.*, 86, 6, 3219–29, 2012.
11. Mahar, J.E., Bok, K., Green, K.Y., Kirkwood, C.D., The importance of intergenic recombination in norovirus GII. 3 evolution. *J. Virol.*, 87, 7, 3687–98, 2013.
12. Jiang, X., Wang, M., Wang, K., Estes, M.K., Sequence and genomic organization of Norwalk virus. *Virology*, 195, 1, 51–61, 1993.
13. Kapikian, A., Norwalk group of viruses, in: *Fields virology*, pp. 783–810, 1996.
14. Green, K., Ando, T., Balayan, M., Berke, T., Clarke, I., Estes, M., *et al.*, Taxonomy of the caliciviruses. *J. Infect. Dis.*, 181, Supplement_2, S322–S30, 2000.
15. Jiang, X., Graham, D.Y., Wang, K., Estes, M.K., Norwalk virus genome cloning and characterization. *Science*, 250, 4987, 1580–3, 1990.

16. Meyers, G., Wirblich, C., Thiel, H.-J., Genomic and subgenomic RNAs of rabbit hemorrhagic disease virus are both protein-linked and packaged into particles. *Virology*, 184, 2, 677–86, 1991.
17. Bertolotti-Ciarlet, A., Crawford, S.E., Hutson, A.M., Estes, M.K., The 3′ end of Norwalk virus mRNA contains determinants that regulate the expression and stability of the viral capsid protein VP1: a novel function for the VP2 protein. *J. Virol.*, 77, 21, 11603–15, 2003.
18. Thorne, L.G. and Goodfellow, I.G., Norovirus gene expression and replication. *J. Gen. Virol.*, 95, 2, 278–91, 2014.
19. McFadden, N., Bailey, D., Carrara, G., Benson, A., Chaudhry, Y., Shortland, A., *et al.*, Norovirus regulation of the innate immune response and apoptosis occurs via the product of the alternative open reading frame 4. *PLoS Pathog.*, 7, 12, e1002413, 2011.
20. Jiang, X., Wang, M., Graham, D.Y., Estes, M.K., Expression, self-assembly, and antigenicity of the Norwalk virus capsid protein. *J. Virol.*, 66, 11, 6527–32, 1992.
21. White, L.J., Hardy, M.E., Estes, M.K., Biochemical characterization of a smaller form of recombinant Norwalk virus capsids assembled in insect cells. *J. Virol.*, 71, 10, 8066–72, 1997.
22. Estes, M., Norwalk virus and other enteric calicivirus, in: *Infections of the gastrointestinal tract*, pp. 1009–34, 1995.
23. White, P., Evolution of norovirus. *Clin. Microbiol. Infect.*, 20, 8, 741–5, 2014.
24. Hardy, M.E., Tanaka, T.N., Kitamoto, N., White, L.J., Ball, J.M., Jiang, X., *et al.*, Antigenic mapping of the recombinant Norwalk virus capsid protein using monoclonal antibodies. *Virology*, 217, 1, 252–61, 1996.
25. Lochridge, V.P., Jutila, K.L., Graff, J.W., Hardy, M.E., Epitopes in the P2 domain of norovirus VP1 recognized by monoclonal antibodies that block cell interactions. *J. Gen. Virol.*, 86, 10, 2799–806, 2005.
26. Tan, M., Huang, P., Meller, J., Zhong, W., Farkas, T., Jiang, X., Mutations within the P2 domain of norovirus capsid affect binding to human histo-blood group antigens: evidence for a binding pocket. *J. Virol.*, 78, 6, 3201–, 2004.
27. Hale, A.D., Tanaka, T.N., Kitamoto, N., Ciarlet, M., Jiang, X., Takeda, N., *et al.*, Identification of an epitope common to genogroup 1 "norwalk-like viruses". *J. Clin. Microbiol.*, 38, 4, 1656–60, 2000.
28. Glass, P.J., White, L.J., Ball, J.M., Leparc-Goffart, I., Hardy, M.E., Estes, M.K., Norwalk virus open reading frame 3 encodes a minor structural protein. *J. Virol.*, 74, 14, 6581–91, 2000.
29. Asanaka, M., Atmar, R.L., Ruvolo, V., Crawford, S.E., Neill, F.H., Estes, M.K., Replication and packaging of Norwalk virus RNA in cultured mammalian cells. *Proc. Natl. Acad. Sci.*, 102, 29, 10327–32, 2005.
30. Chhabra, P., de Graaf, M., Parra, G.I., Chan, M.C.-W., Green, K., Martella, V., *et al.*, Updated classification of norovirus genogroups and genotypes. *J. Gen. Virol.*, 100, 10, 1393–406, 2019. https://doi.org/10.1007/s00705-013-1708-5.

31. De Graaf, M., van Beek, J., Koopmans, M.P., Human norovirus transmission and evolution in a changing world. *Nat. Rev. Microbiol.*, 14, 7, 421–33, 2016.
32. Green, K., Caliciviridae: the noroviruses, in: *Fields virology*, 6th ed, pp. 949–79, Wolters Kluwer Health/Lippincott Williams and Wilkins, Philadelphia, PA, 2013.
33. Ford-Siltz, L.A., Mullis, L., Sanad, Y.M., Tohma, K., Lepore, C.J., Azevedo, M., *et al.*, Genomics analyses of GIV and GVI noroviruses reveal the distinct clustering of human and animal viruses. *Viruses*, 11, 3, 204, 2019.
34. Kroneman, A., Vega, E., Vennema, H., Vinjé, J., White, P.A., Hansman, G., *et al.*, Proposal for a unified norovirus nomenclature and genotyping. *Arch. Virol*, 158, 2059–68, 2013.
35. Bull, R.A., Hansman, G.S., Clancy, L.E., Tanaka, M.M., Rawlinson, W.D., White, P.A., Norovirus recombination in ORF1/ORF2 overlap. *Emerg. Infect. Dis.*, 11, 7, 1079, 2005.
36. Parra, G., Emergence of norovirus strains: a tale of two genes. *Virus Evol.*, 5, vez048, 2019.
37. Lindesmith, L.C., Donaldson, E.F., LoBue, A.D., Cannon, J.L., Zheng, D.-P., Vinje, J., *et al.*, Mechanisms of GII. 4 norovirus persistence in human populations. *PLoS Med.*, 5, 2, e31, 2008.
38. Motomura, K., Yokoyama, M., Ode, H., Nakamura, H., Mori, H., Kanda, T., *et al.*, Divergent evolution of norovirus GII/4 by genome recombination from May 2006 to February 2009 in Japan. *J. Virol.*, 84, 16, 8085–97, 2010.
39. Sanjuán, R., Nebot, M.R., Chirico, N., Mansky, L.M., Belshaw, R., Viral mutation rates. *J. Virol.*, 84, 19, 9733–48, 2010.
40. Shrestha, S., Malla, B., Haramoto, E., Estimation of Norovirus infections in Japan: An application of wastewater-based epidemiology for enteric disease assessment. *Sci. Total Environ.*, 912, 169334, 2024.
41. Hardy, M., Kramer, S., Treanor, J., Estes, M., Human calicivirus genogroup II capsid sequence diversity revealed by analyses of the prototype Snow Mountain agent. *Arch. Virol*, 142, 1469–79, 1997.
42. Jiang, X., Espul, C., Zhong, W., Cuello, H., Matson, D., Characterization of a novel human calicivirus that may be a naturally occurring recombinant. *Arch. Virol*, 144, 2377–87, 1999.
43. Hasing, M.E., Hazes, B., Lee, B.E., Preiksaitis, J.K., Pang, X.L., Detection and analysis of recombination in GII. 4 norovirus strains causing gastroenteritis outbreaks in Alberta. *Infect. Genet. Evol.*, 27, 181–92, 2014.
44. Vakulenko, Y., Orlov, A., Lukashev, A., Patterns and temporal dynamics of natural recombination in noroviruses. *Viruses*, 15, 2, 372, 2023.
45. Rohayem, J., Münch, J., Rethwilm, A., Evidence of recombination in the norovirus capsid gene. *J. Virol.*, 79, 8, 4977–90, 2005.
46. Siqueira, J.A.M., da Silva Bandeira, R., Justino, M.C.A., da Costa Linhares, A., Gabbay, Y.B., Characterization of novel intragenotype recombination events among norovirus pandemic GII. 4 variants. *Infect. Genet. Evol.*, 44, 361–6, 2016.

47. Vinjé, J., Estes, M.K., Esteves, P., Green, K.Y., Katayama, K., Knowles, N.J., *et al.*, ICTV virus taxonomy profile: Caliciviridae. *J. Gen. Virol.*, 100, 11, 1469–70, 2019.
48. Boon, D., Mahar, J.E., Abente, E.J., Kirkwood, C.D., Purcell, R.H., Kapikian, A.Z., *et al.*, Comparative evolution of GII. 3 and GII. 4 norovirus over a 31-year period. *J. Virol.*, 85, 17, 8656–66, 2011.
49. Bok, K., Abente, E.J., Realpe-Quintero, M., Mitra, T., Sosnovtsev, S.V., Kapikian, A.Z., *et al.*, Evolutionary dynamics of GII. 4 noroviruses over a 34-year period. *J. Virol.*, 83, 22, 11890–901, 2009.
50. Tohma, K., Lepore, C.J., Ford-Siltz, L.A., Parra, G.I., Evolutionary dynamics of non-GII genotype 4 (GII. 4) noroviruses reveal limited and independent diversification of variants. *J. Gen. Virol.*, 99, 8, 1027–35, 2018.
51. Mayo, M., A summary of taxonomic changes recently approved by ICTV. *Arch. Virol*, 147, 1655–6, 2002.
52. Bridger, J.C., Hall, G.A., Brown, J.F., Characterization of a calici-like virus (Newbury agent) found in association with astrovirus in bovine diarrhea. *Infect. Immun.*, 43, 1, 133–8, 1984.
53. Oliver, S., Asobayire, E., Dastjerdi, A., Bridger, J., Genomic characterization of the unclassified bovine enteric virus Newbury agent-1 (Newbury1) endorses a new genus in the family Caliciviridae. *Virology*, 350, 1, 240–50, 2006.
54. Smiley, J., Chang, K., Hayes, J., Vinje, J., Saif, L.J., Characterization of an enteropathogenic bovine calicivirus representing a potentially new calicivirus genus. *J. Virol.*, 76, 20, 10089–98, 2002.
55. Green, K., Caliciviridae: the noroviruses: specific virus families, in: *Fields virology*, 6th ed, pp. 582–608, Lippincott Williams & Wilkins, Philadelphia, PA, 2013.
56. Rani, M., Rajyalakshmi, S., Pakalapaty, S., Kammilli, N., Norovirus structure and classification. *Norovirus*, 31, 2021. https://www.sciencegate.app/document/10.5772/intechopen.98216. Norovirus Structure and Classification | ScienceGate
57. Castells, M., Cristina, J., Colina, R., Evolutionary history and spatiotemporal dynamic of GIII norovirus: From emergence to classification in four genotypes. *Transbound. Emerg. Dis.*, 69, 4, 1872–9, 2022.
58. Tohma, K., Lepore, C.J., Martinez, M., Degiuseppe, J.I., Khamrin, P., Saito, M., *et al.*, Genome-wide analyses of human noroviruses provide insights on evolutionary dynamics and evidence of coexisting viral populations evolving under recombination constraints. *PLoS Pathog.*, 17, 7, e1009744, 2021.
59. Krakowka, S. and Ellis, J.A., Evaluation of the effects of porcine genogroup 1 torque teno virus in gnotobiotic swine. *Am. J. Vet. Res.*, 69, 12, 1623–9, 2008.
60. Mauroy, A., Scipioni, A., Mathijs, E., Ziant, D., Daube, G., Thiry, E., Genetic and evolutionary perspectives on genogroup III, genotype 2 bovine noroviruses. *Arch. Virol*, 159, 39–49, 2014.

61. Lopman, B., Atmar, R., Baric, R., Estes, M., Green, K., Glass, R., *et al.*, Global burden of norovirus and prospects for vaccine development, Centers for Disease Control and Prevention, pp. 1–46, 2015.
62. Tran, T.H., Trainor, E., Nakagomi, T., Cunliffe, N.A., Nakagomi, O., Molecular epidemiology of noroviruses associated with acute sporadic gastroenteritis in children: global distribution of genogroups, genotypes and GII. 4 variants. *J. Clin. Virol.*, 56, 3, 269–77, 2013.
63. van Beek, J., de Graaf, M., Al-Hello, H., Allen, D.J., Ambert-Balay, K., Botteldoorn, N., *et al.*, Molecular surveillance of norovirus, 2005–16: an epidemiological analysis of data collected from the NoroNet network. *Lancet Infect. Dis.*, 18, 5, 545–53, 2018.
64. Green, K.Y., Belliot, G., Taylor, J.L., Valdesuso, J., Lew, J.F., Kapikian, A.Z., *et al.*, A predominant role for Norwalk-like viruses as agents of epidemic gastroenteritis in Maryland nursing homes for the elderly. *J. Infect. Dis.*, 185, 2, 133–46, 2002.
65. Vinje, J. and Koopmans, M.P., Molecular detection and epidemiology of small round-structured viruses in outbreaks of gastroenteritis in the Netherlands. *J. Infect. Dis.*, 174, 3, 610–5, 1996.
66. Carlson, K.B., Dilley, A., O'Grady, T., Johnson, J.A., Lopman, B., Viscidi, E., A narrative review of norovirus epidemiology, biology, and challenges to vaccine development. *npj Vaccines*, 9, 1, 94, 2024.
67. Brown, J.R., Roy, S., Ruis, C., Yara Romero, E., Shah, D., Williams, R., *et al.*, Norovirus whole-genome sequencing by SureSelect target enrichment: a robust and sensitive method. *J. Clin. Microbiol.*, 54, 10, 2530–7, 2016.
68. Kundu, S., Lockwood, J., Depledge, D.P., Chaudhry, Y., Aston, A., Rao, K., *et al.*, Next-generation whole genome sequencing identifies the direction of norovirus transmission in linked patients. *Clin. Infect. Dis.*, 57, 3, 407–14, 2013.
69. Chhabra, P., Walimbe, A.M., Chitambar, S.D., Complete genome characterization of Genogroup II norovirus strains from India: Evidence of recombination in ORF2/3 overlap. *Infect. Genet. Evol.*, 10, 7, 1101–9, 2010.
70. Won, Y.-J., Park, J.-W., S-h, H., Cho, H.-G., Kang, L.-H., Lee, S.-G., *et al.*, Full-genomic analysis of a human norovirus recombinant GII. 12/13 novel strain isolated from South Korea. *PLoS One*, 8, 12, e85063, 2013.
71. Nakamura, S., Yang, C.-S., Sakon, N., Ueda, M., Tougan, T., Yamashita, A., *et al.*, Direct metagenomic detection of viral pathogens in nasal and fecal specimens using an unbiased high-throughput sequencing approach. *PLoS One*, 4, 1, e4219, 2009.
72. Batty, E.M., Wong, T.N., Trebes, A., Argoud, K., Attar, M., Buck, D., *et al.*, A modified RNA-Seq approach for whole genome sequencing of RNA viruses from faecal and blood samples. *PLoS One*, 8, 6, e66129, 2013.
73. Wong, T.N., Dearlove, B.L., Hedge, J., Giess, A.P., Piazza, P., Trebes, A., *et al.*, Whole genome sequencing and *de novo* assembly identifies Sydney-like

variant noroviruses and recombinants during the winter 2012/2013 outbreak in England. *Virol. J.*, 10, 1–10, 2013.

74. Brown, A.C., Bryant, J.M., Einer-Jensen, K., Holdstock, J., Houniet, D.T., Chan, J.Z., *et al.*, Rapid whole-genome sequencing of Mycobacterium tuberculosis isolates directly from clinical samples. *J. Clin. Microbiol.*, 53, 7, 2230–7, 2015.
75. Christiansen, M.T., Brown, A.C., Kundu, S., Tutill, H.J., Williams, R., Brown, J.R., *et al.*, Whole-genome enrichment and sequencing of Chlamydia trachomatis directly from clinical samples. *BMC Infect. Dis.*, 14, 1–11, 2014.
76. Depledge, D.P., Palser, A.L., Watson, S.J., Lai, I.Y.-C., Gray, E.R., Grant, P., *et al.*, Specific capture and whole-genome sequencing of viruses from clinical samples. *PLoS One*, 6, 11, e27805, 2011.
77. Fitzpatrick, A.H., Rupnik, A., O'Shea, H., Crispie, F., Cotter, P.D., Keaveney, S., Development of an amplicon-based high throughput sequencing method for genotypic characterisation of norovirus in oysters. *bioRxiv*, 2022.12. 23.521849, 2022. doi: https://doi.org/10.1128/aem.01522-22.
78. Yang, Z., Mammel, M., Wales, S.Q., Near-Complete Genome Sequence of a Human Norovirus GII. 1 [Pg] Strain Associated with Acute Gastroenteritis, Determined Using Long-Read Sequencing. *Microbiol. Resour. Announce.*, 10, 28, 2021. https://doi.org/10.1128/mra.00401-21.
79. Cotten, M., Petrova, V., Phan, M.V., Rabaa, M.A., Watson, S.J., Ong, S.H., *et al.*, Deep sequencing of norovirus genomes defines evolutionary patterns in an urban tropical setting. *J. Virol.*, 88, 19, 11056–69, 2014.
80. CDC, Norovirus Facts and Stats, 2024, [Available from: https://www.cdc.gov/norovirus/data-research/index.html.
81. CDC, NoroSTAT Data, 2024, [Available from: https://www.cdc.gov/norovirus/php/reporting/norostat-data.html.
82. GOV.UK, National norovirus and rotavirus surveillance reports: 2023 to 2024 season, 2024, [Available from: https://www.gov.uk/government/statistics/national-norovirus-and-rotavirus-surveillance-reports-2023-to-2024-season.
83. Government A, Department of Health and Aged Care, 2024, [Available from: https://www.health.gov.au/.
84. CDC, Norovirus Data, 2024, [Available from: https://www.cdc.gov/norovirus/data-research/index.html.
85. Mabasa, V.V., van Zyl, W.B., Ismail, A., Allam, M., Taylor, M.B., Mans, J., Multiple novel human norovirus recombinants identified in wastewater in Pretoria, South Africa by next-generation sequencing. *Viruses*, 14, 12, 2732, 2022.
86. Cantalupo, P.G. and Pipas, J.M., Detecting viral sequences in NGS data. *Curr. Opin. Virol.*, 39, 41–8, 2019.
87. Fonager, J., Stegger, M., Rasmussen, L., Poulsen, M., Rønn, J., Andersen, P., A universal primer-independent next-generation sequencing approach for

investigations of norovirus outbreaks and novel variants. *Sci. Rep.*, 7, 1, 813, 2017.
88. Hasing, M.E., Hazes, B., Lee, B.E., Preiksaitis, J.K., Pang, X.L., A next generation sequencing-based method to study the intra-host genetic diversity of norovirus in patients with acute and chronic infection. *BMC Genomics*, 17, 1–11, 2016.
89. Imamura, S., Haruna, M., Goshima, T., Kanezashi, H., Okada, T., Akimoto, K., Application of next-generation sequencing to investigation of norovirus diversity in shellfish collected from two coastal sites in Japan from 2013 to 2014. *Jpn. J. Vet. Res.*, 64, 2, 113–22, 2016.
90. Cotten, M. and Koopmans, M., *Next-generation sequencing and norovirus*, pp. 719–22, Taylor & Francis, 2016. doi: 10.2217/fvl-2016-0099.
91. Silva, A.J., Yang, Z., Wolfe, J., Hirneisen, K.A., Ruelle, S.B., Torres, A., *et al.*, Application of whole-genome sequencing for norovirus outbreak tracking and surveillance efforts in Orange County, CA. *Food Microbiol.*, 98, 103796, 2021.
92. Bavelaar, H.H., Rahamat-Langendoen, J., Niesters, H.G., Zoll, J., Melchers, W.J., Whole genome sequencing of fecal samples as a tool for the diagnosis and genetic characterization of norovirus. *J. Clin. Virol.*, 72, 122–5, 2015.
93. Kim, Y.E., Song, M., Lee, J., Seung, H.J., Kwon, E.-Y., Yu, J., *et al.*, Phylogenetic characterization of norovirus strains detected from sporadic gastroenteritis in Seoul during 2014–2016. *Gut Pathogens*, 10, 1–15, 2018.
94. Mischenko, VA., Mischenko, A.A., Nikeshina, T.B., Petrova, O.N., Brovko, Y.V., Kushlubaeva, A.I., The problem of norovirus infection in animals (literature review). *Vet. Sci. Today*, 118, 2024. https://doi.org/10.30896/0042-4846.2020.23.11.03-06.
95. Nishida, T., Kimura, H., Saitoh, M., Shinohara, M., Kato, M., Fukuda, S., *et al.*, Detection, quantitation, and phylogenetic analysis of noroviruses in Japanese oysters. *Appl. Environ. Microbiol.*, 69, 10, 5782–6, 2003.
96. Bykov, R., Itani, T., Starikova, P., Skryabina, S., Kilyachina, A., Koltunov, S., *et al.*, Genetic Diversity and Phylogenetic Relationship of Human Norovirus Sequences Derived from Municipalities within the Sverdlovsk Region of Russia. *Viruses*, 16, 7, 1001, 2024.
97. Frydman, C., Miño, S., Iglesias, N.G., Carballeda, J.M., Simari, M., Pisano, M.B., *et al.*, Wastewater surveillance of enteric viruses in eastern Argentina: High rates of detection and first report of NoV GI. 5 and GII. 20. *Environ. Adv.*, 15, 100501, 2024.
98. Vennema, H., De Bruin, E., Koopmans, M., Rational optimization of generic primers used for Norwalk-like virus detection by reverse transcriptase polymerase chain reaction. *J. Clin. Virol.*, 25, 2, 233–5, 2002.
99. Anderson, A.D., Garrett, V.D., Sobel, J., Monroe, S.S., Fankhauser, R.L., Schwab, K.J., *et al.*, Multistate outbreak of Norwalk-like virus gastroenteritis associated with a common caterer. *Am. J. Epidemiol.*, 154, 11, 1013–9, 2001.

100. Jin, M., Wu, S., Kong, X., Xie, H., Fu, J., He, Y., *et al.*, Norovirus outbreak surveillance, China, 2016–2018. *Emerg. Infect. Dis.*, 26, 3, 437, 2020.
101. Mattison, K., Grudeski, E., Auk, B., Charest, H., Drews, S.J., Fritzinger, A., *et al.*, Multicenter comparison of two norovirus ORF2-based genotyping protocols. *J. Clin. Microbiol.*, 47, 12, 3927–32, 2009.
102. Pazdiora, P., Vašíčková, P., Krzyžánková, M., Genotyping of noroviruses from patients of the Pilsen University Hospital in the Czech Republic, 2017-2020. *Epidemiol. Mikrobiol. Imunol.: Casopis Spolecnosti pro Epidemiologii Mikrobiologii Ceske Lekarske Spolecnosti JE Purkyne*, 70, 4, 233–40, 2021.
103. Chhabra, P., Browne, H., Huynh, T., Diez-Valcarce, M., Barclay, L., Kosek, M.N., *et al.*, Single-step RT-PCR assay for dual genotyping of GI and GII norovirus strains. *J. Clin. Virol.*, 134, 104689, 2021.
104. Lee, S.-G., Cho, H.-G., Paik, S.-Y., Molecular epidemiology of norovirus in South Korea. *BMB Rep.*, 48, 2, 61, 2015.
105. Ike, A.C., Brockmann, S.O., Hartelt, K., Marschang, R.E., Contzen, M., Oehme, R.M., Molecular epidemiology of norovirus in outbreaks of gastroenteritis in southwest Germany from 2001 to 2004. *J. Clin. Microbiol.*, 44, 4, 1262–7, 2006.
106. Fankhauser, R.L., Noel, J.S., Monroe, S.S., Ando, T., Glass, R.I., Molecular epidemiology of "Norwalk-like viruses" in outbreaks of gastroenteritis in the United States. *J. Infect. Dis.*, 178, 6, 1571–8, 1998.
107. Lemaitre, E., Allée, C., Vabret, A., Eterradossi, N., Brown, P., Single reaction, real time RT-PCR detection of all known avian and human metapneumoviruses. *J. Virol. Methods*, 251, 61–8, 2018.
108. Li, G., Li, W., Guo, F., Xu, S., Zhao, N., Chen, S., *et al.*, A novel real-time PCR assay for determination of viral loads in person infected with hepatitis B virus. *J. Virol. Methods*, 165, 1, 9–14, 2010.
109. Guo, Z.-N., Liu, J., Xing, Y., Yan, S., Lv, C., Jin, H., *et al.*, Dynamic cerebral autoregulation is heterogeneous in different subtypes of acute ischemic stroke. *PLoS One*, 9, 3, e93213, 2014.
110. Kellner, M.J., Koob, J.G., Gootenberg, J.S., Abudayyeh, O.O., Zhang, F., SHERLOCK: nucleic acid detection with CRISPR nucleases. *Nat. Protoc.*, 14, 10, 2986–3012, 2019.
111. Islam, A., Schulz, S., Afroz, S., Babiuk, L.A., Littel-van den Hurk, S.V.D., Interaction of VP8 with mRNAs of bovine herpesvirus-1. *Virus Res.*, 197210, 116356, 2015.
112. Sarma, M.K., Quadir, M.G.A., Bhaduri, R., Kaushik, S., Goswami, P., Composite polymer coated magnetic nanoparticles based anode enhances dye degradation and power production in microbial fuel cells. *Biosens. Bioelectron.*, 119, 94–102, 2018.
113. Ettayebi, K., Crawford, S.E., Murakami, K., Broughman, J.R., Karandikar, U., Tenge, V.R., *et al.*, Replication of human noroviruses in stem cell–derived human enteroids. *Science*, 353, 6306, 1387–93, 2016.

114. Cortes-Penfield, N.W., Ramani, S., Estes, M.K., Atmar, R.L., Prospects and challenges in the development of a norovirus vaccine. *Clin. Ther.*, 39, 8, 1537–49, 2017.
115. Tan, M., Norovirus vaccines: current clinical development and challenges. *Pathogens*, 10, 12, 1641, 2021.
116. Graziano, V.R., Wei, J., Wilen, C.B., Norovirus attachment and entry. *Viruses*, 11, 6, 495, 2019.
117. Esposito, S. and Principi, N., Norovirus vaccine: priorities for future research and development. *Front. Immunol.*, 11, 1383, 2020.
118. Blazevic, V., Lappalainen, S., Nurminen, K., Huhti, L., Vesikari, T., Norovirus VLPs and rotavirus VP6 protein as combined vaccine for childhood gastroenteritis. *Vaccine*, 29, 45, 8126–33, 2011.
119. Omatola, C.A., Mshelbwala, P.P., Okolo, M.-L.O., Onoja, A.B., Abraham, J.O., Adaji, D.M., *et al.*, Noroviruses: Evolutionary Dynamics, Epidemiology, Pathogenesis, and Vaccine Advances—A Comprehensive Review. *Vaccines*, 12, 6, 590, 2024.
120. Riddle, M.S. and Walker, R.I., Status of vaccine research and development for norovirus. *Vaccine*, 34, 26, 2895–9, 2016.
121. Kitano, M., Hosmillo, M., Emmott, E., Lu, J., Goodfellow, I., Selection and characterization of rupintrivir-resistant Norwalk virus replicon cells *in vitro*. *Antimicrob. Agents Chemother.*, 62, 5, 2018, https://doi.org/10.1128/aac.00201-18.
122. Aktepe TE, D.J., Hyde, J.L., Fritzlar, S., Mead, E.M., Carrera Montoya, J., Hachani, A., Pearson, J.S., White, P.A., Mackenzie, J.M., Norovirus-mediated translation repression promotes macrophage cell death. *PLoS Pathog.*, 20, 9, e1012480, 2024 Sep 3.
123. Santos-Ferreira, N., Van Dycke, J., Chiu, W., Neyts, J., Matthijnssens, J., Rocha-Pereira, J., Molnupiravir inhibits human norovirus and rotavirus replication in 3D human intestinal enteroids. *Antiviral Res.*, 223, 105839, 2024.
124. Carretero-Ledesma, M., Aguilar-Guisado, M., Berastegui-Cabrera, J., Balsera-Manzanero, M., Pachón, J., Cordero, E., *et al.*, Antiviral activity of immunosuppressors alone and in combination against human adenovirus and cytomegalovirus. *Int. J. Antimicrob. Agents*, 63, 5, 107116, 2024.
125. Hayashi, T., Kobayashi, S., Hirano, J., Murakami, K., Human norovirus cultivation systems and their use in antiviral research. *J. Virol.*, 98, 4, e01663-23, 2024.
126. Liu, Y., Li, Q., Shao, H., Mao, Y., Liu, L., Yi, D., *et al.*, CX-6258 hydrochloride hydrate: A potential non-nucleoside inhibitor targeting the RNA-dependent RNA polymerase of norovirus. *Virology*, 595, 110088, 2024.
127. Budicini, M.R., Factors That Influence Murine Norovirus Stability and Tropism, 2024.
128. Rocha-Pereira, J., Neyts, J., Jochmans, D., Norovirus: Targets and tools in antiviral drug discovery. *Biochem. Pharmacol.*, 91, 1–11, 2014.

129. Fua'di, M.T., Er, B., Lee, S., Chan, P.P., Khoo, J., Tan, D., *et al.*, Characteristics of Gastroenteritis Outbreaks Investigated in Singapore: 2018–2021. *Int. J. Environ. Res. Public Health*, 21, 1, 64, 2024.
130. Mouchtouri, V.A., Simou, E., Soteriades, S., Rousou, X., Kontouli, K.M., Kafetsouli, D., *et al.*, Systematic literature review and meta-analysis on preventing and controlling norovirus outbreaks on cruise ships, 1990 to 2020: calling for behaviour change strategies of travellers. *Eurosurveillance*, 29, 10, 2300345, 2024.
131. Sun, Y., Yuan, Y., Mao, H., Su, L., Ge, Q., Gao, J., *et al.*, Molecular Epidemiology of Human Norovirus Variants from Outbreaks in Zhejiang Province, China, during 2021. *Adv. Virol.*, 2024, 1, 7972494, 2024.
132. Hong, X., Xue, L., Cao, Y., Xu, R., Wang, J., Gao, J., *et al.*, The variation of antigenic and histo-blood group binding sites synergistically drive the evolution among chronologically emerging GII. 4 noroviruses. *Heliyon*, 10, 5, 2024. https://doi.org/10.1016/j.heliyon.2024.e26567.
133. Tohma, K., Lepore, C.J., Gao, Y., Ford-Siltz, L.A., Parra, G.I., Population Genomics of GII.4 Noroviruses Reveal Complex Diversification and New Antigenic Sites Involved in the Emergence of Pandemic Strains. *mBio*, 10, 2019. https://doi.org/10.1128/mbio.02202-19.
134. Bartnicki, E., Cunha, J.B., Kolawole, A.O., Wobus, C.E., Recent advances in understanding noroviruses. *F1000Research*, 6, 2017. https://doi.org/10.12688/f1000research.10081.1.
135. Ludwig-Begall, L.F., Mauroy, A., Thiry, E., Noroviruses—The State of the Art, Nearly Fifty Years after Their Initial Discovery. *Viruses*, 13, 2021. https://doi.org/10.3390/v13081541.
136. Zhang, Q., S-r, Z., Zhang, X.-B., Su, L., Ni, J., Zhang, Y., *et al.*, Recent insights into reverse genetics of norovirus. *Virus Res.*, 325, 2023. https://doi.org/10.1016/j.ijbiomac.2025.145355.
137. Karst, S.M., Wobus, C.E., Goodfellow, I.G., Green, K.Y., Virgin, H.W., Advances in norovirus biology. *Cell Host Microbe*, 15, 6, 668–80, 2014.
138. Yang, Z., Mammel, M., Whitehouse, C. A., Ngo, D., Kulka, M., Inter- and Intra-Host Nucleotide Variations in Hepatitis A Virus in Culture and Clinical Samples Detected by Next-Generation Sequencing. *Viruses*, 10, 11, 619, 2018, https://doi.org/10.3390/v10110619.

4

Environmental Factors in Norovirus Outbreaks: Temperature and Rainfall

Neeraj Kumar[1], Akanksha[1], Shriyansh Srivastava[2] and Mohd Mujeeb[1]*

[1]*Department of Pharmacognosy & Phytochemistry, SPER, Jamia Hamdard, New Delhi, India*

[2]*Department of Pharmacology, Delhi Pharmaceutical Sciences and Research University, New Delhi, India*

Abstract

Norovirus outbreaks are significantly influenced by environmental factors, particularly temperature and rainfall. Research indicates that low temperatures, specifically between -6.6°C and 20°C, are conducive to the prevalence of norovirus, as these conditions favor the virus' stability and transmission. This is further supported by a study in South Korea, which found a significant negative correlation between average temperature and the number of norovirus food poisoning outbreaks, with the cold and dry winter season being particularly favorable for the virus spread. Rainfall or precipitation is also of core importance; some studies suggest that precipitation from 1 day to 3 months prior to a norovirus outbreak can either enhance or narrow down the incidence, most likely due to the effects it has on water supply or the potential for pollution. In one of the regions along the coast of China, a predictive model for foodborne norovirus outbreaks indicated that temperature and precipitation are on top of the list of variables with the most sensitive effects on the probability of outbreaks. This model hints of the contribution of both temperature and rainfall in the outbreak of norovirus and states that if these variables are closely monitored, there is an improved control of public health measures against massive outbreak. Collectively, these studies illustrate the complex interplay between environmental conditions and norovirus transmission, emphasizing the need for comprehensive monitoring.

**Corresponding author*: mmujeeb@jamiahamdard.ac.in; ORCID: 0000-0003-3110-6089
Neeraj Kumar: ORCID: 0000-0003-3301-118X
Akanksha: ORCID: 0009-0006-6618-0368
Shriyansh Srivastava: ORCID: 0000-0002-3738-4381

Sachin Kumar, Shriyansh Srivastava, Rishabha Malviya and Ranjit Sah (eds.) Emerging Infectious Disease: Unravelling Norovirus and Pandemic Dynamics, (99–128) © 2026 Scrivener Publishing LLC

***Keywords*:** Norovirus, outbreak, temperature, transmission, evolution, GIV

4.1 Introduction

Norovirus infections are an important global health burden and are the causative factor in approximately 18% of gastroenteritis episodes [1]. Noroviruses are also extremely virulent pathogens as only a handful of particles is enough to infect a person; disease-infected individuals are known to shed large quantities of the virus [2]. It is primarily spread through the fecal–oral route by consuming tainted food or water, touching infected surfaces, or close contact with an infected person. Symptoms may include diarrhea, vomiting, nausea, and stomach cramps, and outbreaks are frequently seen in crowded areas like schools and nursing homes. There is no specific treatment or vaccination, as many disinfectants are unable to destroy the hardest of pathogens; norovirus deteriorates hygiene as much as it does cause it [3]. Commonly referred to as the "winter vomiting bug", norovirus is a seasonal pathogen that leads to viral gastroenteritis, an inflammatory disorder of the stomach and intestinal tract.

First identified after a 1972 outbreak in Norwalk, Ohio, it is now the leading global cause of gastroenteritis. Transmitted primarily through the fecal–oral route, it spreads *via* contaminated food, water, and surfaces or direct contact, particularly in crowded settings like schools, nursing homes, and cruise ships [4]. Noroviruses are classified into multiple genogroups based on genetic diversity, each containing distinct genotypes:

- Genogroups I (GI) and II (GII) primarily affect humans, with GII.4 being the most prevalent and responsible for global outbreaks.
- Genogroups III (GIII) to X (GX) largely involve animal hosts, including bovines (GIII), marine mammals (GVII), bats (GVIII and GX), and cats (GVI).
- Genogroup IV (GIV) affects both humans and animals, while GIX includes viruses previously classified as GII.15.

This classification highlights the diverse hosts and global impact of noroviruses [5].

There are several genetic groupings of norovirus; these are known as genogroups and clusters. The GII.4 cluster is known to be the most widespread and has contributed to many pandemics. Other GII clusters, such as GII.1, GII.2, and GII.3, depict more or less the same impact on public health as those with different clusters. Some recent studies have also revealed genetic types of the GII that had not previously been recognized—for example, GII.13 to GII.17 which have been documented during outbreaks. Additionally, GII.4 genotype has mutated and developed into other variants such as GII.4 Sydney 2012 and GII.4 New Orleans 2009. A classification structure has also been provided, which categorized the strains into three main levels, namely: strain (S), cluster (C), and genogroup (G). Generally, norovirus is characterized by its genogroups and genotypes together with the name of the country where the virus was discovered and its year of discovery. This understanding is important both to the health professionals coping with outbreaks, scientists developing vaccines against the most common strains, and population flows as a way to interfere with mechanisms of norovirus transmission [6]. It is a national municipality belonging to Caliciviridae, consisting of a single species called Norwalk virus which has many genogroups and genotypes. Overall, it can be estimated that GII.4, in particular, is to be accounted for more than 70%–80% during major outbreaks. Generally, individuals infected with the norovirus do not seek treatment, as the virus resolves itself. However, the same cannot be said for the vulnerable, the very young, elderly, and those with weakened immunity, among whom it can have grave consequences [7]. Norovirus remains a global issue, mostly for the potential severe effects it can cause and its high transmissibility [8]. Other hurdles in progress of vaccines and therapies are the multiple virus types. Due to high levels of genetic and antigenic variability and frequent recombination events, new strains are emerging all the time. Effective vaccinations and treatments are becoming more difficult to develop, for instance, because the GII.P17-GII.17 genotype, which recently emerged in Asia, has replaced formerly dominant strains in some locations [9]. It causes approximately 23 million infections annually in the US and is responsible for 60% of global gastroenteritis cases, with incidence peaking during winter. It spreads through contaminated food, water, and surfaces and close contact, often triggering rapid outbreaks in settings like healthcare facilities, schools, and cruise ships. High-risk groups include children, the elderly, and those with weakened immune systems. With its genetic diversity and the emergence of new strains like GII.4, norovirus continues to drive outbreaks. Despite its significant economic and healthcare burden, many cases go unreported, highlighting the importance of robust prevention measures [10]. Norovirus pathogenesis involves a complex interaction with the host immune system, causing gastrointestinal illness. These non-enveloped, single-stranded RNA viruses attach to intestinal cells and replicate within the host, targeting intestinal and immune cells. Histo-blood group antigens (HBGAs) influence susceptibility. The immune response includes both innate and adaptive mechanisms, but the virus' genetic variability allows for reinfections. Symptoms such as nausea, vomiting, diarrhea, and cramps can lead to dehydration and intestinal damage, particularly in vulnerable groups, highlighting the difficulty in managing and preventing infections [11]. Norovirus, a non-enveloped virus lacking a lipid membrane, is structurally characterized by a durable protein shell or capsid measuring 38–40 nm in diameter. The capsid is composed of 180 copies of VP1 (viral protein 1), the major structural protein responsible for its antigenic properties and eliciting an immune response. The virus encloses a single-stranded RNA genome of about 7.5 kb, ordered into three open reading frames (ORFs): ORF1 encodes non-structural proteins that are essential for viral replication, ORF2 encodes VP1 that forms the capsid, and ORF3 encodes VP2, a minor structural protein associated with the capsid. Norovirus' capsid is icosahedral with protrusions made by the P domain of VP1, which plays a significant role in the binding of the virus to the host cell receptor. This virus attacks HBGAs on the epithelial cells lining humans; this enables its penetration and further infection [12]. Noroviruses are also transmitted more because of factors such as population increase, higher international travels, as well as the globalization of food supply [13]. An understanding of the pathogenesis and epidemiology of norovirus will be essential in preventing future outbreaks. More research into the genetics and dynamics of norovirus transmission is required to design comprehensive prevention and treatment programs. This in turn warrants strict monitoring and public health advocacy campaigns in order to defeat this ongoing threat to world health [11]. Surveillance and research are essential to control norovirus infections. Monitoring norovirus strains by surveillance networks like CaliciNet and NoroNet helps discover new variants and assess their effects on public health. More knowledge on the factors that influence norovirus evolution—such as host interactions, environmental conditions, and viral genetics—is necessary for a full understanding of its behaviors. These insights are vital to shape effective public health strategies and enhance prevention as well as control measures [14]. Due to the global burden of acute gastroenteritis, developing a norovirus vaccine is a key public health goal.

Several vaccine candidates, including virus-like particle (VLP) vaccines, VLP vaccine adenovirus vector-based vaccines, and P-particle vaccines, are in clinical trials. However, challenges remain due to the virus' genetic diversity, short-lived immunity, and lack of effective animal models for testing. Strategies to address these issues include developing multivalent vaccines, using adjuvants, and exploring different delivery methods. Ongoing surveillance ensures that vaccines remain effective against emerging strains [12]. VLP vaccines have shown promise in inducing immune responses in adults and children. Some formulations were tested for safety and immunogenicity in phase 1 and 2 clinical trials [15]. Recombinant vaccines: These vaccines employ recombinant DNA technology to produce norovirus proteins, such as VP1, within a vector like recombinant adenovirus, eliciting an immune response. Combination vaccines: Efforts are underway to create vaccines that target multiple pathogens, including norovirus and rotavirus, aiming to provide broader protection against gastroenteritis [16]. Norovirus poses a significant public health challenge, causing widespread outbreaks and straining healthcare systems, particularly during peak seasons. Effective surveillance and reporting are crucial for outbreak management and informing public health strategies [17].

4.2 Evolution of Norovirus

The evolution of norovirus is primarily driven by the following mechanisms and shown in Figure 4.1.

4.2.1 Antigenic Drift

The term "antigenic drift" describes the slow accumulation of mutations that cause immunological responses in the viral genome, particularly in the capsid protein. Over time, the virus is able to evade host defenses thanks to minute modifications to the capsid protein's P2 domain [18]. These alterations produce new strains that can re-infect individuals who have already been exposed to older strains, leading to recurrent infections.

4.2.2 Antigenic Shift

Antigenic shift, on the other hand, is marked by more pronounced genetic mutations, often due to virus strain recombination. The genetic material exchanged in this process results in the creation of new genotypes and those genotypes are often better at infecting and spreading among humans.

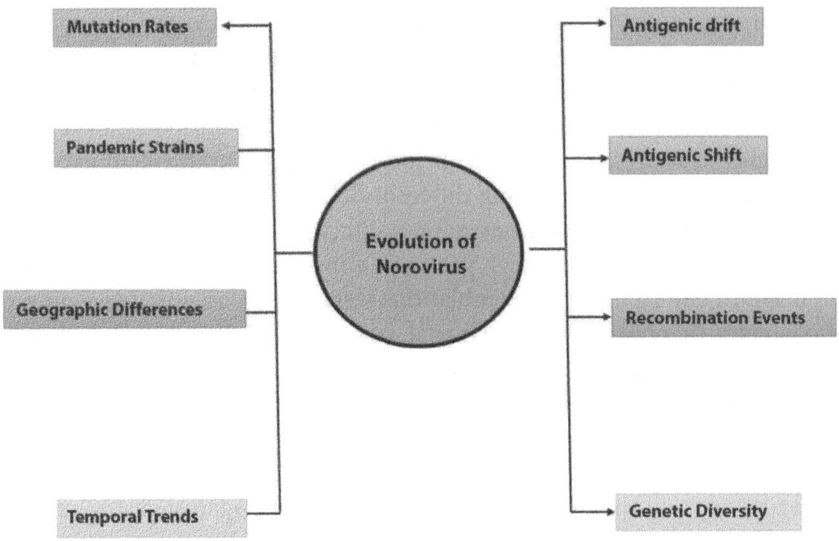

Figure 4.1 Evolution of norovirus.

The most recent GII.4 variants of the norovirus also have features of drift and shift, which makes its management and preventive actions harder [19].

4.2.3 Recombination Events

Recombination, particularly in GII.4 viruses, is the consequence of mixing genetic material from different lineages of the virus. These intragenotype recombination events have been associated with the emergence of pandemic strains, as they enable the virus to adapt rapidly and spread more easily, thereby strengthening its persistence within populations [4].

4.2.4 Genetic Diversity

There are seven genogroups of noroviruses: GI–GVII, with GI and GII being the most commonly associated with human infections. These genogroups encompass a wide range of genotypes, resulting in substantial genetic variability [20].

4.2.5 Mutation Rates

The noroviruses are highly mutable and thus can readily adapt quickly to host immunity changes. This rapid adaptation allows it to evade immune mechanisms established by previous infections or vaccinations, hence the recurrent nature of outbreaks [21].

4.2.6 Pandemic Strains

Certain subtypes, in particular GII.4, are believed to cause pandemics. These GII.4 variants have appeared at different points in time, each bearing slight genetic mutations that enhance their ability for transmission and infection, leading to epidemics [22].

4.2.7 Geographic Differences

The distribution of norovirus strains is spatially and temporally heterogeneous. For example, GII.17 was reported to be prominent strains in Asian outbreaks of some periods, whereas GII.4 continues to be dominant in many other regions.

4.2.8 Temporal Trends

Strain prevalence fluctuates seasonally and annually, driven by factors such as population immunity and patterns of viral transmission [23].

4.3 Fatal Details Regarding the Norovirus

In healthy adults, norovirus infections usually resolve on their own, but they can be quite dangerous for susceptible groups, including the elderly and those with compromised immune systems [24]. Aspiration pneumonia, gastroenteritis, sepsis, and heart problems are among the consequences that can dramatically raise fatality rates during seasonal epidemics in these high-risk groups. It is challenging to determine the exact number of deaths caused by noroviruses, especially in nations with low resources and outdated molecular diagnostic technologies [3]. The cumulative effect on these susceptible groups emphasizes how important it is to better understand and treat norovirus infections. The WHO reports that norovirus is responsible for an estimated 685 million infections and 212,000 deaths annually worldwide, thus indicating its presence as a grave health threat [25]. In the United States, norovirus is estimated to account for around 570 to 800 deaths per year, which forms a small proportion of the 19 to 21 million reported infections.

Globally, it is responsible for approximately 200,000 deaths annually and overwhelmingly impacts the most vulnerable populations, including young children and older adults. These numbers demonstrate the importance of norovirus at a global level and emphasize the continued prevention and management measures focused on those at an increased risk of adverse effects [26]. Generally, norovirus infections have a very low mortality rate, less than 1%, with healthy persons usually recovering uneventfully. Nonetheless, vulnerable groups such as young children, the elderly, and immunocompromised people are more likely to develop severe illness or complications from dehydration. In the United States, norovirus causes a million illnesses, 56,000–71,000 hospitalizations, and 570–800 deaths annually, as seen mostly in older adults and those with underlying conditions. Outbreaks often occur in confined settings like nursing homes and schools, straining healthcare resources. Preventive measures, including proper hygiene, sanitation, and food safety, are essential to reduce infections and complications [27]. The severity of norovirus infections is demonstrated by a number of noteworthy outbreaks: in 2012, Denmark reported 130 cases and one death; in 2018, the United States reported 647 cases and two fatalities; and in 2020, the UK reported 1,500 cases and five deaths. According to Centre for Disease Control and Prevention (CDC) estimates, norovirus kills between 570 and 800 people in the US each year, mostly older adults, whereas it kills 1,000 to 2,000 people in Europe each year, mostly older people. Older adults (65 and above), little children (under 5), and people with long-term illnesses or weakened immune systems are high-risk categories. These groups are most vulnerable to infections, which emphasizes the critical need for improved management, preventive, and treatment plans in order to lower the norovirus burden worldwide [25].

4.4 Mode of Transmission of Norovirus

One of the main causes of gastroenteritis, norovirus, spreads widely and is difficult to control effectively. The short lifetime of immunity, the virus' adaptability to many conditions, and the absence of specific therapies or vaccines are major contributing causes. Norovirus is even harder to control in epidemics on cruise ships because of the close quarters and financial incentives that discourage people from reporting illnesses. The most common ways by which norovirus spreads include the following: fecal–oral routes, exposure to vomit, ingestion of contaminated food and drinks, contact with contaminated surfaces, and close contact with infected individuals [28]. Since the virus might survive in contained environments such as hospitals and schools, having good sanitation practices is of vital importance. For contacts between passengers and crew in close quarters, such as in cruise ships, the spread of the virus is increased by direct person-to-person transmission. In addition, a significant portion of norovirus infections does not manifest symptoms, which makes efforts at control more challenging [29]. The primary transmission of norovirus occurs through the fecal–oral route and may occur through contaminated food, water, and surfaces or direct contact with infected individuals. This means that areas with poor sanitation or contaminated water sources are at a higher risk for outbreaks [30]. Effective management measures are hindered by various challenges, such as the limited cross-protection between different strains and the ephemeral nature of immunity. Some other possible measures for preventing the infection include better environmental cleaning, proper hand hygiene, and separation of contaminated people. It further stresses the importance of using mathematical models to evaluate management techniques and study the norovirus spread. In the end, the best strategy to reduce the spread of the norovirus is to wash your hands thoroughly [31]. Numerous processes contribute to the spread of noroviruses. The most common form of transmission is fecal–oral, in which food, water, or surfaces contaminated by feces are consumed. Furthermore, waterborne transmission is another way that norovirus can spread, especially when contaminated water sources like irrigation water are implicated. Another kind of virus transmission is foodborne, in which the virus contaminates food products, such as fresh produce that has been irrigated with tainted water. Last but not least, person-to-person contact can spread the norovirus, frequently through contaminated hands or surfaces [32]. Hand cleanliness is essential for preventing the spread of the norovirus in nursing facilities, where the illness poses a serious risk, especially to senior citizens. The study evaluated how better hand hygiene procedures could affect the transmission of the norovirus in these institutions using a mathematical modeling approach. The findings suggested a 75% decrease in norovirus incidence if staff members increased their hand hygiene compliance from 15% to 60%. Isolating afflicted residents may also result in 88% reduction in incidence. While increasing the residents' adherence to hand hygiene practices did contribute to a reduction in instances, its impact was not as great [33].

4.4.1 Environmental Source

Various environmental factors can trigger norovirus outbreaks in closed and semi-closed environments. Restrooms play a significant role, as the virus has frequently been found on surfaces, faucets, and doorknobs, suggesting that these spaces are key in the spread of outbreaks. In kitchens, for instance, during foodborne outbreaks, food contact surfaces and work areas may be expected hotspots where infection is probable. Additionally, relatively universal places like TV remote controls, elevator buttons, and dining tables appear to harbor the virus, thereby contributing to its transmission. Handrails, toilet handles, and elevator buttons are some high-risk surfaces bearing the virus, which, according to the researchers, calls for directed cleaning and disinfection activities [34]. It can survive on surfaces around houses, including countertops, kitchen utensils, bathroom fixtures, and door handles for days and even up to weeks, thereby also raising the possibility of transmission by direct contact. Also, fomites—the objects that harbor viruses through such things as shared towels or utensils—are a notable factor that transmits this virus. When an infected person touches these surfaces or items, the virus can be passed on to others who subsequently come into contact with them [35].

4.4.2 Influence of Temperature on Norovirus Outbreaks

Climate has a major influence on the spread of the norovirus, particularly on temperature and precipitation. The principal observations consist of the following:

(a) Temperature effects: During the winter months in the United States, the cold, dry weather is associated with an increase in norovirus activity. Usually, outbreaks reach their zenith between October and May during these cooler months. The effective temperature range for norovirus outbreaks is reported to be between -6.6°C and 20°C [36].
(b) Rainfall contributions: Increased rainfall levels can potentially increase the norovirus transmission rate, suggesting that weather patterns play a critical role in its spread. Rainfall significantly influences norovirus transmission and outbreaks, with heavy precipitation linked to increased hospitalizations, especially in children. Mechanisms include waterborne transmission through contaminated runoff, sewer overflows releasing untreated wastewater, and seasonal variations in rainfall, particularly in warmer months. Additionally, the effects of rainfall on norovirus infections may have a delayed impact, with hospitalizations rising several days after heavy rain events [37].

(c) Behavioral influences: When the weather is colder and wetter, individuals tend to cluster closer together, which makes indoor gatherings more frequent and a behavioral element in the virus' transmission.
(d) Geographic variations: Data from 2009 to 2019 show distinct geographic variations in the number of norovirus outbreaks. While the south central and western regions display less noticeable seasonal shifts, the midwest and northeast demonstrate strong winter seasonality. In several states, there is even no discernible seasonal trend for outbreaks. Over the past decade, 24,995 norovirus outbreaks were reported, with a significant number occurring in winter, indicating a link between cold, dry weather and increased outbreaks. Emerging variants, particularly GII.4 strains, may also influence seasonal outbreaks, potentially exacerbated by environmental factors. Ongoing surveillance is essential to understand how climate change affects norovirus spread and inform public health strategies. These findings highlight the importance of tailoring prevention plans to local climates, launching seasonal education during peak periods, and adjusting public health measures based on local temperature and humidity to mitigate outbreaks and improve public health responses [38]. Norovirus outbreaks are seasonal due to several environmental reasons. While its precise impact is yet unknown, temperature is an important component. Rainfall is also important since precipitation differences can impact the virus' persistence and transmission. The state of water sources and watersheds may also affect the transmission. Higher population density accelerates the norovirus transmission within communities. Some studies have been done regarding the impact of economic factors such as GDP on the prevalence of the virus. Regional differences are also very interesting since patterns of the seasonal outbreaks vary between northern and southern countries. For instance, in the northern hemisphere, countries like Australia and New Zealand, whose schools operate throughout the year, are less likely to experience outbreaks compared with those with traditional school calendars; this points out that maybe school structures influence the dynamics of norovirus transmission differently at various seasons [39].

4.5 Correlation Between Temperature Fluctuations and Outbreaks

Norovirus genomes are consistently recombinant and prone to frequent mutations. Notably, specific genomic mutations in the GII.4 strain have been linked to increased infectivity and diverse outbreak patterns.

Factors such as prolonged viral shedding in certain individuals, the virus' resilience outside the host, the potential impact of rainfall on outbreaks, and the likelihood of zoonotic transmission from animal reservoirs play crucial roles in norovirus transmission and outbreak dynamics [40].

4.6 Effect of Norovirus on Humidity

Particularly when it comes to murine norovirus (MNV), humidity significantly affects the infectivity and persistence of the virus. At low relative humidity (RH) levels, especially around 10%, MNV is best preserved. This is also the time when human norovirus virus-like particles (VLPs) bind to saliva and MNV is most infectious. On the other hand, MNV infectivity is significantly reduced at medium humidity levels (35%–55% RH). At 55% and 85% RH at 25°C, total inactivation occurs in 12 h, while VLP binding completely disappears after a week. Excessive humidity (85%–100% RH) somewhat reduces MNV infectivity and VLP binding, suggesting that residual liquid may contribute to viral persistence. Absolute humidity (AH) of less than 0.007 kg water/kg air is ideal for MNV survival; values higher than 0.01 kg are detrimental. Furthermore, MNV survival is further enhanced by lower temperatures (9°C) at all humidity levels, indicating that the ideal environment for norovirus survival is one that combines low temperature and humidity. Low AH conditions (96.3% of days <0.007 kg water/kg air) throughout the winter likely contribute to the environmental persistence and transmission of noroviruses, according to meteorological data from Paris [41].

4.7 The Impact of Rainfall on the Spread of the Norovirus

The research shows a significant correlation between rainfall and norovirus outbreaks in Victoria, Australia, between 2002 and 2007. Average monthly rainfall and the frequency of norovirus-associated gastroenteritis outbreaks (NAGO) were found to be statistically significantly correlated ($p < 0.05$), with an average 3-month lag between peak rainfall and subsequent epidemic increases. According to studies, rainfall may increase the water sources' turbidity, which may facilitate the spread of waterborne norovirus from environmental reservoirs. These findings suggest that rainfall contributes to the norovirus' persistence since outbreaks typically peak toward the end of the year after times of higher rainfall earlier in the year [42].

4.8 The Impact of Water on the Spread of the Norovirus

About 8,000 people were affected by a drinking water outbreak in Finland in 2007 due to water supply poisoning. Three of the 67 schoolchildren and teachers involved in the 2011 incident at the water park in Korea had excrement and groundwater contaminated with noroviruses. Five outbreaks in recreational water settings in the United States in 2007 and 2008 resulted in 121 cases overall, in both treated and untreated water sources. In 2007, there was another epidemic of drinking water in Sweden that resulted in 35 cases. The patient samples and well water were found to have noroviruses. During a 2007 outbreak in Denmark that claimed 453 lives, noroviruses were discovered in patient samples together with other infections. A 2010 epidemic in Rio de Janeiro, Brazil, revealed noroviruses in wastewater treatment facility influent and effluent samples. Moreover, norovirus outbreaks occurred in long-term care institutions in the United States in 2010–2011; these outbreaks accounted for 59% of cases that were recorded in those settings [43]. A waterborne norovirus outbreak that affected 130 individuals and had a 51% attack rate among 368 homes happened in Kalundborg, Denmark, in December 2012. The norovirus GII.4 New Orleans 2009 variant, which was found in both stool and water samples and had the same virus sequences in both, was linked to the outbreak. The virus was made easier to spread after a sewage pipe leak contaminated a broken tap water line. On December 12 and 13, tap water consumption was associated with illness, with a relative risk of 6.0, indicating a dose–response connection based on the amount of tap water drank. A water ban and a boil water advisory were issued as a precaution on December 15 and 16, respectively, coupled with attempts to clean up the water system and provide clean water [44].

4.9 Norovirus Combined Effects of Temperature and Rainfall

The study found that the incidence of norovirus GII was negatively correlated with temperature, precipitation, sunlight, and other environmental parameters. Temperature was found to be an important factor after adjusting for sunshine and rainfall; it showed a strong negative link with the incidence of norovirus (β = -0.8773, $p < 0.0001$). This suggests that norovirus GII tends to spread less when temperature levels rise. Norovirus outbreaks in Taiwan exhibited clear seasonal trends, with the majority of genotypes reporting more than half of their cases in the winter and spring. Temperature responses varied between norovirus genotypes; among them, GII.6 was reported to occur more frequently in the summer and at higher temperature levels (especially above 25°C). Compared to GII.2 or GII.17, a notably higher percentage of GII.6 cases was detected in the 30%–35°C range, but GII.6 infections were less common in the 20%–25°C range. The study acknowledged that rainfall was a relevant component, but it did not clarify how rainfall and temperature worked together. The multivariate linear regression model indicated that rainfall did not significantly affect the incidence of norovirus ($p = 0.6303$) on its own. In addition, the prevalence of norovirus GII in Taiwan's northern, western, and southern regions was distributed geographically according to similar meteorological conditions. Lastly, the study found that Taiwan's humidity levels were consistently high (70–80%) all year round [45].

4.10 Stability of Norovirus

Control strategies are greatly hampered by the norovirus' extraordinary resilience and capacity to spread infection in a variety of environmental settings. Its capacity to withstand low pH levels, which allows it to endure the stomach's acidic environment and reach target cells in the small intestine, is one of the main factors influencing its stability. Norovirus can also persist as an infectious agent in food and surfaces for prolonged periods of time, which makes it more likely to cause outbreaks and makes measures to stop its spread more difficult. Understanding the virus' ability to survive in various environments depends on its stability over a broad temperature range. Moreover, it is common practice to compare the stability of norovirus to other viruses, such as murine norovirus 1 (MNV-1) and feline calicivirus (FCV), highlighting the importance of these viruses as models for investigating the stability and inactivation of human norovirus [46]. Norovirus is resilient across a broad temperature range, capable of surviving freezing conditions and tolerating some heat. However, it is vulnerable to high temperatures, and cooking food to proper temperatures (e.g., 70°C or 160°F) can effectively deactivate the virus [47]. Norovirus remains stable in environments with a neutral to slightly acidic pH, typically between pH 6 and 7, allowing it to survive in the gastrointestinal tract where pH levels vary. The virus is also resistant to acidic conditions, such as those in the stomach (pH 1.5 to 3.5), although extended exposure to highly acidic environments may lead to some degradation. However, norovirus is more susceptible to damage in alkaline conditions (pH above 8). Research indicates that the virus' capsid may break down in alkaline environments, reducing its infectivity and hindering its ability to cause illness [48].

4.11 Case Study

There were 1,130 norovirus outbreaks in the US between 2009 and 2013, which resulted in 21,109 sickness cases that were documented, 1,044 hospital admissions, and 150 deaths. California (143), Texas (93), Florida (83), New York (73), and Illinois (63) were the states with the highest number of outbreaks. Summertime (June to August) saw the fewest norovirus occurrences, with winter months—especially between December and April—usually seeing the highest number of outbreaks. Children under the age of 5 years constituted the largest risk category, making up 34% of cases; seniors 65 years and older accounted for 23% of cases. Of the two strains, GII.4 caused 70% of outbreaks and GII.2 15% of them. The GII.4 strain was the most prevalent [49]. The details are presented in Table 4.1.

A norovirus outbreak that affected 332 students at a university in Michigan in 2016 resulted in a high attack rate of 14.4%. Rapid transmission, severe symptoms, and a major disruption to university operations were the hallmarks of this outbreak. To manage the problem, more thorough cleaning and disinfection, isolating the sick, and postponing group activities were all done. The pandemic had significant financial consequences with an estimated cost of $159,644 [41]. Over 200 students and staff members were impacted by a norovirus outbreak at a US institution in February 2016, severely disrupting academic programs. In order to examine the outbreak, public health officials gathered information on symptoms, demography, and potential exposures in addition to sampling the environment. Stool testing was used to identify the norovirus strain GII. P16-GII.2 as the cause of the incident. Contaminated food prepared by

Table 4.1 Case study related to the norovirus.

Region/year	Outbreak setting	Cases	Key features	Main cause	Response measures	References
US (2009–2013)	General outbreaks	21,109	Highest outbreaks in CA, TX, FL, NY, and IL; winter peak; GII.4 strain caused 70% of cases	Person-to-person, contaminated food	Isolation, hygiene reinforcement	[49]
US (Feb 2016)	University cafeteria	217	14.4% attack rate; vomiting (85%), diarrhea (83%)	Contaminated food by food handler	Disinfection, isolation, outbreak investigation	[50]
US (2016)	Michigan university	332	High attack rate (14.4%); rapid transmission disrupted operations	Unknown source	Enhanced cleaning, isolation, activity suspension	[41]

(Continued)

Table 4.1 Case study related to the norovirus. (*Continued*)

Region/year	Outbreak setting	Cases	Key features	Main cause	Response measures	References
England (2012–2013)	Nationwide	~1,116,146	Sydney 2012 strain introduced *via* travel; peak in January 2013; hygiene lapses exacerbated spread	Person-to-person, environment	Hygiene education, containment measures	[50]
England (1990–2012)	Hospitals, schools, care homes	12,116	Winter peak; mostly adults and children under five	Person-to-person, food, water	Hygiene reinforcement	[51]
Africa (Various)	Pediatric settings	~13.5%	Genogroup II (84.1%) prevalent; higher rates in children under two	Contaminated water, zoonotic factors	Improved surveillance	[52]
Cruise Ship (2002)	Cruise ship	>400	Contaminated food and water	Food/water contamination	Enhanced hygiene	[40]

(*Continued*)

Table 4.1 Case study related to the norovirus. (*Continued*)

Region/year	Outbreak setting	Cases	Key features	Main cause	Response measures	References
Hospital (2006)	Hospital	>100	Spread by a contaminated food handler	Poor infection control	Infection control measures	[40]
School (2007)	School	>200	Linked to contaminated food and water	Poor sanitation	Hygiene improvements	[40]
Restaurant (2008)	Restaurant	>100	Illness traced to contaminated food	Unsafe food handling	Food safety practices reinforced	[40]
Nursing home (2009)	Nursing home	>50	Contaminated food and water	Poor infection control	Infection control measures	[40]
Community (2010)	Community water supply	>500	Contaminated water source	Poor water treatment	Improved water treatment	[40]
Hotel (2011)	Hotel	>200	Illness linked to contaminated food and water	Hygiene lapses	Hygiene practices enforced	[40]

(*Continued*)

Table 4.1 Case study related to the norovirus. (*Continued*)

Region/year	Outbreak setting	Cases	Key features	Main cause	Response measures	References
Food plant (2012)	Food processing plant	>100	Outbreak traced to food contamination	Poor sanitation	Hygiene protocols updated	[40]
University (2013)	University	>500	Contaminated food and water led to widespread illness	Sanitation lapses	Improved cleaning, outbreak investigation	[40]
Australia (2015–2019)	Hospitals, schools, care homes	~100+ per outbreak	Winter peaks; significant disruptions in institutions	Person-to-person, food contamination	Containment and cleaning measures	[53]
UK (2015–2019)	Hospitals, schools, care homes	~100+ per outbreak	Recurrent outbreaks forced closures and heightened infection control	Food and environmental contamination	Enhanced hygiene and containment measures	[53]

an infected food handler and served in the university cafeteria was identified as the source. With 217 instances reported overall, the attack rate was 14.4%. Vomiting (85%), diarrhea (83%), cramping in the abdomen (82%), nausea (80%), fever (45%), and headache (35%) were the most frequent symptoms [50]. The newly identified Sydney 2012 strain caused a major norovirus outbreak in England in 2012, affecting over a million individuals and seriously interrupting healthcare systems. The outbreak peaked in January 2013, with 1,116,146 cases reported nationwide, mostly in Yorkshire and the Humber and the North West. It started in October 2012. The introduction of the Sydney 2012 strain was through international travel, quick person-to-person transmission in confined areas like hospitals and schools, and environmental contamination made worse by poor hygiene and cleaning techniques, which were all found to be contributing factors in an epidemiological investigation [50]. There were 235 norovirus outbreaks in England and Wales between 1990 and 2012, in addition to 15 outbreaks in general practices, hospitals, care facilities, and schools. These outbreaks mainly affected adults, older people, and children under the age of 5 years. A total of 12,116 cases were documented, while 145 additional cases were unreported. Frequent signs and symptoms were fever, vomiting, diarrhea, and stomach pain. Hospitals and assisted living facilities were particularly impacted, where epidemics peaked in the winter, from December to February. The primary method of transmission was person-to-person contact; however, tainted food and water also had a role. Usually lasting a few days to weeks, the epidemics resulted in a significant number of cases and deaths in medical facilities [51]. Based on 19 research from 14 African nations, norovirus is a prominent cause of gastroenteritis in children, with an average frequency of 13.5% in unwell individuals and 9.7% in asymptomatic individuals. Children under the age of 5 years (11.8%) are less likely to contract the virus than those up to 2 years old (18%). In most studies, genogroup II strains accounted for 84.1% of cases, with NoV GII.4 being the predominant genotype. Different places had different seasonal patterns; some saw peaks during the cool, dry seasons, while others saw them during the wet ones. For example, South Africa showed a 14.3% frequency in children under 13, peaking in spring/early summer, while Ghana reported a 15.9% prevalence in hospitalized children under 2 years. Studies on the environment also revealed high levels of norovirus contamination in sewage and surface water, with notable and unique strains being identified. These findings underline the importance of norovirus in pediatric gastroenteritis in Africa and the necessity of more thorough surveillance to comprehend the virus' distribution throughout the region [52]. Changes in viral infectivity and epidemic features are correlated with the numerous mutations and recombinations that occur in the norovirus genome, especially within the GII.4 genotype. Long-term viral shedding by some infected individuals, the virus' ability to survive in the environment, the potential impact of rainfall on outbreaks, and the potential for zoonotic transmission from animal reservoirs are all important environmental factors in the transmission of noroviruses. These elements are highlighted in several case studies, such as the following:

- Cruise ship outbreak (2002): Over 400 individuals were affected due to contaminated food and water, underscoring the importance of hygiene.
- Hospital outbreak (2006): More than 100 patients and staff contracted the virus from a contaminated food handler, emphasizing the necessity of infection control in healthcare settings.
- School outbreak (2007): More than 200 individuals were impacted due to contaminated food and water, highlighting the importance of sanitation.
- Restaurant outbreak (2008): Over 100 customers became ill from contaminated food, stressing the need for safe food handling practices.
- Nursing home outbreak (2009): More than 50 residents were affected by contaminated food and water, indicating the importance of infection control in long-term care facilities.
- Community outbreak (2010): Over 500 individuals were impacted due to contaminated water, emphasizing the need for effective water treatment practices.
- Hotel outbreak (2011): More than 200 guests fell ill due to contaminated food and water, reinforcing the importance of hygiene.
- Food processing plant outbreak (2012): Over 100 workers were affected, with the outbreak traced back to food contamination.
- University outbreak (2013): More than 500 students and staff were impacted by contaminated food and water, again emphasizing the need for sanitation practices.
- Hospital outbreak (2014): Over 200 individuals contracted the virus due to contaminated food and water, reaffirming the necessity for effective infection control measures [40].

There have been successive large-scale norovirus outbreaks in Australia in recent years, which involved multiple institutions. The hospital outbreaks were so widespread that they affected over 100 patients and staff members in both 2016 and 2019. During the outbreaks in 2015 and 2018, it affected around 100 students and staff that made the affected schools close for a certain period of time. Over 50 residents and staff members were affected by a care facility outbreak in 2017, prompting containment measures. These cases epitomize the ongoing challenges faced in managing norovirus in clinical, educational, and residential settings across Australia [53]. Major norovirus outbreaks have been reported in the UK at different institutions over the years. A hospital reported over 100 cases involving patients and staff in 2019, prompting the establishment to put in control measures. Similar reports of the outbreak at the school last year led to more than 100 students and staff being affected and causing it to close. Efforts to contain the virus began after an outbreak that involved over 50 people, residents, and staff members in a care facility in 2017. In another hospital outbreak in 2016, about 100 people were infected. It thus had to enforce additional controls to curb its spread. The schools had to close after an outbreak that infected over 100 pupils and staff members in 2015. These attacks reflect the ongoing challenges with norovirus in numerous settings [53].

4.12 Global Patterns in Temperature- and Rainfall-Related Outbreaks

Other environmental factors that influence the survival and transmission of the virus greatly impact global patterns of norovirus outbreaks associated with rainfall and temperature. Seasonal trends indicate that the outbreaks occur more frequently in cool season months, and this can be understood as the virus survives longer in the environment and in food products when there is less temperature. For instance, a study indicated that temperature extremes play a significant role in the survival of noroviruses. While very low temperatures can facilitate a greater survival of the virus in contaminated food and water, high temperatures may lower it because of such things as UV exposure and solar radiation. Events with high rainfall increase the amount of runoff from contaminated sources, like sewage and agricultural regions, which in turn helps the norovirus spread. This source of runoff has the potential to increase norovirus levels within irrigation and recreational waters, leading to an outbreak. Flooding also causes issues for sanitary systems, contaminating drinking water sources and increasing the danger of transmission within the community. A feature of remarkable interest is the impact of geographic heterogeneity on outbreak patterns. There are several locations that have incidence rates unique to each location, which appear to depend upon the climate of the respective location. For instance, seasonal patterns may vary between a tropical versus temperate location because the climate locally influences the virus' behavior. Moreover, urban areas often have higher rates of outbreaks due to the likelihood of sewage contamination and greater population density. However, it may be that agricultural runoff has a greater effect in rural locations, leading to greater virus concentrations in irrigation water supplies [54].

The continuing impacts of global climate change are likely to alter temperature ranges and precipitation patterns, with more frequent, heavy downpouring events. Norovirus is likely to increase in areas already at risk due to these changes, and weather patterns may become more variable, making outbreaks unpredictable, thus increasing the difficulty of preparedness and public health measures.

There is a complex interconnection between temperature, rainfall, and viral transmission that must be appreciated in order to control the risk related to norovirus outbreaks. For public health officials creating better monitoring systems and contingency plans, this knowledge is essential. When there is heavy rainfall, it is particularly important to regulate the water quality to halt the spread of the virus to non-contaminated sources of water. By and large, seasonal patterns, regional variations, and effects of climate change are just a few variables that dictate the times of norovirus epidemics. Monitoring these elements is crucial in guiding public health initiatives successfully to avoid significant repercussions of epidemics. Health officials can better prepare for and respond to possible outbreaks by understanding the environmental factors that drive norovirus transmission, thereby protecting public health [54].

4.13 Mitigation Strategies and Public Health Interventions

The study of norovirus distribution, frequency, and contributing variables is the main goal of norovirus epidemiology. Although it affects people of all ages, little children and elderly people are more at danger from this extremely contagious virus. The CDC estimates that norovirus causes 570–800 fatalities, 56,000–71,000 hospital admissions, and 1.7–2.1 million outpatient visits yearly in the US. It is the most common cause of foodborne infections in the country. Wintertime is usually when infection rates are at their highest, particularly in December and January [55].

(a) Personal hygiene: Handwashing well is crucial to stopping the norovirus from spreading. Hands should be cleaned for at least 20 s with soap and water, especially after using the restroom, before eating, and after meeting ill people. Hand sanitizer is an alternative that can be used in situations where soap and water are not accessible. Wearing gowns, masks, and gloves are examples of personal protective equipment (PPE) that caregivers for norovirus-infected patients should wear.
(b) Environmental cleaning and disinfection: It is important to clean and sanitize every surface, particularly in locations where sick people have been. A 1:10 diluted bleach solution should be used for this. High-touch areas such as light switches, counters, and doorknobs require extra care. Disinfectants that exhibit efficacy against norovirus should be registered with the EPA.
(c) Food safety: It is important to follow safe food handling procedures, particularly when handling ready-to-eat items. Foods must be cooked to a minimum safe internal temperature of 165°F (74°C), and it is crucial to keep surfaces and food items separate from one another.
(d) Control of the outbreak: To stop the spread of the norovirus, affected people must be located and isolated. Places like schools and hospitals should be shuttered to stop the spread of an infection. It is recommended to use infection control measures, such as improved cleaning and disinfection practices.
(e) Immunization: Norovirus vaccination should be taken into consideration, particularly for those who are more vulnerable, such as healthcare personnel and people with weakened immune systems.
(f) Additional measures: Reducing the concentration of airborne norovirus can be achieved by improving ventilation in locations with ill people. UV light can also be used to disinfect surfaces and lower the number of viruses in the air. Last but not least, enacting sick leave regulations is critical to motivate the ill to remain at home and prevent the norovirus from spreading [56].
(g) Special measures for food service settings: Food service establishments should ensure that strict food safety measures, for instance, proper hand washing and safe food handling, are in place to reduce the incidences of norovirus contamination. Cleanliness and sanitizing of facilities where foods are prepared especially after the provision of meals to infected people is necessary. Food service providers should also not provide high-risk

foods to susceptible individuals such as the elderly and small children, such as raw oysters [57].

(h) Isolation during outbreaks: Symptomatic individuals should remain at home and avoid contact with others for at least 48 h after their symptoms have resolved to prevent further transmission.

(i) Safe handling of contaminated items: Care should be taken for those infected when handling their laundry and dishes. Use gloves and wash items in hot water to eradicate the virus.

(j) Public awareness: Increasing awareness about norovirus, including its symptoms, transmission methods, and preventive measures, is vital for reducing the frequency of outbreaks [58].

(k) Patient cohorting and isolation precautions: Minimize exposure to vomitus or diarrhea. Place symptomatic patients in single-occupancy rooms under contact precautions. If single rooms are unavailable, ensure that symptomatic patients are kept separate from those who are asymptomatic. Continue contact precautions for at least 48 h after symptoms have resolved.

(l) Communication: Establish clear, written policies for outbreak communication to ensure that timely updates are provided to all stakeholders, including local health authorities.

(m) Active case-finding: Conduct active case-finding when clusters of acute gastroenteritis are identified. Use defined case criteria to monitor exposed and symptomatic individuals effectively.

(n) Diagnostics: Rely on clinical and epidemiological evidence to identify potential norovirus outbreaks. Submit stool samples for laboratory testing to confirm suspected cases [59].

4.14 Conclusion

Norovirus is an extremely contagious virus and a primary cause of acute gastroenteritis worldwide, characterized by symptoms such as vomiting, diarrhea, and stomach pain. It spreads swiftly through contaminated food, water, and surfaces and direct contact, particularly in enclosed environments like schools, healthcare facilities, and cruise ships. The virus shows seasonal patterns, with outbreaks peaking during the dry season in tropical regions and in colder months in temperate areas. Vulnerable populations, including young children, the elderly, and individuals with weakened immune systems, face a higher risk of severe infection. Norovirus' ability to adapt and persist makes it a significant public health challenge, leading to widespread illness and substantial economic impact globally. Its high rate of mutation through antigenic drift and recombination has made it difficult to contain the virus because it evades immune defenses and constantly produces new variants, especially GII.4. This complicates the building of long-term immunity or the development of effective vaccines. Preventive measures are mainly through hygiene maintenance practices such as frequent hand washing, proper handling of food, and effective environmental sanitation. Public health campaigns concentrate on increasing awareness to curb transmission and manage outbreaks, yet controlling the virus especially proves difficult in high-risk environments like healthcare facilities and communal living spaces. To combat norovirus, ongoing research focuses on developing vaccines and antiviral treatments that reduce infection severity and spread. Scientists are studying its mechanisms of infection and evasion of immune response to advance preventive and therapeutic measures. Nevertheless, the complexity of the virus and its high transmissibility stand as challenging barriers for vaccine and treatment development. Meaningful reduction of the global burden of norovirus can only be attained through a comprehensive preventive measure combined with public health education and continuous surveillance. The measures in place to monitor food safety, improve hygiene practices, and advance scientific research are important in minimizing the impact of the virus. Nonetheless, creation of effective medical interventions will remain fundamental to reducing societal and economic costs from norovirus outbreaks.

AI disclosure statement: The author acknowledges the use of AI-based tools for minor language editing or grammatical correction. However, all intellectual contributions, analyses, and conclusions presented in this chapter are entirely the author's own.

References

1. Ahmed, S.M., Hall, A.J., Robinson, A.E., Verhoef, L., Premkumar, P., Parashar, U.D., *et al.*, Global prevalence of norovirus in cases of gastroenteritis: A systematic review and meta-analysis. *Lancet Infect. Dis. [Internet]*, 14, 8, 725–30, 2014, Available from: http://dx.doi.org/10.1016/S1473-3099(14)70767-4.
2. Ushijima, H., Fujimoto, T., Müller, W.E., Hayakawa, S., Norovirus and Foodborne Disease: A Review. *Food Saf.*, 2, 3, 37–54, 2014.
3. Babaei, A., Rafiee, N., Taheri, B., Sohrabi, H., Mokhtarzadeh, A., Recent Advances in Early Diagnosis of Viruses Associated with Gastroenteritis by Biosensors. *Biosensors*, 12, 7, 499, 2022.

4. White, P.A., Evolution of norovirus. *Clin. Microbiol. Infect.*, 20, 8, 741–5, 2014.
5. Chhabra, P., de Graaf, M., Parra, G.I., Chan, M.C.W., Green, K., Martella, V., *et al.*, Updated classification of norovirus genogroups and genotypes. *J. Gen. Virol.*, 100, 10, 1393–406, 2019.
6. Zheng, D.P., Ando, T., Fankhauser, R.L., Beard, R.S., Glass, R.I., Monroe, S.S., Norovirus classification and proposed strain nomenclature. *Virology*, 346, 2, 312–23, 2006.
7. Mattison, C.P., Cardemil, C.V., Hall, A.J., Progress on norovirus vaccine research: public health considerations and future directions. *Expert Rev. Vaccines [Internet]*, 17, 9, 773–84, 2018, Available from: https://doi.org/10.1080/14760584.2018.1510327.
8. Pérez, D., A bibliometric analysis of the literature on trade unionism in geography. *Rev. Geogr. Venez.*, 63, 1, 158–73, 2022.
9. Zhang, P., Hao, C., Di, X., Chuizhao, X., Jinsong, L., Guisen, Z., *et al.*, Global prevalence of norovirus gastroenteritis after emergence of the GII.4 Sydney 2012 variant: a systematic review and meta-analysis. *Front. Public Health*, 12, 1–17, June 2024.
10. Atmar, R.L. and Estes, M.K., The Epidemiologic and Clinical Importance of Norovirus Infection. *Gastroenterol. Clin. North Am.*, 35, 2, 275–90, 2006.
11. Wobus, C.E., Thackray, L.B., Virgin, H.W., Murine Norovirus: a Model System To Study Norovirus Biology and Pathogenesis. *J. Virol.*, 80, 11, 5104–12, 2006.
12. Cortes-Penfield, N.W., Ramani, S., Estes, M.K., Atmar, R.L., Prospects and Challenges in the Development of a Norovirus Vaccine. *Clin. Ther.*, 39, 1537–1549, 2017.
13. King, T., Cole, M., Farber, J.M., Eisenbrand, G., Zabaras, D., Fox, E.M., *et al.*, Food safety for food security: Relationship between global megatrends and developments in food safety. *Trends Food Sci. Technol.*, 68, 160–75, 2017.
14. De Graaf, M., Van Beek, J., Koopmans, M.P.G., Human norovirus transmission and evolution in a changing world. *Nat. Rev. Microbiol.*, 14, 7, 421–33, 2016.
15. Zhang, M., Fu, M., Hu, Q., Advances in human norovirus vaccine research. *Vaccines*, 9, 7, 1–11, 2021.
16. Esposito, S. and Principi, N., Norovirus Vaccine: Priorities for Future Research and Development. *Front. Immunol.*, 11, 1–9, July 2020.
17. Debbink, K., Lindesmith, L.C., Baric, R.S., The state of norovirus vaccines. *Clin. Infect. Dis.*, 58, 12, 1746–1752, 2014 Jun.
18. Peña-gil, N., Santiso-bellón, C., Gozalbo-rovira, R., Buesa, J., Monedero, V., Rodríguez-díaz, J., The role of host glycobiology and gut microbiota in rotavirus and norovirus infection, an update. *Int. J. Mol. Sci.*, 22, 24, 13438, 2021.
19. Sitabkhan, A., Molecular surveillance of norovirus and study of the evolutionary mechanisms of antigenic drift and shift, 2017, Available from:

http://unsworks.unsw.edu.au/fapi/datastream/unsworks:46662/SOURCE02?view=true.
20. Wei, N., Ge, J., Tan, C., Song, Y., Wang, S., Bao, M., *et al.*, Epidemiology and evolution of Norovirus in China. *Hum. Vaccin. Immunother. [Internet]*, 17, 11, 4553–66, 2021, Available from: https://doi.org/10.1080/21645515.2021.1961465.
21. Parra, G.I., Emergence of norovirus strains: A tale of two genes. *Virus Evol.*, 5, 2, 1–9, 2019.
22. Rocha-Pereira, J., Van Dycke, J., Neyts, J., Norovirus genetic diversity and evolution: implications for antiviral therapy. *Curr. Opin. Virol.*, 20, 92–8, 2016.
23. Siqueira, J.A.M., Bandeira, R.D.S., Oliveira, D.D.S., Dos Santos, L.F.P., Gabbay, Y.B., Genotype diversity and molecular evolution of noroviruses: A 30-year (1982-2011) comprehensive study with children from Northern Brazil. *PLoS One*, 12, 6, 1–21, 2017.
24. Ruvoën-Clouet, N., Belliot, G., Le Pendu, J., Noroviruses and histo-blood groups: The impact of common host genetic polymorphisms on virus transmission and evolution. *Rev. Med. Virol.*, 23, 6, 355–66, 2013.
25. Lucero, Y., Matson, D.O., Ashkenazi, S., George, S., O'ryan, M., Norovirus: Facts and reflections from past, present, and future. *Viruses*, 13, 12, 1–18, 2021.
26. Burke, R.M., Shah, M.P., Wikswo, M.E., Barclay, L., Kambhampati, A., Marsh, Z., *et al.*, The Norovirus Epidemiologic Triad: Predictors of Severe Outcomes in US Norovirus Outbreaks, 2009-2016. *J. Infect. Dis.*, 219, 9, 1364–72, 2019.
27. Goswami, R., Pavon, C.G., Miller, I.G., Berendam, S.J., Williams, C.A., Rosenthal, D., *et al.*, Prenatal Immunization to Prevent Viral Disease Outcomes During Pregnancy and Early Life. *Front. Virol.*, 2, 1–23, May 2022.
28. Sukhrie, F.H.A., Teunis, P., Vennema, H., Copra, C., Thijs Beersma, M.F.C., Bogerman, J., *et al.*, Nosocomial transmission of norovirus is mainly caused by symptomatic cases. *Clin. Infect. Dis.*, 54, 7, 931–7, 2012.
29. Isakbaeva, E.T., Widdowson, M.A., Beard, R.S., Bulens, S.N., Mullins, J., Monroe, S.S., *et al.*, Norovirus transmission on cruise ship. *Emerg. Infect. Dis.*, 11, 1, 154–7, 2005.
30. Koo, E.S. and Jeong, Y.S., Comparative study of the geographical spread of genogroup II porcine norovirus and human norovirus. *J. Microbiol.*, 59, 7, 644–50, 2021.
31. Towers, S., Chen, J., Cruz, C., Melendez, J., Rodriguez, J., Salinas, A., *et al.*, Quantifying the relative effects of environmental and direct transmission of norovirus. *R. Soc. Open Sci.*, 5, 3, 170602, 2018.
32. Wang, Z., Shin, H., Jung, S., Yeo, D., Park, H., Shin, S., *et al.*, Effects of weather and environmental factors on the seasonal prevalence of foodborne viruses in irrigation waters in Gyeonggi province, Korea. *Microorganisms*, 8, 8, 1–19, 2020.

33. Assab, R. and Temime, L., The role of hand hygiene in controlling norovirus spread in nursing homes. *BMC Infect. Dis. [Internet]*, 16, 1, 1–10, 2016, Available from: http://dx.doi.org/10.1186/s12879-016-1702-0.
34. Rico, E., Pérez, C., Belver, A., Sabaté, S., Razquin, E., de Benito, J., *et al.*, Norovirus detection in environmental samples in norovirus outbreaks in closed and semi-closed settings. *J. Hosp. Infect.*, 105, 1, 3–9, 2020.
35. Bitler, E.J., Matthews, J.E., Dickey, B.W., Eisenberg, J.N.S., Leon, J.S., Norovirus outbreaks: A systematic review of commonly implicated transmission routes and vehicles. *Epidemiol. Infect.*, 141, 8, 1563–71, 2013.
36. Shamkhali Chenar, S. and Deng, Z., Environmental indicators for human norovirus outbreaks. *Int. J. Environ. Health Res.*, 27, 1, 40–51, 2017.
37. Wang, P., Goggins, W.B., Chan, E.Y.Y., A time-series study of the association of rainfall, relative humidity and ambient temperature with hospitalizations for rotavirus and norovirus infection among children in Hong Kong. *Sci. Total Environ.*, 643, 414–22, 2018.
38. Kambhampati, A.K., Calderwood, L., Wikswo, M.E., Barclay, L., Mattison, C.P., Balachandran, N., *et al.*, Spatiotemporal Trends in Norovirus Outbreaks in the United States, 2009-2019. *Clin. Infect. Dis. [Internet]*, 76, 4, 667–73, 2023, Available from: https://doi.org/10.1093/cid/ciac627.
39. Kraut, R.Y., Snedeker, K.G., Babenko, O., Honish, L., Influence of School Year on Seasonality of Norovirus Outbreaks in Developed Countries. *Can. J. Infect. Dis. Med. Microbiol.*, 2017, 9840632, 2017.
40. Marshall, J.A. and Bruggink, L.D., The dynamics of norovirus outbreak epidemics: recent insights. *Int. J. Environ. Res. Public Health*, 8, 4, 1141–9, 2011.
41. Steele, M.K., Wikswo, M.E., Hall, A.J., Koelle, K., Handel, A., Levy, K., *et al.*, Characterizing norovirus transmission from outbreak data, United States. *Emerg. Infect. Dis.*, 26, 8, 1818–25, 2020.
42. Bruggink, L.D. and Marshall, J.A., The incidence of norovirus-associated gastroenteritis outbreaks in victoria, Australia (2002-2007) and their relationship with rainfall. *Int. J. Environ. Res. Public Health*, 7, 7, 2822–7, 2010.
43. Yates, M,V., Norovirus, in: *Microbiology of Waterborne Diseases: Microbiological Aspects and Risks.* S.L. Percival, M.V. Yates, D.W. Williams, R.M. Chalmers, N.F. Gray (Eds.), pp. 515–522, 2nd ed, Amsterdam, Academic Press, 2013.
44. Van Alphen, L.B., Dorléans, F., Schultz, A.C., Fonager, J., Ethelberg, S., Dalgaard, C., *et al.*, The application of new molecular methods in the investigation of a waterborne outbreak of norovirus in Denmark, 2012. *PLoS One*, 9, 9, e105603, 2014.
45. Chiu, S.C., Hu, S.C., Liao, L.M., Chen, Y.H., Lin, J.H., Norovirus Genogroup II Epidemics and the Potential Effect of Climate Change on Norovirus Transmission in Taiwan. *Viruses*, 14, 3, 518, 2022.
46. Cannon, J.L., Papafragkou, E., Park, G.W., Osborne, J., Jaykus, L.A., Vinjé, J., Surrogates for the study, of norovirus stability and inactivation in the environment: A comparison of murine norovirus and feline calicivirus. *J. Food Prot.*, 69, 11, 2761–5, 2006.

47. Pogan, R., Dülfer, J., Uetrecht, C., Norovirus assembly and stability. *Curr. Opin. Virol.*, 31, 59–65, 2018.
48. Cuellar, J.L., Meinhoevel, F., Hoehne, M., Donath, E., Size and mechanical stability of norovirus capsids depend on pH: A nanoindentation study. *J. Gen. Virol.*, 91, 10, 2449–56, 2010.
49. Hall, A.J., Lopman, B.A., Payne, D.C., Patel, M.M., Gastañaduy, P.A., Vinjé, J., *et al.*, Norovirus disease in the united states. *Emerg. Infect. Dis.*, 19, 8, 1198–205, 2013.
50. Ondrikova, N., HE, C., NA, C., Iturriza-Gomara, M., Vivancos, R., Harris, J.P., Understanding norovirus reporting patterns in England: a mixed model approach. *BMC Public Health*, 21, 1, 1–9, 2021.
51. Inns, T., Harris, J., Vivancos, R., Iturriza-Gomara, M., O'Brien, S., Community-based surveillance of norovirus disease: A systematic review. *BMC Infect. Dis.*, 17, 1, 1–10, 2017.
52. Mans, J., Armah, G.E., Steele, A.D., Taylor, M.B., Norovirus epidemiology in Africa: A review. *PLoS One*, 11, 4, 1–15, 2016.
53. Lopman, B.A., Adak, G.K., Reacher, M.H., Brown, D.W.G., Two Epidemiologic Patterns of Norovirus Outbreaks: Surveillance in. *Emerg. Infect. Dis.*, 9, 1, 71–7, 2003.
54. Chenar, S.S. and Deng, Z., Development of genetic programming-based model for predicting oyster norovirus outbreak risks. *Water Res.*, 128, 20–37, 2018 October.
55. Harris, J.P., Norovirus Surveillance: An Epidemiological Perspective. *J. Infect. Dis.*, 213, S8–S11, 2016.
56. Barclay, L., Park, G.W., Vega, E., Hall, A., Parashar, U., Vinjé, J., *et al.*, Infection control for norovirus. *Clin. Microbiol. Infect.*, 20, 8, 731–40, 2014.
57. Malone, L., Cm, M.B.A., Grigorenko, E., Stalons, D., Id Week 2015, 2, September, 2633851, 2017.
58. Winder, N., Gohar, S., Muthana, M., Norovirus: An Overview of Virology and Preventative Measures. *Viruses*, 14, 12, 2811, 2022.
59. MacCannell, T., Umscheid, C.A., Agarwal, R.K., Lee, I., Kuntz, G., *et al.*, Guideline for the Prevention and Control of Norovirus Gastroenteritis Outbreaks in Healthcare Settings. *Infect. Control Hosp. Epidemiol.*, 32, 10, 939–69.s, 2011.

5

Zoonotic Transmission of Norovirus: Current Understanding and Future Implications

Vini Mehta[1]*, Sapna Negi[1], Snehasish Tripathy[1], Sarvesh Rastogi[2] and Ankita Mathur[1]

[1]*Department of Dental Research Cell, Dr. D. Y. Patil Dental College and Hospital, Dr. D. Y. Patil Vidyapeeth, Pune, India*
[2]*School of Applied and Life Science, Uttaranchal University, Dehradun, Uttarakhand, India*

Abstract
Countries worldwide have made substantial progress in managing and preventing gastrointestinal illnesses. However, no progress has been made in relation to gastroenteritis associated with norovirus infection. Norovirus is responsible for causing the majority of non-bacterial-associated gastrointestinal infections. It can spread through food, water, and inanimate surfaces, eventually resulting in outbreaks. Similar to symptomatic cases, asymptomatic patients consistently release the virus with similar patterns of viral quantity and shedding period, which also may lead to norovirus outbreaks. The virus has 10 genogroups ranging from GI to GX and each genogroup consists of genotypes that include viral strains within it. GIII to GVII, GX, and GNA1 have been found in various animals. On the other hand, genogroups I, II, IV, VIII, and IX infect humans. While cases of human infection through animal norovirus strains have not been documented so far, the close genetic relationship and unknown origins of new human genotypes suggest possible inter-species transmission. In this chapter, we discuss the emerging evidence that has reported domestic as well as wildlife animal species as reservoirs for norovirus infection and evidence on reverse transmission from animal to human. We further elaborated on existing possible mechanisms of zoonotic transmission of norovirus. Ultimately, addressing noroviruses' zoonotic potential necessitates a multidimensional approach that involves expanded research, increased

**Corresponding author*: vini.mehta@statsense.in

Sachin Kumar, Shriyansh Srivastava, Rishabha Malviya and Ranjit Sah (eds.) Emerging Infectious Disease: Unravelling Norovirus and Pandemic Dynamics, (129–150) © 2026 Scrivener Publishing LLC

surveillance, interdisciplinary collaboration, and strong public health policies to better prevent and respond to norovirus outbreaks.

***Keywords*:** Norovirus, zoonoses, infectious disease transmission, One Health, public health, disease reservoir, genotypes, gastroenteritis

5.1 Brief Overview of Norovirus

Countries worldwide have achieved substantial progress in the management and prevention of gastrointestinal illnesses. There has been a significant decrease in diarrhea-related mortality rates from 42 per 100,000 in the year 2000 to 15.4 per 100,000 people in the year 2021 [1], particularly in <5 years age group of children, because of improvements in water, sanitation, and hygiene practices. Additionally, the advent of rotavirus vaccination has played a crucial role in controlling parasitic infections. Nevertheless, even with this significant enhancement, fatalities resulting from diarrheal disease remain the fourth most prevalent cause of death overall. Specifically, no progress has been made in relation to gastroenteritis caused by norovirus. Norovirus is a non-segmented RNA virus of the *Caliciviridae* family and is responsible for causing the majority of all non-bacterial-associated gastrointestinal infections [2]. Although it resolves in a few days with no long-term damage, the virus is known for its highly infectious nature, low infection dose, and ability to persist in the environment, which make it a significant public health concern. Norovirus is composed of 7.5 kb of genetic material arranged into three open reading frames (ORF)—1, 2, and 3 [3]. ORF1 encodes non-structural proteins that aid in viral replication, whereas ORF2 encodes the main capsid protein (VP1), which forms an outer protective shell around the virus and serves as the basis for its antigenic traits. ORF-3 encodes the minor capsid protein that helps in the stabilization of viral particles. ORF1 and 2 play critical roles in the formation of new strains *via* diverse recombination events, although their origin remains unknown [4, 5]. The virus has 10 genogroups ranging from GI to GX and each genotype consists of genotypes that further include viral strains within it [6]. GIII to GVII, GX, and GNA1 have been found in varied animals including sheep, cattle, dogs, cats, rodents, bats, and harbor porpoises [4]. On the other hand, genogroups I, II, IV, VIII, and IX infect humans. Currently, GI and GII are frequently identified in humans, with GII.4 being the most common strain and the primary cause of acute gastroenteritis (AGE) epidemic among both children and adults globally [5, 6]. These norovirus strains account for 70%–80% of reported outbreaks for the last 10 years. They are known for their antigenic evolution, forming a new variant of norovirus at a regular interval of 2–4 years [7] with modifications in transmission and pathogenesis mechanisms.

5.2 Epidemiology of Norovirus

Acute gastroenteritis is common across the globe with 19% of diarrheal cases associated with norovirus infection [8] and could potentially lead to nearly 685 million morbidities and 2 lakh diarrheal deaths annually [9, 10]. In the United States, norovirus is the primary cause of medically attended gastroenteritis with nearly 2,500 outbreaks reported each year with 58% of these outbreaks due to foodborne infection. The risk of acquisition and norovirus infections is higher in children (<5 years of age), the elderly, and individuals with low immunity and poor socioeconomic status [11]. According to CDC data [11], norovirus-related infection results in nearly 200 million cases and 50,000 deaths among children at the global level. At the same time, it is associated with 1 million pediatric in-patient admissions and 900 elderly mortalities in the United States. A systematic review and meta-analysis reported South America as the continent with the highest prevalence of norovirus-associated gastroenteritis, followed by Europe, Asia, Africa, and Oceania [12]. Another recent meta-analysis reported a higher positivity rate of norovirus in underdeveloped nations than in developed nations [8]. However, despite this visibly high burden of norovirus-associated gastroenteritis, the true estimate might exceed the reported cases since patients with mild symptoms may not seek medical treatment [13]. Liao *et al.* [12] found a higher number of cases in the community compared to healthcare settings, suggesting that people often do not seek care until the symptoms of norovirus do not resolve. These findings across the available evidence indicate that norovirus-associated gastroenteritis affects all age groups and socioeconomic classes, while occurrence differs substantially across these groups. The virus outbreak has been reported more from enclosed or semi-enclosed establishments, such as chronic care facilities [14], nursing homes [15], childcare centers, schools [16], restaurants, and hotels [17].

5.3 Symptoms of Norovirus-Associated Gastroenteritis

Malaise, nausea and vomiting, diarrhea, headache, and fever are among the most prevalent symptoms of norovirus-associated gastroenteritis [3, 6, 13, 18]. Additionally, recent research has documented asymptomatic norovirus cases. The global pooled prevalence of asymptomatic/subclinical norovirus infections is nearly 21.8% during outbreaks with the GII genotype being prevalent [19].

Similar to symptomatic cases, asymptomatic patients consistently release the virus with similar patterns of viral quantity and shedding period, which also may lead to norovirus outbreaks. Therefore, neglecting to identify asymptomatic persons during an outbreak could extend its length and amplify its severity [19]. Infection with norovirus can be triggered by a relatively small amount of 18 viral particulates as well [2]. Hence, both asymptomatic and symptomatic norovirus cases are crucial in disease transmission. Thus, it is essential to comprehend the transmission patterns and related elements to enhance the empirical foundation for interventions targeted at mitigating the impact of norovirus outbreaks.

5.4 Mode of Transmission of Norovirus

Transmission is the process by which an infectious agent is spread from one individual to another. Viruses are commonly spread from person to person or from animals to humans when an infected person releases the virus into the immediate surroundings. Transmission can occur through several modes. Norovirus exhibits high environmental resilience, heat up to 60°C, withstanding freezing temperatures, and disinfectants involving ethanol or chlorine [20]. This endurance enables the norovirus to spread through various mediums including inanimate surfaces, food, or drinking water, resulting in disease outbreaks, particularly in close-knit environments (Figure 5.1).

Norovirus is primarily spread from sick individuals to healthy ones through the means discussed below.

5.4.1 Vomitus or Stool

Transmission of the virus, particularly GII, can occur directly or indirectly via exposure to stool or vomitus of infected patients while giving care.

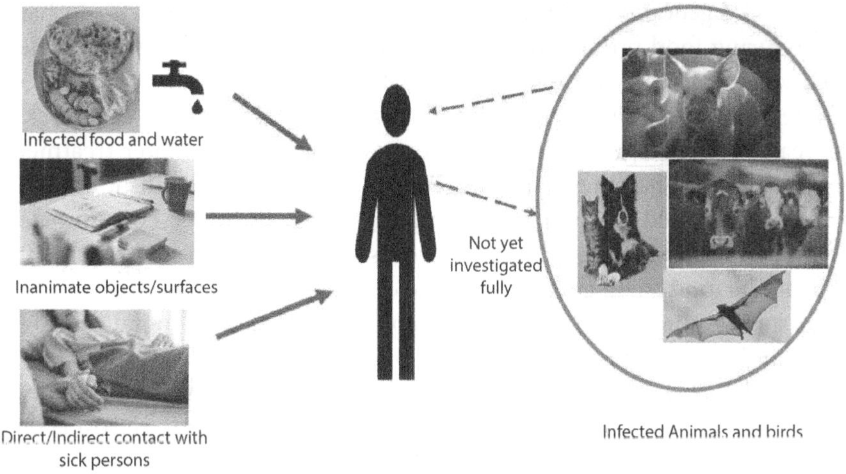

Figure 5.1 Transmission mode of norovirus.

Infected individuals produce tons of virus particles in every gram of vomit or feces, reaching the highest concentrations within the first 24 to 48 h after the onset of symptoms [2, 21]. This shedding of viral particles typically lasts for an average of 4 weeks. In immunodeficient hosts, the virus particle excretion in feces can continue for several months after infection and may persist for years, leading to a long-term, persistent illness [10]. In immunocompetent people, the feces can contain virus RNA for weeks after the symptoms disappear. Asymptomatic people can excrete an enormous amount of viruses which have a role in transmission [22]. Similarly, vomiting may contaminate the environment and spread by fomites and bioaerosols [2]. Bioaerosol particle size affects both airborne duration and inhalation probability. Disease-related microbial aerosols typically range from 0.1 to 20 μm in diameter [2]. Norovirus often transmits via bioaerosols and direct contact with infected individuals or via contaminated environments.

5.4.2 Through Contaminated Food and Water

Norovirus-associated gastroenteritis can also be caused by consuming contaminated food or water. Food and water are the most prevalent means of norovirus transmission, and food-borne norovirus infection is one of the most common types of infectious disease worldwide [6, 9]. GI [3, 6, 7] and GII [3, 6, 10] are more frequently linked to foodborne illness. Conversely, the GI genogroup is linked to the transmission through water, as opposed to the GII genogroup [23, 24]. Foodborne transmission, which can happen during food production or preparation or food handling by infected individuals, is a key route for the global propagation of noroviruses [25]. Water source contamination with norovirus-containing feces can quickly spread to a wide population in regions with insufficient water treatment and sanitation facilities. Leafy vegetables, fresh fruits, shellfish, etc., can be contaminated during the irrigation process by water contaminated with sewage or by harvesters and processors infected with the virus [9]. Given the food chain globalization, foodborne outbreaks are common and can spread norovirus strains. These epidemics may also contain norovirus combinations, promoting viral recombination. The worldwide breadth of foodborne norovirus outbreaks remains hard to determine because countries track outbreak epidemiology separately. However, retrospective research found that 7% of outbreaks occur during foreign travel and had a common source [26].

5.5 Zoonotic Transmission

Although norovirus infection is mostly linked with transmission between humans, new evidence indicates that zoonotic transmission also plays an integral part in the ecology of norovirus [9]. The term "zoonotic transmission" describes the spread of infectious pathogens from humans to other vertebrate animals [8]. Zoonotic viruses continue to pose a global health risk to both humans and animals. This covers infections that humans contract from animals either directly or indirectly, for example, through vector-borne infections, environmental diseases, or food-borne pathogens [7]. Since the 1970s, zoonotic diseases have gained popularity as a result of increased human activity [10]. The encroachment of humans and livestock into wildlife natural habitats, whether through urbanization, tourism, wildlife hunting, or intensified agricultural practices that reduce biodiversity, has led to greater incidences and closer interaction between humans and domestic and wild animals. This has drastically influenced 75% of land habitats and 66% of aquatic ecosystems, resulting in the emergence of zoonoses [27]. Zoonotic diseases have profound effects on both human and animal well-being, as well as on livelihoods and economies [28]. Over the past 20 to 30 years, viral pathogens causing zoonotic illnesses have become dominant in the global infection landscape. These diseases have resulted in various forms of illness and death, as well as cost and declines in productivity. Approximately one billion episodes of disease and millions of deaths occur annually worldwide due to zoonoses. Zoonoses account for

61% of all infectious illnesses, 75% of newly emerging infectious diseases, and 80% of potential biological warfare pathogens that are reported worldwide. A total of more than 30 novel human infections have been identified during the past 30 years, with 75% of them having originated in animals [29]. RNA viruses present significant zoonotic hazards due to their ability to quickly emerge and spread. A study among livestock revealed the cytoplasmic replication capacity of the virus, without entering the nucleus, as the most influential factor in determining its ability to cross-species transmission between animals and humans [30]. Therefore, understanding the zoonotic transmission of norovirus is crucial for public health, epidemiology, and global health security. It involves a multidisciplinary approach that considers the complex interactions between humans, animals, and the environment. By studying zoonotic transmission, we can develop effective strategies to predict, prevent, and respond to norovirus outbreaks, ultimately protecting human health and promoting sustainable ecosystems [3].

While cases of human infection through animal norovirus strains have not been documented so far, the close genetic relationship and unknown origins of new human genotypes suggest possible inter-species transmission [4]. Some evidence for the domestic as well as wildlife animal species as reservoirs for norovirus infection has been reported across several studies [4, 31, 32].

5.5.1 Animal Reservoirs

Domestic Animals

Norovirus first emerged in 1968 as a gastroenteritis outbreak among schoolchildren in Ohio, USA. It was later classified in 1972 as a small, round-shaped virus with a 27–35-nm diameter based on immune electron microscopy observation [10]. Similar viruses were subsequently detected in the fecal material of domestic cows and pigs, confirming the genetically similar norovirus pathogen in both domestic animals and humans [10, 31].

(a) Pigs: With the global demand for pork rising, diverse pig farming centers and practices worldwide such as breeding centers, pig farms, live animal markets, abattoirs, and trade exhibitions create numerous contact points between pigs and humans. These interactions increase the likelihood of viral transmission. Moreover, pigs are being utilized more and more in xenotransplantation and as research subjects for human diseases because of their physiological, genetic, and immunological resemblances to humans. As a result, pigs have frequently served as intermediate and amplification hosts in various disease epidemics of humans. Porcine noroviruses, particularly the GII genotype, have been initially discovered in domestic pigs in the United States. Subsequently, many GII genotypes [11, 18, and 19] were documented in both healthy and sick pigs in surveillance studies conducted across the globe. Three GII genotypes and a possible recombinant were found in fecal samples from adult swine in the United States. One genotype, GII.18, was found to be closely linked to human noroviruses and able to replicate in gnotobiotic pigs [19]. In line with this, a study conducted in Italy on 225 swine fecal samples discovered a high prevalence of norovirus (GII.P11, P18) in older pigs with significant genetic diversity, including different genetic subgroups and unique amino acid variation in norovirus, indicating the co-circulation of different swine norovirus subgroups [31]. Pigs have also been found to develop antibodies against the human norovirus, with prevalences ranging from 52% to 70%. Virus-like particles of various genotypes were identified by these antibodies, but their ability to attach to pig mucin was unaffected. Cross-reactivity investigations revealed that antibodies to GII.1 and GII.3 genotypes exhibited cross-reactivity with the porcine GII.11 genotype; however, no such findings were noted with GI genotypes. These findings call into doubt the co-circulation/reverse transmission of some human norovirus genotypes from humans to pigs, particularly in areas with a higher risk of pig–human interaction [3, 4].

(b) Calves: Norovirus identified in stool samples from domestic cows or calves is called bovine norovirus. GIII.1 and GIII.2 were the first norovirus genotypes identified in Germany and England, respectively [33], and later in Africa, America, and Asia. Many published articles reported discovering norovirus in the intestinal samples of diarrheic and asymptomatic cattle [32, 34–37]. In a research investigation carried out in Konya, RT-PCR was used to analyze the feces of 80 cattle from a slaughterhouse. In addition, 7.5% of the samples tested positive for norovirus [32]. Another Argentine study found norovirus sequences related to genotypes GIII.1 and GIII.2 in diarrheal bovine herds from the nation's main dairy regions. Furthermore, one of the sequences (B4881) identified by the genetic analysis could correspond to a novel genotype of the bovine norovirus GIII, which is currently being provisionally classified GIII.4 [38]. In a study conducted in Uruguay [35], RT-qPCR testing of 761 fecal or intestinal samples of dairy and beef calves revealed a high incidence (66%) of bovine norovirus, with more cases in dairy calves and higher infection rates in those under 2 weeks old. Two main genotypes—GIII.1 and GIII.2—and four hybrid variants were identified, indicating the global spread of these strains [35]. Numerous experimental investigations have revealed that norovirus sheds both before and after animals exhibit symptoms of diarrhea. Although asymptomatic calves can also transmit the virus, very little virus identification has been done in this field.

(c) Dogs: Norovirus in dogs, commonly known as canine norovirus, was originally discovered in 2-month-old pups in Italy [39] and was later reported to other countries worldwide [40]. Seroprevalence of norovirus in dogs across 14 European nations showed extensive circulation of GVI norovirus in dogs throughout Europe, ranging from 0% to 64% in Hungary and Portugal, respectively [41]. Furthermore, GIV, GV, and GVII norovirus genotypes are frequently identified in both diarrheal and healthy dogs, particularly the younger ones. In addition, a few investigations have highlighted concerns about human–dog transmission of canine norovirus due to the discovery of animal-specific antibodies (GIV.2) norovirus strains in serological specimens [42, 43]. One study in Uruguay [44] has shown the presence of a canine norovirus GVII strain, marking the first occurrence of this strain in the American continent. The study also raises concerns about potential zoonotic transmission, as canine strains were found in domestic waste. Nevertheless, a study successfully developed a method based on full-genome amplicons to assist in the genotyping of canine noroviruses that are now prevalent in the United States. The study revealed that exclusively animal strains generated GVI/GIV chimeric viruses, suggesting constraints in impeding transmission between other species [44].

(d) Cats: Some studies have also been carried out on feline norovirus seen in cats. The first case was reported in an Italy based investigation when 16% of cats tested positive for GIV.2-specific antibodies [45]. Another study conducted among cats in shelter houses in the United States detected feline norovirus in 42.8% of diarrheic fecal samples [46]. Upon genome sequencing, the norovirus among cats was found to be similar to the norovirus found in other canine animals, suggestive of interspecies circulation of the virus [45, 46]. Takano *et al.* (2015) conducted an experiment in which healthy cats were administered a fecal emulsion containing feline norovirus and observed for clinical symptoms [47]. The results showed that three out of four cats developed diarrhea, and one cat began vomiting, demonstrating that feline norovirus (FNoV) can cause infectious gastroenteritis in cats. In a subsequent study, Takano *et al.* [48] explored whether cats that had previously been infected with genogroup GVI norovirus were immune to reinfection with the same strain. They found that higher levels of anti-FNoV IgG antibodies in the cats' blood were associated with lower viral loads in their stool samples and milder clinical symptoms [48]. Nonetheless, complete protection against reinfection was not observed (Table 5.1).

Table 5.1 Norovirus genotypes found among different hosts.

Host	Norovirus genotype	Location reported
Human	GI, GII	Global [19]
Pigs	GII.1, GII.11, GII.18, GII.19	America, Asia, Europe, Africa [18, 31, 49]
Cattles	GIII.1, GIII.2, GIII.4	America, Asia, Europe, Africa [33, 34, 34, 36]
Dogs	GIV, GVI, GV, GVII	America, Asia, Europe [43, 50, 51]
Cats	GIV.2	America, Europe [45, 46, 48]
Rats	GV	America, Europe [52]
Bats	GX.1	Asia [52]
Sea Lion	GII, GIV2	America, Asia [52]
Fox	GVII	Asia [53]
Wild birds	GI, GII.3, GII,4	Europe [52]
Baltic Sea mussels	GI, GII	Europe [54]

5.5.2 Wild Animals

Besides domestic animals, instances of norovirus infection have also been reported among a few wild animals and birds. Summa *et al.* [52] found that wild birds, especially gulls and crows, as well as rodents, can spread human noroviruses (HuNoVs) in the environment, with 27% of bird samples and 2% of rat samples testing positive. Another study found murine norovirus (MNV) in 54.6% of fecal samples from healthy mice in an Italian facility, with two co-circulating MNV variants of GV norovirus detected despite strict animal care guidelines [55]. A novel GVII norovirus was identified in fox fecal samples from Xinjiang, China, showing close genetic relation to canine GVII norovirus strains but with distinct differences in his blood group antigen binding sites and patterns [53]. Bigoraj *et al.* found human noroviruses (NoV GI: 18.3%, NoV GII: 23.3%) in Baltic Sea mussels from the Polish coast, suggesting their use as bioindicators of seawater contamination [54]. Besides that, norovirus strains have been found in harbor porpoises (*Phocoena phocoena*), California sea lions (*Zalophus californianus*), and bat species (*Rhinolophus pusillus*, *Rhinolophus sinicus*, and *Rhinolophus affinis*) [4] (Figure 5.2).

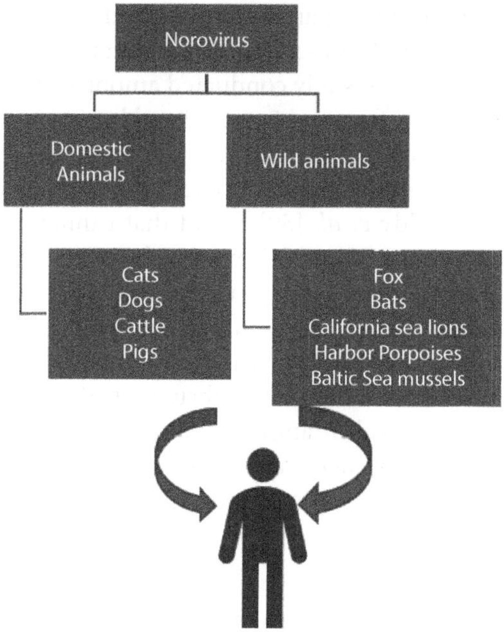

Figure 5.2 Animal reservoir of norovirus.

5.6 Zoonotic Transmission Mechanisms of Norovirus

5.6.1 Animal-to-Human Transmission

So far, there has been no conclusive evidence to support the presence of a zoonotic reservoir for norovirus that can infect humans. However, the discovery of norovirus (NoV) in animals that have close contact with infected humans suggests the possibility of zoonotic or cross-species transmission as well as the presence of an animal reservoir for human-infective norovirus [4]. There is serological evidence suggesting that people may be able to contract norovirus from animals, specifically from cows and dogs. Widdowson *et al.* [56] assessed the potential for human infection by bovine noroviruses using enzyme immunoassays (EIAs) to detect antibodies in 210 veterinarians and 630 controls. The study found that 28% of veterinarians had IgG antibodies to bovine norovirus compared to 20% of controls, indicating possible infection. The study revealed a substantial correlation between the amount of time that the youth spent on farms and their participation in the bovine practitioners' association with the presence of antibodies to bovine norovirus. One study [41] tested the serum samples of small animal veterinarians for IgG antibodies to canine norovirus and found that 22.3% of the veterinarians had antibodies to CaNoV compared to 5.8% of the controls, indicating a higher potential for canine norovirus to cause disease in humans. A study conducted among birth cohorts in India showed higher positivity for norovirus GIII and bovine norovirus antibodies in children and their mothers [57]. Another study [58] found that a quarter of blood donors in Sweden possessed immunoglobulins against bovine norovirus. Caddy *et al.* [59] found that canine norovirus (CNV) belonging to genogroups IV and VI can attach to histo-blood group antigens (HBGAs) present in human saliva. This finding demonstrates the first stage of canine norovirus invading human cells and suggests the possible risk of norovirus transmission from canine animals to humans. This offers insights into norovirus evolution. Villabruna *et al.* [4] analyzed human norovirus sequences found in animal stool samples, revealing that most animal strains are nearly identical to human strains, particularly GII.4 variants, suggesting close genetic relationships and potential for interspecies transmission.

5.6.2 Human-to-Animal Transmission

Various studies have tried to find a possible link to norovirus transmission from humans to animals. One case–control study among 92 dogs in Finland found four dogs positive for norovirus in the household. Two individuals experienced symptoms of severe gastroenteritis a few days before the collection of samples from the canines. Three variants of GII.4 and one variant of GII.12 genotype were identified. Among them, one GII.4 variation was the same as the virus discovered in the owner's fecal sample [60]. Another study in Sydney found that dogs were infected with human norovirus GII.4 during an outbreak where two children on the premises were hospitalized with norovirus 2 weeks before. Among the dogs, 11.11% of the 18 adult dogs and 83.33% of the six puppies exhibited clinical manifestations of norovirus. Genetic and phylogenetic investigations revealed that the whole genotypes of canine and human noroviruses were correlated with human norovirus GII.Pe-GII.4 [50]. In a Turkey-based study, human norovirus (NoV) GII was detected in 3.91% of stool samples from 128 pet dogs with diarrhea, with the virus most commonly found in crossbreed, female dogs aged 24 months or older, highlighting the importance of personal hygiene and proper care practices for dog owners to prevent the transmission of NoV GII to pets [51]. Caddy *et al.* [43] found that canine tissues can bind to multiple genotypes of human norovirus-like particles, and serological evidence from 43 of 325 canine serum samples suggests previous human norovirus exposure, indicating potential zoonotic transmission. Similar findings have been noted among pigs as well. In a longitudinal study in Japan, 11 out of 354 intestinal samples from 6-month-old healthy pigs tested positive for human GII norovirus, with strains classified as GII.4, GII.3, and GII.13 all linked to human outbreaks during that season [49]. Liu *et al.* [61] report the detection and genome sequencing of a GII.17 norovirus from a rhesus monkey at a monkey farm in southwestern China, revealing a virus that matches 99% to human norovirus strains. Another study found a high prevalence (51%–61%) of anti-norovirus antibodies in rhesus macaques at a primate research center, suggesting frequent exposure and possible transmission between human and non-human primates [62]. Besides that, an initial RT-PCR screening in Copenhagen's sewer system detected a recombinant human norovirus GI.b/GI.6 strain in one of 11 Norway rats [63]. Overall, these investigations point to the possibility that human noroviruses could infect animals, albeit additional research is required to determine the exact mechanisms and processes.

5.7 Zoonotic Transmission Mechanism

Understanding the mechanisms of norovirus attachment is crucial for assessing the risk and mechanism of zoonotic transmission. Noroviruses (NoVs) are significant pathogens that can potentially cross species barriers. The ability of certain animal noroviruses to bind to human HBGAs suggests that these viruses can potentially infect humans, posing a public health risk. Conversely, bovine GIII.2 and murine GV noroviruses depend on receptors not found in humans, limiting their zoonotic potential. In mice, GV infection relies on terminal sialic acids and glycoproteins, with the CD300 receptor playing a significant role, although it does not function similarly in humans. The presence of HBGAs, influenced by the FUT1 and FUT2 genes, is crucial for host susceptibility [4]. Understanding these interactions helps assess zoonotic risks and informs public health strategies to prevent cross-species norovirus transmission, emphasizing the need for continued surveillance and research. Enhanced surveillance and diagnostic tools are needed to monitor and identify such cross-species infections.

5.8 Future Implications and Research Directions

Cross-species contamination is becoming more likely due to the increased connection between people and animals. The direct evidence of animal-to-human transmission is limited and heavily depends on serological data; yet, the probability of such instances, albeit rare, cannot be dismissed. These transmissions may be asymptomatic or infrequent, making them difficult to detect without advanced surveillance systems specifically developed to detect noroviruses in human stool samples. However, reported occurrences of viral transmission from people to animals and vice versa indicate the need to explore whether human noroviruses in animals or recombinant human–animal norovirus genomes can be transferred back to humans, potentially affecting the re-emergence of noroviruses. Emerging norovirus strains could infect humans through zoonotic transmission, making prevention, diagnosis, and treatment more difficult. These novel strains have the potential to circumvent current human immunity, resulting in widespread disease and higher healthcare costs. The high rates of mutation in noroviruses may cause them to evolve quickly, giving rise to increasingly virulent types and a serious risk to the security of global health. However, because of the specificity of the primers used in present diagnostic procedures, animal noroviruses are frequently missed. While generic primers are broad, they may not be sensitive enough to detect animal strains, and protocols aimed at human-specific norovirus genotypes may fail to identify animal strains. As a result, the possibility of norovirus zoonotic transmission presents serious issues for public health in the future that call for an all-encompassing strategy.

To overcome these challenges, we must fill several knowledge gaps regarding the processes, reservoirs, and cross-species adaptability of noroviruses. There is currently a dearth of thorough surveillance data on the incidence of norovirus in animal populations and its potential to spread to humans. Targeted surveillance is necessary to solve these issues. This entails using precise detection techniques and gathering samples from humans and animals that are in close proximity, especially during outbreaks. Our knowledge of norovirus evolution and emergence will be improved by an understanding of norovirus reservoirs and the dynamics of their movement between species, which will help in the creation of efficient preventive and control measures. Besides that, future studies should utilize cutting-edge tools including metagenomics to investigate the virome of animal populations, systems to study gene functions associated with host vulnerability and viral replication, and next-generation sequencing (NGS) for in-depth genomic analysis.

Multidisciplinary methods are essential. By combining the fields of environmental, animal, and human health, the One Health initiative may investigate the mechanisms of transmission to develop preventative measures. To advance norovirus research and create novel treatments, academic institutions, governmental organizations, and the corporate sector must collaborate through international research networks and cross-sector partnerships. The main focus of policy and public health recommendations should be on routine testing and reporting as well as improved surveillance to track the frequency of norovirus in both human and animal populations. The development and dissemination of public health standards that prioritize hygiene practices, animal handling protocols, and food safety precautions are crucial in the prevention of zoonotic transmission. Vaccines against animal and human noroviruses can be developed to lower the danger of cross-species transmission. Public health campaigns should encourage preventative actions among high-risk populations, such as farmers, veterinarians, and pet owners, and increase knowledge about the dangers of zoonotic norovirus transmission.

5.9 Conclusion

Although norovirus is primarily linked to human-to-human transmission, new research emphasizes the significance of zoonotic transmission for the virus' ecology. The finding that norovirus may infect both domestic and wild animals raises the possibility that these animals could act as reservoirs for diseases that affect humans. Besides that, the close genetic affinities between animal and human norovirus strains suggest the possibility of interspecies transmission from animal to human and vice versa. However, there is a lack of direct evidence on animal-to-human transmission. Fewer serological research has shown that those who work closely with animals—such as farmers and veterinarians—have greater prevalences of antibodies against animal noroviruses, which may indicate exposure to the virus and potential infection. This emphasizes the importance of targeted surveillance and epidemiological research to better understand the scope of zoonotic transmission and its implications for public health. The discovery of unique histo-blood group antigens (HBGAs) and proteins that aid in norovirus attachment and entrance into host cells emphasizes the complexities of these interactions and the importance of precise molecular research. Furthermore, noroviruses' high mutation rates lead to their rapid evolution and enhanced virulence, highlighting the significance of ongoing surveillance and research. Addressing noroviruses' zoonotic potential necessitates a multidimensional approach that involves expanded research, increased surveillance, interdisciplinary collaboration, and strong public health policies. Implementing these methods will allow us to better forecast, prevent, and respond to norovirus outbreaks, preserving public health and improving global health security.

Abbreviations

RNA	ribonucleic acid
CDC	Centre for Disease Control
ORF	open reading frames
AGE	acute gastroenteritis
HBGAs	histo-blood group antigens
RT-PCR	reverse transcriptase polymerase chain reaction
FNOV	feline norovirus
CNOV	canine norovirus
HuNoVs	human noroviruses
MNV	murine norovirus
EIA	enzyme immunoassays
FUT	fructosyltransferase
NGS	next-generation sequencing

AI disclosure statement: The author acknowledges the use of AI-based tools for minor language editing or grammatical correction. However, all intellectual contributions, analyses, and conclusions presented in this chapter are entirely the author's own.

References

1. Dattani, S., Spooner, F., Ritchie, H., Roser, M., *Diarrheal Diseases*, Our World Data, 2023 Dec 28, [cited 2024 Jul 18]; https://ourworldindata.org/diarrheal-diseases.
2. Tan, M., Tian, Y., Zhang, D., Wang, Q., Gao, Z., Aerosol Transmission of Norovirus. *Viruses*, 16, 1, 151, 2024 Jan.
3. de Graaf, M., van Beek, J., Koopmans, M.P.G., Human norovirus transmission and evolution in a changing world. *Nat. Rev. Microbiol.*, 14, 7, 421–33, 2016 Jul.
4. Villabruna, N., Izquierdo Lara, R.W., Szarvas, J., Koopmans, M.P.G., Graaf, M., Phylogenetic Investigation of Norovirus Transmission between Humans and Animals. *Viruses*, 12, 11, 1287, 2020 Nov.

5. Carlson, K.B., Dilley, A., O'Grady, T., Johnson, J.A., Lopman, B., Viscidi, E., A narrative review of norovirus epidemiology, biology, and challenges to vaccine development. *npj Vaccines*, 9, 1, 1–9, 2024 May 29.
6. Bitler, E.J., Matthews, J.E., Dickey, B.W., Eisenberg, J.N.S., Leon, J.S., Norovirus outbreaks: a systematic review of commonly implicated transmission routes and vehicles. *Epidemiol. Infect.*, 141, 8, 1563–71, 2013 Aug.
7. Misumi, M. and Nishiura, H., Long-term dynamics of Norovirus transmission in Japan, 2005–2019. *PeerJ.*, 9, e11769, 2021 Jul 12.
8. Zhang, P., Hao, C., Di, X., Chuizhao, X., Jinsong, L., Guisen, Z., *et al.*, The global prevalence of norovirus gastroenteritis after the emergence of the GII.4 Sydney 2012 variant: a systematic review and meta-analysis. *Front. Public Health*, 12, 1373322, 2024 Jun 27.
9. Omatola, C.A., Mshelbwala, P.P., Okolo, M.L.O., Onoja, A.B., Abraham, J.O., Adaji, D.M., *et al.*, Noroviruses: Evolutionary Dynamics, Epidemiology, Pathogenesis, and Vaccine Advances—A Comprehensive Review. *Vaccines*, 12, 6, 590, 2024 Jun.
10. Lucero, Y., Matson, D.O., Ashkenazi, S., George, S., O'Ryan, M., Norovirus: Facts and Reflections from Past, Present, and Future. *Viruses*, 13, 12, 2399, 2021 Dec [cited 2024 Jul 16.
11. CDC., Norovirus Facts and Stats. *Norovirus*, 2024 [cited 2024 Jul 17], https://www.cdc.gov/norovirus/data-research/index.html.
12. Liao, Y., Hong, X., Wu, A., Jiang, Y., Liang, Y., Gao, J., *et al.*, Global prevalence of norovirus in cases of acute gastroenteritis from 1997 to 2021: An updated systematic review and meta-analysis. *Microb, Pathog.*, 161, 105259, 2021 Dec 1.
13. Hofmann, F.M., Olawumi, E., Michaelis, M., Stößel, U., Hofmann, F., Significance of norovirus in occupational health: a review of published norovirus outbreaks in Central and Northern Europe. *Int. Arch. Occup. Environ. Health*, 93, 8, 911–23, 2020 Nov 1.
14. Calderwood, L.E., Wikswo, M.E., Mattison, C.P., Kambhampati, A.K., Balachandran, N., Vinjé, J., *et al.*, Norovirus Outbreaks in Long-term Care Facilities in the United States, 2009–2018: A Decade of Surveillance. *Clin. Infect. Dis.*, 74, 1, 113–9, 2022 Jan 1.
15. Rico, E., Pérez, C., Belver, A., Sabaté, S., Razquin, E., de Benito, J., *et al.*, Norovirus detection in environmental samples in norovirus outbreaks in closed and semi-closed settings. *J. Hosp. Infect.*, 105, 1, 3–9, 2020 May 1.
16. Jin, M., Wu, S., Kong, X., Xie, H., Fu, J., He, Y., *et al.*, Norovirus Outbreak Surveillance, China, 2016–2018. *Emerg. Infect. Dis.*, 26, 3, 437–45, 2020 Mar.
17. Fanaselle, W., Pouillot, R., Papafragkou, E., Liggins, G., Williams, L., Doren, J.M.V., Evaluation of the Impact of Compliance with Mitigation Strategies and Frequency of Restaurant Surface Cleaning and Sanitizing on Control of Norovirus Transmission from Ill Food Employees Using an Existing Quantitative Risk Assessment Model. *J. Food Prot.*, 85, 8, 1177–91, 2022 Aug 1.

18. Ibrahim, Y.M., Werid, G.M., Zhang, H., Fu, L., Wang, W., Chen, H., *et al.*, Potential zoonotic swine enteric viruses: The risk ignored for public health. *Virus Res.*, 315, 198767, 2022 Jul 2.
19. Wang, J., Gao, Z., Liu, K., Zhang, H., Global prevalence of asymptomatic norovirus infection in outbreaks: a systematic review and meta-analysis. *BMC Infect. Dis.*, 23, 1, 595, 2023 Sep 12.
20. Cromeans, T., Park, G.W., Costantini, V., Lee, D., Wang, Q., Farkas, T., *et al.*, Comprehensive Comparison of Cultivable Norovirus Surrogates in Response to Different Inactivation and Disinfection Treatments. *Appl. Environ. Microbiol.*, 80, 18, 5743–51, 2014 Sep 15.
21. Hagbom, M., Lin, J., Falkeborn, T., Serrander, L., Albert, J., Nordgren, J., *et al.*, Replication in Human Intestinal Enteroids of Infectious Norovirus from Vomit Samples. *Emerging Infect. Dis. J. CDC*, 27, 8, 2212–4, August 2021. [cited 2024 Jul 25]; https://wwwnc.cdc.gov/eid/article/27/8/21-0011_article.
22. Phattanawiboon, B., Nonthabenjawan, N., Boonyos, P., Jetsukontorn, C., Towayunanta, W., Chuntrakool, K., *et al.*, Norovirus transmission is mediated by asymptomatic family members in households. *PLoS One*, 15, 7, e0236502, 2020 Jul 23.
23. Khamrin, P., Kumthip, K., Thongprachum, A., Sirilert, S., Malasao, R., Okitsu, S., *et al.*, Genetic diversity of norovirus genogroup I, II, IV and sapovirus in environmental water in Thailand. *J. Infect. Public Health*, 13, 10, 1481–9, 2020 Oct 1.
24. Matthews, H.S., Palmer, R.L., Baynam, G.S., Quarrell, O.W., Klein, O.D., Spritz, R.A., *et al.*, Large-scale open-source three-dimensional growth curves for clinical facial assessment and objective description of facial dysmorphism. *Sci. Rep.*, 11, 1, 12175, 2021 Jun 9.
25. Derrick, J., Hollinghurst, P., O'Brien, S., Elviss, N., Allen, D.J., Iturriza-Gómara, M., Measuring transfer of human norovirus during sandwich production: Simulating the role of food, food handlers and the environment. *Int. J. Food Microbiol.*, 348, 109151, 2021 Jun 16.
26. Verhoef, L., Kouyos, R.D., Vennema, H., Kroneman, A., Siebenga, J., van Pelt, W., *et al.*, An Integrated Approach to Identifying International Foodborne Norovirus Outbreaks. *Emerg. Infect. Dis.*, 17, 3, 412–8, 2011 Mar.
27. Alimi, Y. and Wabacha, J., Strengthening coordination and collaboration of one health approach for zoonotic diseases in Africa. *One Health Outlook*, 5, 1, 10, 2023.
28. Tomori, O. and Oluwayelu, D.O., Domestic Animals as Potential Reservoirs of Zoonotic Viral Diseases. *Annu. Rev. Anim. Biosci.*, 11, 33–55, 2023 Feb 15.
29. Said, M.S., Tirthani, E., Lesho, E., Animal Zoonotic Related Diseases, in: *StatPearls*, StatPearls Publishing, Treasure Island FL, 2024 [cited 2024 Jul 20], http://www.ncbi.nlm.nih.gov/books/NBK570559/.
30. Pulliam, J.R.C. and Dushoff, J., Ability to replicate in the cytoplasm predicts zoonotic transmission of livestock viruses. *J. Infect. Dis.*, 199, 4, 565–8, 2009 Feb 15.

31. Cavicchio, L., Tassoni, L., Laconi, A., Cunial, G., Gagliazzo, L., Milani, A., *et al.*, Unrevealed genetic diversity of GII Norovirus in the swine population of North East Italy. *Sci. Rep.*, 10, 1, 9217, 2020 Jun 8.
32. Dik, I., Bulut, O., Avci, O., Hasoksuz, M., Palanci, H.S., Aslim, H.P., *et al.*, Molecular Detection and Characterization of Bovine Noroviruses from Cattle in Konya, Turkey. *Pak Vet. J.*, 2023.
33. Di Felice, E., Mauroy, A., Pozzo, F.D., Thiry, D., Ceci, C., Di Martino, B., *et al.*, Bovine noroviruses: A missing component of calf diarrhea diagnosis. *Vet. J.*, 207, 53–62, 2016 Jan 1.
34. Symes, S., Allen, J., Mansell, P., Woodward, K., Bailey, K., Gilkerson, J., *et al.*, First detection of bovine noroviruses and detection of bovine coronavirus in Australian dairy cattle. *Aust. Vet. J.*, 96, 6, 203–8, 2018.
35. Castells, M., Caffarena, R.D., Casaux, M.L., Schild, C., Castells, F., Castells, D., *et al.*, Detection, risk factors and molecular diversity of norovirus GIII in cattle in Uruguay. *Infect. Genet. Evol.*, 86, 104613, 2020 Dec 1.
36. Thomas, C., Jung, K., Han, M.G., Hoet, A., Scheuer, K., Wang, Q., *et al.*, Retrospective serosurveillance of bovine norovirus (GIII.2) and nebovirus in cattle from selected feedlots and a veal calf farm from 1999 to 2001 in the United States. *Arch. Virol*, 159, 1, 83–90, 2014.
37. Jung, K., Scheuer, K.A., Zhang, Z., Wang, Q., Saif, L.J., Pathogenesis of GIII.2 bovine norovirus, CV186-OH/00/US strain in gnotobiotic calves. *Vet. Microbiol.*, 168, 1, 202–7, 2014 Jan 10.
38. Ferragut, F., Vega, C.G., Mauroy, A., Conceição-Neto, N., Zeller, M., Heylen, E., *et al.*, Molecular detection of bovine Noroviruses in Argentinean dairy calves: Circulation of a tentative new genotype. *Infect. Genet. Evol.*, 40, 144–50, 2016 Jun 1.
39. Martella, V., Lorusso, E., Decaro, N., Elia, G., Radogna, A., D'Abramo, M., *et al.*, Detection and Molecular Characterization of a Canine Norovirus. *Emerg. Infect. Dis.*, 14, 8, 1306–8, 2008 Aug.
40. Bodnar, L., Lorusso, E., Di Martino, B., Catella, C., Lanave, G., Elia, G., *et al.*, Identification of a novel canine norovirus. *Infect. Genet. Evol.*, 52, 75–81, 2017 Aug 1.
41. Mesquita, J.R., Costantini, V.P., Cannon, J.L., Nascimento, M.S.J., Vinjé, J., Presence of Antibodies against Genogroup VI Norovirus in Humans. *Virol. J.*, 10, 1, 176, 2013 Jun 4.
42. Di Martino, B., Di Profio, F., Ceci, C., Di Felice, E., Green, K.Y., Bok, K., *et al.*, Seroprevalence of Norovirus Genogroup IV Antibodies among Humans, Italy, 2010–2011. *Emerg. Infect. Dis.*, 20, 11, 1828–32, 2014 Nov.
43. Caddy, S.L., de Rougemont, A., Emmott, E., El-Attar, L., Mitchell, J.A., Hollinshead, M., *et al.*, Evidence for Human Norovirus Infection of Dogs in the United Kingdom. *J. Clin. Microbiol.*, 53, 6, 1873–83, 2015 May 14.
44. Ford-Siltz, L.A., Mullis, L., Sanad, Y.M., Tohma, K., Lepore, C.J., Azevedo, M., *et al.*, Genomics Analyses of GIV and GVI Noroviruses Reveal the Distinct Clustering of Human and Animal Viruses. *Viruses*, 11, 3, 204, 2019 Mar.

45. Di Martino, B., Di Profio, F., Melegari, I., Sarchese, V., Cafiero, M.A., Robetto, S., *et al.*, A novel feline norovirus in diarrheic cats. *Infect. Genet. Evol.*, 38, 132–7, 2016 Mar 1.
46. Pinto, P., Wang, Q., Chen, N., Dubovi, E.J., Daniels, J.B., Millward, L.M., *et al.*, Discovery and Genomic Characterization of Noroviruses from a Gastroenteritis Outbreak in Domestic Cats in the US. *PLoS One*, 7, 2, e32739, 2012 Feb 28.
47. Takano, T., Kusuhara, H., Kuroishi, A., Takashina, M., Doki, T., Nishinaka, T., *et al.*, Molecular characterization and pathogenicity of a genogroup GVI feline norovirus. *Vet.Microbiol.*, 178, 3, 201–7, 2015 Aug 5.
48. Takano, T., Hiramatsu, K., Matsuyama, M., Mutoh, K., Matsumoto, Y., Fukushima, T., *et al.*, Viral shedding and clinical status of feline-norovirus-infected cats after reinfection with the same strain. *Arch. Virol*, 163, 6, 1503–10, 2018 Jun 1.
49. Nakamura, K., Saga, Y., Iwai, M., Obara, M., Horimoto, E., Hasegawa, S., *et al.*, Frequent Detection of Noroviruses and Sapoviruses in Swine and High Genetic Diversity of Porcine Sapovirus in Japan during Fiscal Year 2008. *J. Clin. Microbiol.*, 48, 4, 1215–22, 2010 Apr.
50. Charoenkul, K., Nasamran, C., Janetanakit, T., Tangwangvivat, R., Bunpapong, N., Boonyapisitsopa, S., *et al.*, Human Norovirus Infection in Dogs, Thailand. *Emerg. Infect. Dis.*, 26, 2, 350–3, 2020 Feb.
51. Sokel, S. and Kale, M., Investigation of Norovirus genogroups (GI, GII, and GIV) in the stool of pet dogs with diarrhea. *Pesqui Veterinária Bras.*, 39, 402–8, 2019 Jul 18.
52. Summa, M., Henttonen, H., Maunula, L., Human noroviruses in the feces of wild birds and rodents—new potential transmission routes. *Zoonoses Public Health*, 65, 5, 512–8, 2018.
53. Wang, J., Li, L., Xu, Y., Mao, T., Ma, Y., Sun, X., *et al.*, Identification of a novel norovirus species in fox. *Infect. Genet. Evol.*, 98, 105214, 2022 Mar 1.
54. Bigoraj, E., Kwit, E., Chrobocińska, M., Rzeżutka, A., Occurrence of Norovirus and Hepatitis A Virus in Wild Mussels Collected from the Baltic Sea. *Food Environ. Virol.*, 6, 3, 207–12, 2014 Sep 1.
55. Tofani, S., Ianiro, G., De Sabato, L., Monini, M., Angeloni, G., Ponterio, E., *et al.*, Detection and whole genome sequencing of murine norovirus in animal facility in Italy. *Anim. Biotechnol.*, 33, 6, 1142–9, 2022 Nov 18.
56. Widdowson, M.A., Rockx, B., Schepp, R., van der Poel, W.., Vinje, J., YT, .D., *et al.*, Detection of serum antibodies to bovine norovirus in veterinarians and the general population in the Netherlands. *J. Med. Virol.*, 76, 1, 119–28, 2005 [cited 2024 Jul 25], https://onlinelibrary.wiley.com/doi/abs/10.1002/jmv.20333.
57. Menon, V.K., George, S., Shanti, A.A., Saravanabavan, A., Samuel, P., Ramani, S., *et al.*, Exposure to Human and Bovine Noroviruses in a Birth Cohort in Southern India from 2002 to 2006. *J. Clin. Microbiol.*, 51, 7, 2391–5, 2020

Dec 21 [cited 2024 Jul 25], https://journals.asm.org/doi/full/10.1128/jcm.01015-13.

58. Vildevall, M., Grahn, A., StefanL, O., JC, B., Charpilienne, A., Poncet, D., *et al.*, Human antibody responses to bovine (Newbury-2) norovirus (GIII.2) and association to histo-blood group antigens. *J. Med. Virol.*, 82, 7, 1241–6, 2010 [cited 2024 Jul 25], https://onlinelibrary.wiley.com/doi/abs/10.1002/jmv.21776.
59. Caddy, S., Breiman, A., le Pendu, J., Goodfellow, I., Genogroup IV and VI Canine Noroviruses Interact with Histo-Blood Group Antigens. *J. Virol.*, 88, 18, 10377–91, 2014 Sep 15 [cited 2024 Jul 25], https://journals.asm.org/doi/full/10.1128/jvi.01008-14.
60. Summa, M., von Bonsdorff, C.H., Maunula, L., Pet dogs–a transmission route for human noroviruses? *J. Clin. Virol. Off. Publ. Pan. Am. Soc. Clin. Virol.*, 53, 3, 244–7, 2012 Mar.
61. Liu, B., Tao, Y., Li, C., Li, X., Liu, J., He, Z., *et al.*, Complete Genome Sequence of a GII.17 Norovirus Isolated from a Rhesus Monkey in China. *Genome Announc.*, 4, 5, 2016 Sep 8. 10.1128/genomea.00904-16.
62. Farkas, T., Dufour, J., Jiang, X., Sestak, K., Detection of norovirus-, sapovirus- and rhesus enteric calicivirus-specific antibodies in captive juvenile macaques. *J. Gen. Virol.*, 91, 3, 734–8, 2010.
63. Wolf, S., Reetz, J., Johne, R., Heiberg, A.C., Petri, S., Kanig, H., *et al.*, The simultaneous occurrence of human norovirus and hepatitis E virus in a Norway rat (Rattus norvegicus). *Arch. Virol.*, 158, 7, 1575–8, 2013 Jul 1.

Part 2

TRANSMISSION, DYNAMICS AND ENVIRONMENTAL IMPACTS

6

Genetic Changes in Norovirus and Their Impact on Infectivity/Virulence

Soumya Patil[1], Vijayalaxmi Bangra Manjeshwara[2], Uday Venkat Mateti[1] and Raushan Kumar Chaudhary[3*]

[1]*Department of Pharmacy Practice, NGSM Institute of Pharmaceutical Sciences (NGSMIPS), Nitte (Deemed to be University), Mangalore, Karnataka, India*
[2]*Department of Physiology, KS Hegde Medical Academy (KSHEMA), Nitte (Deemed to be University), Mangalore, Karnataka, India*
[3]*Centre for Integrative Omics Data Science (CIODS), Yenepoya (Deemed to be University), Mangalore, Karnataka, India*

Abstract

Norovirus (NoV) exhibits significant genetic diversity with high mutation rate and major contributors of acute gastroenteritis outbreaks all over the world. This chapter delves into the genetic changes in noroviruses and their significant impact on infectivity and virulence. The NoV GII.4 genotype is the most prevalent and emerge every 2–5 years due to the accumulation of mutations at sites such as VP1 protein, enabling the virus to evade herd immunity. The main processes influencing norovirus' genomic evolution are point mutations and recombination events, which have had a major influence on their propensity to spread like wildfire. Furthermore, the primary breakpoint that propels NoV development is recombination at the ORF1/ORF2 overlap region, which encodes RdRp and VP1, respectively. Recombination occurring at the P2 domain of the VP1 capsid protein affects the epitopes crucial for immune recognition and contributes to the development of new strains. These genetic changes result from selective pressure exerted by host immune responses, impacting both immune evasion and vaccine effectiveness. Key findings highlight the evolutionary pressures exerted by host immune responses, which drive both antigenic drift (point mutation) and antigenic shift (recombination), facilitating

**Corresponding author*: chaudharyrosen91@gmail.com; ORCID: https://orcid.org/0000-0002-5862-3636
Soumya Patil: ORCID: https://orcid.org/0000-0002-5804-3498
Uday Venkat Mateti: ORCID: https://orcid.org/0000-0001-8149-2067

Sachin Kumar, Shriyansh Srivastava, Rishabha Malviya and Ranjit Sah (eds.) Emerging Infectious Disease: Unravelling Norovirus and Pandemic Dynamics, (153–186) © 2026 Scrivener Publishing LLC

the re-infection of previously exposed individuals. To create efficient surveillance, diagnostic, and treatment plans against the ever-evolving threat posed by noroviruses, it is imperative to comprehend these genetic dynamics.

Keywords: Genogroup, genotype, norovirus, open reading frames, point mutation, recombination

6.1 Introduction

A primary cause of gastroenteritis is norovirus (NoV), which impacts people of every age and substantially adds to the overall disease burden. Nearly 685 million cases are reported globally every year, with an estimated 1.5 million deaths [1, 2]. Despite its self-limiting nature, NoV infection can be severe in infants, geriatrics, and patients with impaired immune systems. In developing nations, deaths from NoV infections are common among children resulting from severe diarrhea and dehydration, whereas in developed countries, it is more often observed in elderly and immunocompromised patients [3]. It was first documented in the year 1929 by Zahorsky as "winter vomiting disease", which was associated with abrupt symptoms of vomiting and diarrhea [4]. The name "winter vomiting disease" reflects the higher incidence of the illness during the winter months. Previously, this virus was called the "Norwalk virus" after the 1968 gastroenteritis outbreak in "Norwalk", Ohio [5]. In 1972, Kapikian identified NoV as the main culprit behind this outbreak [6]. These highly transmittable viruses can spread quickly in crowded environments such as healthcare facilities, educational institutions, etc. They are mainly disseminated through the fecal–oral pathway, frequently contaminated food, water, or surfaces [7]. Surprisingly, there was decline in the incidence of NoV infection during COVID-19 due to pandemic lockdown restrictions [8]. However, post-COVID, the NoV infection rates have returned to pre-pandemic levels [9, 10]. The NoV capacity to escape the host immune system largely depends on their genetic diversity, which contributes to the major public health burden. The NoV genome undergoes frequent genetic recombination and mutation events that promote the evolution of novel variants that are frequently linked to enhanced infectivity, modified virulence, and the capacity to bypass pre-existing immunity. To develop effective prevention and control measures, it is essential to comprehend the genetic mechanisms driving the evolution of NoV and how they affect virulence and infectivity. Recent advances in genomic sequencing and molecular epidemiology have provided insights into the patterns of NoV genetic diversity and their association with frequent outbreaks. This chapter explores the relationship between genetic changes in Noroviruses and their implications for infectivity and virulence, shedding light on the ongoing challenges in managing NoV infections and highlighting potential avenues for future research and public health interventions.

6.2 Classification of NoV

NoV belongs to the non-enveloped Caliciviridae family, which consists of 11 genera, namely, *Norovirus*, *Sapovirus*, *Vesivirus*, *Lagovirus*, *Bavovirus*, *Valovirus*, *Nacovirus*, *Nebovirus*, *Minovirus*, *Recovirus*, and *Sal virus* (Figure 6.1). Among these, only NoV and Sapovirus are capable to infect both human and animals. In the mid-1990s, the nucleotide sequencing of the RdRp gene led to its classification in various genogroups and genotypes [11]. In the early 2000s, complete sequencing of the VP1 amino acids became the standard for assigning noroviruses to different genogroups and genotypes, initially using a 15% sequence difference threshold [12]. The organization of NoV data was impacted by classification errors and conflicts caused by the absence of a standard classification scheme. As a result, a single, dual nomenclature system that specifies the genotypes of capsid protein (C) and RNA-dependent RNA polymerase (P) was provided by the Norovirus Classification Working Group (NCWG) [13]. This system enabled the classification of six genogroups (GI–GVI) and over 40

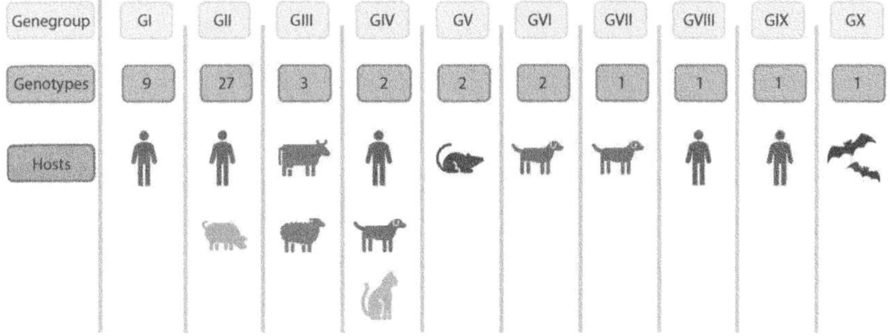

Figure 6.1 Norovirus genera classification with host of origin.

genotypes. Subsequent advancements expanded the classification to 10 genogroups (GI-GX) and 49 genotypes (Figure 6.1). Partly, RdRp sequencing further identified 60 P-types and several tentative P-groups and P-types. Phylogenetic analysis segregates noroviruses into genogroups and P-groups, with genotypes and P-types defined by VP1 and RdRp sequence diversity, respectively [14]. Recent reports of new noroviruses necessitate updates to the classification scheme, incorporating complete genome sequences. The international NoV classification working group continues to coordinate and disseminate these updates, ensuring a robust and universally accepted system [15].

6.3 Genomic Structure and Functions

Norovirus exhibits significant environmental persistence due to its single-stranded RNA genome's inherent stability. Encapsulated within a robust icosahedral capsid, the genome can withstand temperatures up to 60°C and a broad pH range (pH 3–7), contributing to the virus' high infectivity and transmission efficiency [16]. There are three open reading frames which make up the norovirus genome, each coding for proteins necessary for viral replication, structural integrity, and pathogenesis (Figure 6.2).

6.3.1 ORF1: The Replication Engine

ORF1 is responsible for coding a polyprotein which is essential for viral replication by modulating host cell machinery; it consists of six non-structural proteins ("NS1/2 to NS7"). These proteins participate in various functions, including the disassembly of cellular structures, such as the Golgi apparatus (NS1/2), the hydrolysis of nucleotide triphosphates (NS3),

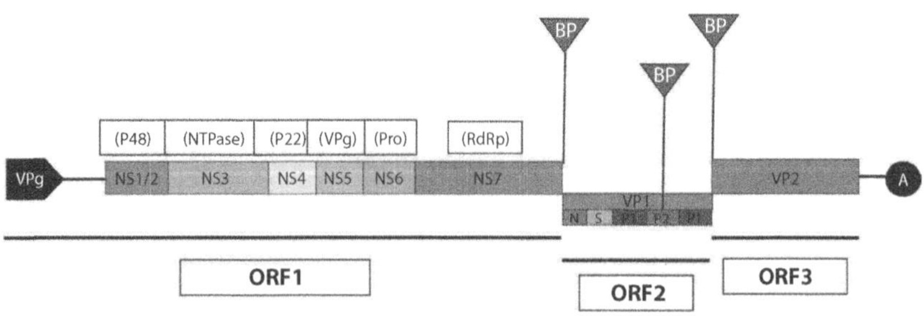

Figure 6.2 Norovirus genome structure. Major breakpoints are represented at the ORF1–ORF2 junction, within ORF2, and at the ORF2–ORF3 junction.

Table 6.1 Norovirus genomic organization and function.

Open reading frame	Gene	Protein name	Location	Function
ORF1	NS1/2	N terminal protein (p48)	Golgi apparatus	Disassembles Golgi into discrete aggregates; interacts with VAP-A, possibly anchoring membrane-bound replication complexes
	NS3	Nucleotide triphosphate -NTPase (p41)	Cytoplasm	Exhibits NTPase activity and contains a helicase domain; hydrolyze NTPs for unwinding
	NS4	3A-like protein or (p22)	Cellular membranes	Suggested to be involved in cellular membrane trafficking and replication complexes
	NS5	Viral VpG protein	Viral RNA	Essential for infectious calicivirus production; binds to translational initiation factors, may recruit ribosomes to viral RNA; involved in RNA replication, linked to RNA *via* a phosphodiester bond, and uridylylated by RdRp
	NS6	Viral protease - 3C-like protease (3CLpro)	Cytoplasm	Crucial for proteolytic processing of ORF1 polyprotein into six non-structural proteins; contains catalytic triad essential for enzymatic activity; suggested to cleave host poly[A]-binding protein, inhibiting cellular translation

(*Continued*)

Table 6.1 Norovirus genomic organization and function. (*Continued*)

Open reading frame	Gene	Protein name	Location	Function
	NS7	"RNA-dependent RNA polymerase (RdRp)"	Cytoplasm	"Involved in norovirus genome replication; functions as a bifunctional enzyme with both protease and RdRp activities during viral replication"
ORF2	VP1	Viral protein 1 (VP1)	Viral capsid	Major capsid protein; forms the icosahedral capsid shell with S domain; P domain involved in immune and cellular recognition, with P2 subdomain being the most exposed and hypervariable
ORF3	VP2	Viral protein 2 (VP2)	Viral capsid	Minor structural protein; upregulates VP1 expression, stabilizes virus structure, and protects VP1 from disassembly and degradation; may interact with RNA and contribute to the formation of infectious viral particles
ORF4[a]	VF1	Virulence factor 1 (VF1)	Mitochondria	Antagonize innate response to infection

[a]Present in murine norovirus. Adapted from [16].

and the formation of replication complexes on cellular membranes (NS4). The viral "VPg protein" (NS5) is essential for both translating and replicating viral RNA, while the main function of 3C-like protease (NS6) is to cleave the polyprotein into its functional components. Furthermore, the ability to replicate is solely dependent on mediated RdRp (NS7) [16, 17] (Table 6.1).

6.3.2 ORF2: The Master Builder

The viral capsid consists of structural protein "VP1" which is encoded by ORF2. This capsid is icosahedral and consists of 180 copies of VP1 arranged as dimers [18]. The VP1 protein consists of two domains, namely, a shell (S) that encapsulates the viral RNA and a protruding (P) domain. Furthermore, the P domain consists of subdomains, namely, P1 and P2. The P2 subdomain is critical for interactions with host cell receptors present on the virion's outer surface, evading the host immune response and major hotspot for antigenic drift [19, 20] (Table 6.1).

6.3.3 ORF3: The Stabilizer

The ORF3 encodes for the minor structural protein VP2, which significantly stabilizes the viral structure. VP2 is involved in upregulating VP1 expression and protecting the capsid from disassembly and degradation. Although present in minute quantity it may also interact with viral RNA and contribute to the formation of infectious viral particles [16, 19, 20] (Table 6.1).

6.3.4 ORF4: Murine Norovirus

Exclusive to murine noroviruses, ORF4 encodes Virulence Factor 1, a protein localized to mitochondria. VF1 does not play any role in viral replication but is thought to suppress the host's first line of immunity [21, 22] (Table 6.1).

6.4 NoV Life Cycle: Molecular Mechanism of NoV Genome Replication

The process of norovirus replication is triggered after the viral particles attach to the host cell receptors, particularly the ABO histo-blood group antigens (HBGAs). These carbohydrates are expressed superficially on the enterocytes. The expression of HBGAs is facilitated by fucosyl-transferase-2 (FUT2) gene. The HBGAs required for viral attachment are expressed by those with a wild-type FUT2 gene (also known as "secretors"), whereas those without a functional FUT2 gene (also known as "non-secretors") do not express the proper HBGA, making them resistant to specific genotypes, such as GII.4 and GI.1 [23, 24]. Non-secretors account for around 20% of the European population, with a more significant frequency in South Asia [25]. The attachment facilitates the invasion of the virus into the enterocytes of the host *via* endocytosis. Following entry, the viral capsid disassembles, releasing the virus' genetic material into the cytoplasm. This uncoating is critical for the viral genome's translation and replication. Upon release into the cytoplasm, the VPg protein linked with positive-sense RNA genome at the 5′ end is immediately recognized and translated by the host cell's ribosomes. Non-structural proteins, particularly NS1/2 and NS4, create a replication complex necessary for synthesizing new viral RNA during the replication process [26]. The viral RdRp initiates the replication process by binding to the positive-sense viral RNA genome as the initial template. RdRp catalyzes the synthesis of negative-sense RNA, which serves as an intermediate in the replication process [27]. Positive-sense RNA is synthesized using the newly formed negative-sense RNA as a template, which is then packaged into new virions or used for further replication rounds. Subgenomic RNA (SG RNA) is also synthesized by RdRp, which facilitates the replication of subgenomic RNA, derived from specific regions of the genome. *De novo* initiation is proposed as the model for norovirus RNA synthesis, which involves the direct addition of nucleotides without requiring a primer. The newly synthesized positive-sense RNA genomes are either encapsidated into new virus particles or used as templates for additional cycles of replication and translation. Structural proteins and viral RNA assemble into complete virions for the release and infection of new cells, ensuring efficient viral replication and propagation within the host [28]. The replication process involves reorganization of the Golgi apparatus and the endoplasmic reticulum. The membrane vesicles form 12 h post-infection [29]. Noroviruses utilize primer-dependent initiation mechanisms for RNA synthesis. *De novo* initiation requires an RNA template, an initiating nucleotide, and a second nucleotide, whereas primer-dependent initiation can involve a protein-linked oligonucleotide or a capped oligonucleotide from host mRNA. Some viral RdRps also exhibit terminal transferase activity, initiating RNA replication *via* a "copy-back" synthesis mechanism [30–32]. Once new genomic RNA and structural proteins (mainly VP1) are synthesized, they are assembled into new virions within the cytoplasm. The VP1 and VP2 proteins form the capsid, encapsulating the newly synthesized positive-sense RNA genomes. The assembled virions accumulate in the host cell and are eventually released through cell lysis or exocytosis, ready to infect neighboring cells or new hosts [32].

6.5 Transmission

Human noroviruses (HuNoVs) are extremely infectious viruses that are waterborne and foodborne and can spread through the environment and fecal–oral pathways as well as *via* aerosolized vomitus particles. The large-scale outbreaks are caused by multiple factors, including the low infectious dose of 18–1,000 virus particles, extended shedding by infected persons irrespective of the development of symptoms, insufficient long-lasting immunity, and viral stability across a variety of environmental circumstances [33]. High population density and frequent social interactions are key risk factors for transmission, with symptomatic individuals in homes being significant predictors of disease in both children and adults. Foreign travel, particularly when accompanied by behavioral changes or exposure to new strains, increases the infection risk [34]. Transmission through food remains a major route of the spread of norovirus. It often occurs due to sick employees' handling while food is being prepared and served. Contamination with ready-to-eat meals like noodles, sandwiches, salads, and uncooked food like soft fruits and vegetables or raw seafood that has not been fully cooked is common. Feces and sewages could contaminate frozen and fresh berries. Moreover, shellfish, mussels, clams, and oysters collect NoV strains by filter feeding that leads to high chances of viral co-infection and recombination [35, 36]. Globalization of the food chain facilitates genetic exchange between NoV strains and complicates outbreak recognition. Approximately 14% of HuNoV outbreaks are foodborne. It has been reported that foodborne outbreaks often involve mixed strains, increasing the risk of viral recombination. Nearly 7% of foodborne outbreaks are international events with a common source, highlighting the challenge of tracking outbreaks globally [37]. Additionally, nosocomial transmission in healthcare facilities is a major risk, with attack rates of 53.2% for patients and 24.5% for healthcare providers. Enhanced hand hygiene, thorough cleaning, patient movement restrictions, and exclusion of ill staff are recommended to control outbreaks [38]. Long-term care facilities (LTCFs) are particularly susceptible to outbreaks due to shared living spaces and the presence of higher-risk individuals, contributing to higher infection rates, hospitalization, and mortality. LTCFs accounted for 62.5% of US outbreaks from 2009 to 2013, with other common settings including restaurants, schools, and cruise ships [39–41]. NoV outbreaks frequently occur in communal dining settings and institutions like hospitals and care homes, where close quarters and poor hygiene facilitate the virus spread [42]. Moreover, genogroups like "GI, GII, GIV, GVII, and GIX" are capable of infecting humans (Figure 6.1). However, recent studies have identified zoonotic transmission by birds and rodents as potential routes, though it remains to be determined if these animals can support HuNoV replication. Genome copies of various HuNoVs (GI and GII) have been detected in the feces of gulls, crows, and rats, suggesting that these animals might act as mechanical carriers, spreading the virus in the environment and possibly contaminating food sources [43]. Additionally, people with impaired immune systems may harbor these viruses for months or even years, serving as breeding grounds for new strains [41, 44]. Viral diversity arises during transmission events in healthy populations. Immunocompromised patients can harbor numerous variants that may escape herd immunity [45]. Surveillance networks like CaliciNet and NoroNet have shown that GII.4 strains primarily spread through person-to-person contact which contributes to most of the outbreaks (approximately 80% of outbreaks). Additionally, the great survival capability of "GI strains" in water facilitates its waterborne transmission while non-GII.4 strains spread through food [46, 47]. The transmission ability of NoV in community settings depends on the host factors such as age of the person, health status, history of prior exposure, and immunity toward these viruses. Noroviruses are among the top causes of foodborne diseases worldwide. Noroviruses are resilient against various decontamination methods and virucidal treatments. They may survive in ambient waters for weeks or months and have been found up to 10 km from their discharge location [48]. Traditional depuration methods are ineffective against NoV contamination in bivalve molluscs due to the viruses' binding to molluscan carbohydrate structures. Bivalve molluscs can act as hotspots for multiple NoV strains, potentially facilitating viral recombination and introducing novel strains to humans [49]. These NoV strains can linger in the environment for a long time and GII.4 strains predominate in the majority of outbreaks.

6.6 Molecular Diversity

Gastroenteritis outbreaks have been primarily caused by NoV for over 20 years. GII.4 strains have been at the center of multiple outbreak incidences across the globe. Some of the widely mentioned GII.4 strains outbreaks are "GII.4 Camberwell", "GII.4 Farmington Hills", "GII.4 Den Haag", "GII.4 Hunter", "GII.4 New Orleans", and "GII.4 Sydney" which have become the norm. Of the abovementioned strains, GII.4 Sydney strain is the most recently reported and is quite frequently associated with outbreaks caused by NoV [50]. New strains such as "GII.13[P16]" and "GII.3[P12]" are associated with the recombination of NoV have been a major hindrance to the stability of the GII.4 strain. Strains that are caused by the above-mentioned genogroups are classified as GII and GI, which constitute 90% of NoV cases registered [50, 51]. There has been a relative increase in the parts of the world that have been reporting increasing cases of acute gastroenteritis due to the GII.4 strain [53]. Further corroborating this, a report from Ethiopia reported that the "GII.3" genotype was more common than GII.4 in recent cases. There has been a relative switch in the genotype GII.4 and now there is a surge in the GII.P16/GII.2 genotype since 2017. GII.17 also experienced a rapid transition in genotype during the 2014–2015 outbreaks with GII.4 being the primary virus. A significant rise in acute gastroenteritis cases was observed in Europe and Asia during the 2016–2017 epidemic season due to the GII.2[P16] variant strain. The replacement of dominant strains with less prevalent ones is mostly due to the recurrent mutation and recombination events both within and between genotypes [54].

6.7 Evolutionary Mechanisms and Dominance

Due to limited cross-protection among different NoV genotypes, individuals can experience multiple infections throughout their lives. According to phylogenetic analysis and epidemiological observations, closely related genotypes might offer a certain level of cross-protection, lowering the lifetime risk of infection for a person. Even though there are several genotypes, "GII.4" is still the most commonly circulating around the world [1, 45].

There are no clear mechanisms behind the stable and dominant nature of GII.4 One hypothesis suggests that new GII.4 variants emerge in succession, evading immune responses generated against earlier variants [1, 45]. These variants can remain prevalent for years with minimal phenotypic alterations until they are eventually supplanted by the next variant [1].

Phylogenetic analyses of the major structural protein encoding gene VP1 suggest that the GII.4 variant has a different origin, with significant differences mapped to specific antigenic sites. Immune pressure drives the evolution of GII.4, with antibodies targeting these antigenic sites playing a crucial role in obstructing the interaction between HBGAs and GII.4 virus-like particles (VLPs). Recent studies have connected amino acid alterations in epitopes A, epitope C, and epitope G to the occurrence of novel GII.4 variants [55].

Improved surveillance systems and extensive sequencing are needed to identify low-prevalence pre-pandemic virus. Unlike GII.4, other NoV genotypes have fewer variants and can persist with minimal changes in their VP1 protein for decades. Despite their limited overall prevalence due to evolutionary constraints, non-GII.4 NoV strains can still trigger epidemics and occasionally surpass "GII.4 strains" in dominance [56]. The recent appearance of "GII.2" and "GII.17" in several countries underscores their potential to cause large-scale outbreaks. These viruses are antigenically distinct from GII.4, and the rise of epidemic GII.2 strains seems to be associated with alterations in the RdRp, emphasizing the importance of non-structural proteins in the evolution and emergence of dominant viruses [57].

6.7.1 Point Mutation

Point mutations in the NoV RNA genome have been extensively reported in healthy and immunocompromised individuals. RNA viruses lack proofreading systems, which is the main cause of the high mutation rates ranging from 10^{-3} to 10^{-5} per nucleotide during replication. Specifically, mutations in the NoV genome have been documented at 1.9 to 9.0×10^{-3} substitutions per nucleotide per year [51, 58]. The RdRp plays a central role in this process, with strain-dependent differences in mutation rates contributing to the rapid evolution of certain genotypes, particularly the "GII.4" variants. The higher mutation rates observed in "GII.4 strains", compared to less common genotypes, facilitate antigenic drift and allow the virus to evade immune detection, thereby maintaining its infectivity and prevalence [59]. Comparative genomics have demonstrated an inverse correlation between the viral fidelity of RdRp in certain NoV strains. For instance, the GII.4 and GII.b/GII.3 recombinants NoV types with lower RdRp fidelity than the GII.7 strain have been more commonly encountered domestically and globally. The cost in fidelity associated with the presence of antigenic diversity might enhance the virus' fitness potential in the face of immune response. Moreover, changes in regions other than the evolutionary conserved RdRp due to antigenic drift or shift because of accumulated point mutations would also enhance immune evasion [60]. For example, a molecular epidemiology study found many mutations in the P2 subdomain of the ORF2 region is believed to harbor two key epitopes—P1 and P2—targeted by antibodies. This is further supported by recent findings from a protein mutation analysis of NoV sequences from a long-term diarrheal surveillance study, which linked amino acid sequence changes in the P2 domain of the major capsid protein (VP1) to antigenic evolution [60, 61]. P2 domain mutations are particularly significant in GII.4 strains due to the change in HBGA binding specificity, consequently influencing the prevalence of the virus in the infection of acute diarrhea. Numerous amino acid variations in the epitope P2 domain have been reported in GII.2, GII.7, and recombinant GII.b/GII.3 non-GII.4 strains [61]. It is also possible that some of the changes are caused by GII.4 strains containing new mutations in the HBGA-binding P2 domain that would minimize the vulnerability of the strains to immune recognition [62]. This suggests an evolutionary strategy by the virus to either increase or decrease antibody binding capacity to escape herd immunity. In immunocompetent hosts, NoV typically sheds for 1 to 4 weeks following the onset of diarrhea. However, in prolonged infections, sequencing analysis has demonstrated that genomic mutations leading to amino acid alterations can result in neutralization escape, reduced virus clearance, and extended viral shedding [63]. The existence of time-dependent mutations in the non-structural replication protein RdRp and structural viral protein "VP1" and "VP2" in a host which was infected for 3 months with the GII.14[P7] strain without developing any form of immunity explains the phenomenon of long-term shedding [64]. The GII.4 NoV strains which have been dominant in the past may develop more distinct characteristics due to intra-host emergence of the pandemic structural variants which lead to strong immunity. This is furthered by the evidence that long-term infections result in the evolution and development of multiple characteristics of the virus. For example, the globoside present in the Fut2 gene mutations, in turn, leads to the development of strong resistance toward increased NoV infections [65]. Moreover, the variations in particular NoV receptors including "VP1" across NoV sub-groups are a strong determinant of the differential reception of a particular viral strain by various individuals [66]. Many people carry non-secretory HBGAs, exploring that HBGAs' immunity mechanisms could paint a clearer picture as to why people have varying degrees of resistance in relation to geographic location. There is active NoV replication along the epithelial layer; human enterocytes are co-localized sites for the viral non-structural and structural proteins [67]. The two viral proteins mentioned, point mutations in capsid proteins while RdRp, are critical in the viral evolutionary diversification process. Mutations in the VP1 capsid protein accumulate during the emergence of new GII. 4 variants, whereas other genotypes such as GII. 2, GII. 17 and isolates with mutations are associated with alterations in RdRp. Being on the verge of switch and replicative-bottleneck, the experimental evidence indicates that single RdRp point mutations have a strong impact on the nucleic acid replication dynamic of the virus and that both the non-structural and structural genomic region undergoes the evolutionary step in NoV replication. For example, emerging GII. 17[P17] viruses are characterized by a mutation in the RdRp that corresponds to a similar mutation found in the GII. P16 strains which predominated in 2016 [68]. Though their actual impact is still being studied, these changes are probably going to offer new biological characteristics that make infection and dissemination easier. Certain single-point mutations at crucial locations can alter the RdRp's metabolic characteristics, influencing the virus' fitness and propensity for epidemic spread [69]. Human noroviruses (HuNoV) have high mutation rates primarily dictated by RdRp, which possesses functional similarities to other RNA virus polymerases and is a major contributor to NoV genetic diversity. *In vitro*, mutation rates for HuNoVs are consistent with those of other RNA viruses; however, they vary greatly between strains. For example, the "GII. 4" strains have five to 36 times the mutation rates of rarer strains like "GII. P21 and GII. 7". A mutation rate of 1.5×10^{-4} substitutions per nucleotide cycle is estimated by a recent study on Norwalk virus replication in human cells [70, 71]. U-to-C and A-to-G substitutions make up a significant percentage of spontaneous mutations. These changes can be caused by host-induced hypermutation, namely, double-strand RNA-dependent adenosine deaminases (ADARs) editing adenosine to inosine. This implies that intrinsic RdRp fidelity and host-driven mutations both contribute to NoV diversity [71]. Nonsynonymous mutations in "GII.4" and other genotypes tend to cluster in specific regions of the viral capsid, particularly in the hypervariable P2 domain, indicating that these regions are under strong immune selection. However, evolutionary rates across different NoV genotypes can be similar, suggesting that factors beyond RdRp fidelity, such as recombination events, also play a role in NoV evolution. Moreover, studies on murine NoV (MNV) support the importance of point mutations in NoV evolution. Mutation rates in MNV are similar to those in HuNoV for most genome regions but are higher in the capsid and other structural proteins [72]. Single-point mutations can drastically alter MNV behavior, such as changing a nonpersistent strain to a persistent one or attenuating viral pathogenicity. Additionally, mutations affecting RdRp fidelity can influence MNV replication and transmission, highlighting the potential of targeting RdRp in antiviral therapies [73]. Point mutations in norovirus can arise due to a variety of factors, including viral factors such as reduced RdRp fidelity and the absence of a proofreading mechanism in RNA viruses, host factors like adenosine deaminase, as well as environmental influences such as UV radiation, chemical mutagenesis, and susceptibility in certain populations (as shown in Figure 6.3).

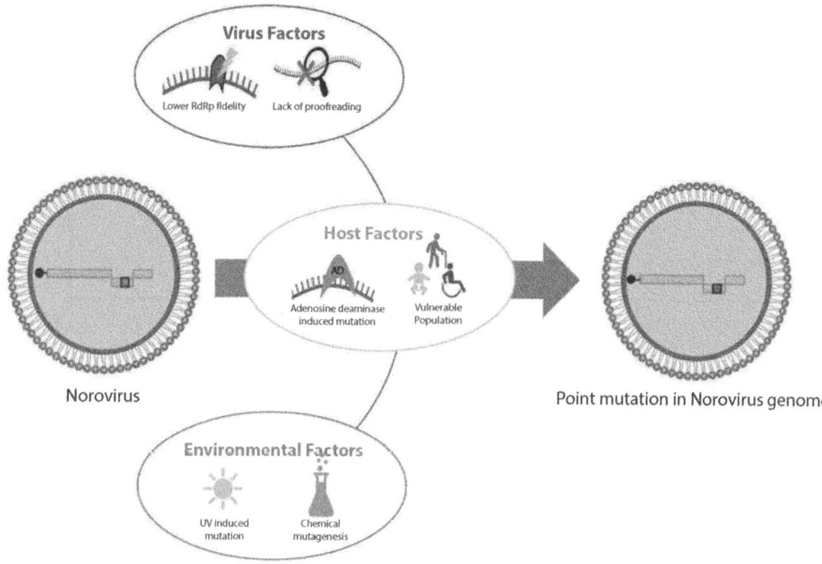

Figure 6.3 Point mutation (antigenic drift). Various factors like virus factors, host factors, and environmental factors are responsible for point mutation in the norovirus genome.

6.7.2 Recombination

Recombination is a fundamental mechanism driving genetic diversity in noroviruses, playing a critical role in their evolution and adaptation. During co-infection, recombination in noroviruses usually entails the exchange of genetic material between strains, giving rise to new variants that may have improved virulence and immune capacities. This recombination process frequently takes place at specific locations within the viral genome, especially in the overlapping junction of ORF1 and ORF2, and to a lesser extent at the overlap between "ORF2 and ORF3" [51, 74]. The primary mechanism involves RdRp, which can switch templates during replication, especially when encountering subgenomic promoters. This template switching can generate recombinant viruses with novel combinations of "ORF1" and "ORF2/ORF3" sequences. Two types of recombination are observed in noroviruses, namely, intragenotypic and intergenotypic. Intragenotypic recombination occurs within the same genotype, often near the junctions of "ORF1 and ORF2" and occasionally at the "ORF2–ORF3" junction. This type of recombination is associated with antigenic shifts that help the virus escape immune detection [75]. For instance, the rise of new "GII.4 variants" like "New Orleans 2009" and "Sydney 2012" is linked to intragenotypic recombination events (see Figure 6.4) [76]. On the other hand,

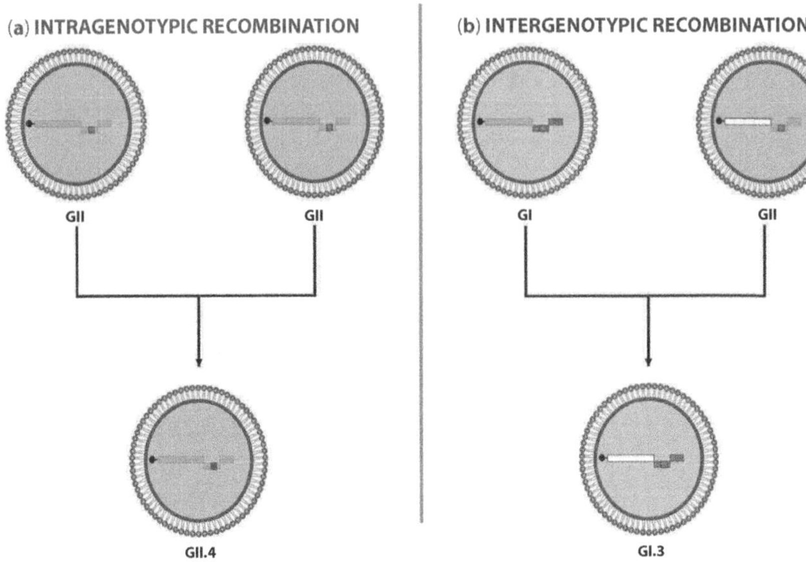

Figure 6.4 Genetic recombination: (a) Intragenotypic recombination occurs within the same genotype. (b) Intergenotypic recombination occurs between different genotypes and can lead to the blending of different genome portions.

intergenotypic recombination occurs between different genotypes and can lead to the blending of different genome portions, commonly at the ORF1–ORF2 junction. This type can create novel virus strains with enhanced fitness and virulence [75]. A novel recombinant strain was identified with a polymerase gene fragment is like the NoV "GI.3 virus" and a capsid gene resembling the "GII.4 strain". Recombination occurred at nucleotide positions 790 in the GI strain and 808 in the GII strain (see Figure 6.4) [77].

6.7.2.1 Mechanisms of Recombination

The mechanisms underlying NoV recombination involve several key steps (illustrated in Figure 6.4), namely:

i. Co-infection

- Process: Recombination in norovirus primarily occurs when two distinct strains or variants simultaneously infect the host cell. This co-infection provides the necessary conditions for combining genetic material from different viral genomes.

- Facilitators: Conditions that favor co-infection include food-originated outbreaks with multiple NoV strains, especially in people with chronic infections, impaired immune system, and infections in malnourished individuals.

ii. Recombination breakpoints

- Process: Once the host cell is coinfected, the viral RdRp initiates replication of the viral RNA. During this process, recombination break points may occur at specific regions within the viral genome, resulting in the generation of distinct recombinant strains.
- Location: Recombination typically occurs at some special sites in the viral genome, such as the "ORF1–ORF2 junction", which encode the NS1/2 to NS7 proteins and the VP1 protein. This region is considered a hotspot for recombination events. Less frequently, recombination break points may also occur at the overlap between "ORF2" and "ORF3", which encodes VP2 protein (see Figure 6.2).

iii. Template switching

- Process: The RdRp creates a complementary negative-sense RNA strand during viral RNA replication. If the RdRp encounters a region of sequence similarity between the two co-infecting viral genomes, it can switch templates. This template switching allows the polymerase to incorporate sequences from one viral strain into the genome of another, resulting in a recombinant virus.
- Outcome: The creation of a recombinant virus with a unique combination of ORF1 and ORF2/ORF3 due to template swapping may have changed antigenic characteristics and enhanced fitness [75, 78–80].

Recombination in noroviruses is a selective process where only those events that confer a fitness advantage or alter pathogenicity are likely to persist in the viral population. This selection ensures that recombination events which enable the virus to evade host immune responses or adapt to new environments are maintained, contributing to the ongoing evolution of noroviruses. Such recombination may contribute to the birth of new variants with altered antigenic properties, complicating efforts in vaccine development and molecular surveillance. Homologous recombination in the capsid gene has been linked to increased virulence and challenges in genotyping, as observed in recent studies.

The ongoing evolution of noroviruses, particularly the GII.4 variants, is a result of both point mutations and recombination events, with significant implications for viral fitness, immune evasion, and the appearance of novel epidemic strains. The interplay between intrinsic viral factors, such as RdRp fidelity, and extrinsic host factors, such as immune pressure, continues to shape the genetic diversity and evolutionary trajectory of Noroviruses [81].

6.8 Recent Epidemiological Trends

The molecular epidemiology and evolution of NoV have been extensively studied across various regions, revealing a wide range of genogroups and genotypes as well as the appearance of rare and novel strains. We have provided a comprehensive overview of NoV prevalence and the genetic diversity that underpins its transmission and evolution from the literature published between 2020 and 2024 (Table 6.2).

In Nigeria, Arowolo *et al.* identified a prevalence of NoV of 2.3% and detected the rare "GII.P30" genotype [82]. Similarly, Badjo *et al.* reported a NoV prevalence of 10.5% and identified multiple genotypes, including GII.P31, which originated from the co-infection of GI and GII strains [83].

Furthermore, in Brazil, novel recombinant "GIX.1[GII.P15]" [84], "GII.6[P7]" [85], and unconventional recombinants from the Amazon region such as GI.P13/GI.3 and GII.P30/GII.3 were reported [86].

In diverse locations like Russia and Ghana, a significantly higher NoV prevalence of 38% [87] and 36.2% [88], respectively, has been reported, where GII.17 and GII.4 were found to be co-dominant capsid genotypes. Chen *et al.* observed that 91.9% of NoV samples in China were non-GII.4 strains [89], while in Mozambique, Chilaule *et al.* found a prevalence of 14.2% for GII.4 Sydney 2012[P31] [90].

In Germany, studies have identified rare and novel NoV strains, including "GII.13, GII.14, and GII.26". These genotypes' low incidence but notable presence was highlighted by their detection across a wide range of strains [92]. Additionally, "GII.P16-GII.4 Sydney" and "GII.P31-GII.4 Sydney" have been observed to be dominant genotypes, indicating the emergence and persistence of specific strains in recent years [92].

Recently, in 2024, Khamrin *et al.* identified a very low prevalence (0.4%) of "GI.3[P13]" and "GI.4[P4]" in Thailand [93]. Similarly, in Bangladesh,

Table 6.2 Molecular epidemiological data from recent publications (2020–2024).

Author	Study period	Location	No. of patients	Genogroups/ genotype	Major findings
Arowolo *et al.*, 2023 [82]	2015–2017	Nigeria	175	GII	Prevalence of norovirus 2.3% Rare GII.P30 genotype was detected in Nigeria
Badjo *et al.*, 2024 [83]	2018–2021	Burkina Faso	1,295	"GI, GII.P16, GII.P31, GII. P16, GII.P7, GII.P8, GII. P16, GII.P7, GII.PNA7, GII.P16, GII. P7, GII.P30, GII.P31"	Prevalence of norovirus 10.5% GII.P31 genotypes detected to originate from the co-infection of GI and GII strains
Bykov *et al.*, 2024 [87]	2022–2023	Russia (Sverdlovsk Region)	510	GI, GII	Prevalence of norovirus 38%. GII.4 and GII.17 were the co-dominant capsid genotypes
Chen *et al.*, 2020 [89]	2014–2018	China (Huzhou)	450	"GII.2[P16], GII.3[P12], GII.17[P17]"	91.9% samples were non-GII.4 strains
Chilaule *et al.*, 2022 [90]	2014–2015	Mozambique	204	GII	Prevalence of norovirus 14.2% GII.4 Sydney 2012[P31] was most prevalent
Costa *et al.*, 2024 [86]	2018–2022	Brazil (Amazon)	695	GI, GII	Prevalence of norovirus 25.8% GII.P30/GII.3 and GI.P13/GI.3 were the unconventional recombinants detected
Degiuseppe *et al.*, 2020 [97]	2010–2017	Argentina	523	GI, GII	Prevalence of norovirus infection 12% to 29%
Dey *et al.*, 2021 [94]	2014–2019	Bangladesh	466	GII.7, GII.3	Prevalence of norovirus infection 5%

(*Continued*)

Table 6.2 Molecular epidemiological data from recent publications (2020–2024). (*Continued*)

Author	Study period	Location	No. of patients	Genogroups/ genotype	Major findings
Ennuschat *et al.*, 2021 [91]	2013–2018	Germany (Leipzig)	7509	"GI.1, GI.2, GI.3, GI.4, GI.5, GI.6, GII.2, GII.3, GII.4, GII.6, GII.7, GII.13, GII.14, GII.17, GII.26, GI.3, GII.4, GII.2, GII.4"	Prevalence of norovirus 8.1% Rare genotypes such as GII.13, GII.14, and GII.26 were detected
Hungerford *et al.*, 2020 [100]	2012–2015	Malawi	684	GII.4	Prevalence of norovirus infection 9.3% to 11.4%
Ji *et al.*, 2020 [95]	2018	China (Huzhou)	551	GI, GII	Prevalence of norovirus 18.1% Predominantly GII genogroup was identified (majorly GII.2[P16] and GII.17[P17])
Kebe *et al.*, 2022 [99]	2007–2010	Senegal	599	GI, GII	Prevalence of norovirus 13.2% Most prevalent were GII.4 and GII.6 genotypes
Khamrin *et al.*, 2024 [93]	2015–2021	Thailand	2829	"Recombinant strains GI.3[P13], GI.5[P4], GI.6[P11] Non-recombinant strains GI.4[P4], GI.5[P5], GI.6[P6]"	0.4% were positive for NoV GI

(*Continued*)

Table 6.2 Molecular epidemiological data from recent publications (2020–2024). (*Continued*)

Author	Study period	Location	No. of patients	Genogroups/ genotype	Major findings
Kirdar *et al.*, 2022 [102]	2017–2019	Turkey	92	GII	Prevalence of norovirus 18.4%
Lartey *et al.*, 2020 [88]	2008–2017	Ghana	1337	"GI.[P4], GI.7GII.4, GII.3, GII.6, GII.17, GII.5"	Prevalence of norovirus 36.2% GII.4 was the predominant capsid genotype
Li *et al.*, 2023 [96]	2017–2019	China (Hubei)	1216	"GII.4 Sydney [P31], GII.3[P12], GII.2[P16], GII.4 Sydney[P16], GII.17[P17], GII.6[P7], GII.3[P16], GII.17[P17]"	Prevalence of norovirus 14.64% Novel recombinant of GII.4 Sydney 2012 and GII.4 Sydney 2016 was detected
Lo *et al.*, 2021 [101]	2018–2019	India (Kolkata)	2812	GII.4, GII.3, GII.13, GII.17	Prevalence of norovirus 8.2% GII.4 Sydney 2012 was the dominant norovirus capsid genotype
Niendorf *et al.*, 2020 [92]	2018	Germany	66	GI, GII	"GII.P16-GII.4 Sydney and GII.P31-GII.4 were Sydney predominant genotypes"
Sarmento *et al.*, 2021[84]	2017–2018	Brazil	1546	"GI, GII, GII.4[P16], GII.2[P16], GII.4[P31], GII.6[P7], GIX.1[GII. P15]"	Prevalence of norovirus 32.1% First time intergenogroup recombinant genotype GIX.1[GII.P15] was detected
Tegegne *et al.*, 2020 [52]	May 2021–November 2021	Ethiopia	519	"GII.3, GII.6, GII.10, GII.17, GII.21, GI.3, GI.5"	Prevalence of norovirus 8.9% GII.21 was reported for the first time

(*Continued*)

Table 6.2 Molecular epidemiological data from recent publications (2020–2024). (*Continued*)

Author	Study period	Location	No. of patients	Genogroups/ genotype	Major findings
Tinker *et al.*, 2020 [85]	2010–2016	Brazil	251	"GII.4_ Sydney[P31], GII.6[P7], GII.17[P17], GII.1[P33], GII.3[P16], GII.2[P16], GII.2[P2], GII.4_New Orleans[P4]"	Prevalence of norovirus 19.9% Distinct GII.6[P7] was identified
Utsumi *et al.*, 2020 [98]	2015–2019	Indonesia (East Java)	966	GII	Prevalence of norovirus 12.3% GII.[P31] was most prevalent

Dey *et al.* identified a lower prevalence of 5%, with GII.7 and GII.3 being the dominant genotypes [94]. However, GII genogroup strains continued to be dominant globally, with novel recombinants like a new GII.4 Sydney strain detected in China [95, 96] and GII.21 from Ethiopia [52].

These studies highlight the predominance of the GII genogroup [98, 99] and specifically the GII.4 genotype across different regions [99, 100]. The high prevalence of GII.4 Sydney 2012 and co-infection with GI strains have given rise to novel and rare genotypes [101, 102]. However, they also highlight the dynamic character of NoV evolution, as evidenced by the appearance of unusual strains and geographical variations in genotype prevalence. Persistent genetic surveillance is required to monitor the evolution of dominant and emerging genotypes and to better understand the virus' epidemiology.

6.9 Global Impact and Challenges

Norovirus genetic changes have a profound global impact, posing significant challenges to public health, vaccine development, and outbreak management. NoV infecting humans are responsible for substantial morbidity and mortality in developing countries, which significantly contribute to the economic burden in developed countries [103]. Each year, NoVs result in 0.6 billion cases and approximately 219,000 deaths. The economic impact is considerable, with 60,300 million dollars in societal costs and 4,200 billion dollars in direct healthcare costs due to productivity losses from absenteeism or mortality [104]. The virus' genetic diversity is driven by mechanisms like mutation and recombination, which complicate the efforts to control its spread and mitigate its effects. Notably, GII.4 infections are responsible for 55%–85% of previous HuNoV outbreaks which are associated with higher hospitalization and death rates. Until 2012, GII.4 NoV has appeared every 2 to 3 years, making this NoV the most common genotype in humans for more than 20 years. Since then, the GII.4 Sydney strain has remained dominant. This dominance is ascribed to several exclusive advantages, such as bypassing of collective population immunity against dominant genotypes and improved adaptation to host receptors [105]. The survival and dominance of "GII.4" noroviruses are caused by genetic alterations, including intragenotypic recombination, unique missense mutation in capsid sequences, and alteration in HBGA binding patterns. Norovirus genetic diversity has been greatly impacted by the accumulation of point mutations and intricate recombination patterns within the GII.4 lineage [106]. However, the emergence and reappearance of different intra- and intergenotype recombinants is challenging the position of dominant "GII.4" variations, changing long-term global norovirus diversity patterns [26]. These genetic changes can lead to new strains with enhanced virulence and transmission capabilities, resulting in more severe outbreaks and increased disease burden, particularly among susceptible populations such as young children, geriatric population, and immunocompromised individuals. Additionally, NoV's ability to evolve enables it to evade the host immune system, complicating vaccine development as natural immunity may not protect against new strains. The constant evolution of the virus can trigger global outbreaks with significant public health and economic consequences, straining healthcare systems, disrupting economies, and increasing mortality rates. The challenges posed by norovirus genetic changes extend to vaccine development, where the virus' genetic diversity makes it difficult to create broadly protective vaccines. The variability complicates diagnostic accuracy, with current tests potentially failing to detect emerging strains, leading to delays in outbreak response. Although antivirals for norovirus are not widely available, the potential for resistance due to genetic changes remains a concern. Furthermore, public health strategies must continuously adapt to monitor emerging strains and respond effectively, requiring ongoing international collaboration and updates to diagnostic tools and preventive measures. In summary, the genetic changes in norovirus present significant global challenges, impacting the effectiveness of vaccines, diagnostics, and public health responses and underscoring the need for ongoing research and innovation in prevention and control strategies.

6.10 Conclusion

The genetic variability of noroviruses plays a pivotal role in their ability to cause widespread gastroenteritis outbreaks. A high mutation rate, driven by RdRp, and frequent recombination events, particularly at the "ORF1–ORF2" junction, result in a diverse population of quasi-species within a single host. This genetic diversity enables the rapid evolution of noroviruses, leading to the appearance of new strains that can evade host immunity, posing significant challenges for public health. The GII.4 genotype exhibits a concerning trend of rapid evolution with variants emerging every few years due to selective pressures from the host immune response. This evolutionary trend is attributed to the genotype's ability to evolve through both antigenic drift and shift, driven by mutations in the capsid protein and RdRp. Persistent infections, particularly in individuals with weakened immune systems, facilitate the buildup of mutations that can result in antigenically different strains that are able to invade herd immunity. Additionally, host factors like HBGAs play a critical role in mediating viral adherence and susceptibility to infection, further complicating NoV epidemiology. The ongoing evolution of noroviruses not only complicates vaccine development but also necessitates the continuous monitoring of viral strains to anticipate and mitigate outbreaks. Understanding the mechanisms behind these genetic changes is critical for developing effective surveillance and intervention strategies. Future research should explore specific pathways of NoV evolution and their implications for infectivity and virulence as well as exploring innovative approaches to vaccine design that can accommodate the virus' adaptive capabilities. Given the significant public health threat posed by noroviruses, especially in vulnerable populations, a comprehensive understanding of their genetic dynamics is essential for informing effective prevention and control measures. The insights gained underscore the urgent need for collaborative efforts in research, surveillance, and public health policy to address the challenges presented by this highly adaptable virus.

Abbreviations

ADARs	adenosine deaminases
FUT2	fucosyl-transferase-2
HBGAs	histo-blood group antigens
HuNoVs	human noroviruses
LTCFs	long-term care facilities
MNV	murine norovirus
NCWG	Norovirus Classification Working Group
NoV	norovirus
ORF1	open reading frame 1
ORF2	open reading frame 2
ORF3	open reading frame 3
ORF4	open reading frame 4
RdRp	RNA-dependent RNA polymerase
VLPs	virus-like particles
VP1	viral protein 1
VP2	viral protein 2

AI disclosure statement: The author acknowledges the use of AI-based tools for minor language editing or grammatical correction. However, all intellectual contributions, analyses, and conclusions presented in this chapter are entirely the author's own.

References

1. Zhang, X., Chen, C., Du, Y., Yan, D., Jiang, D., Liu, X., Yang, M., Ding, C., Lan, L., Hecht, R., Yang, S., Global burden and trends of norovirus-associated diseases from 1990 to 2019: an observational trend study. *Front. Public Health*, 7, 10, 905172, 2022.
2. Institute for Health Metrics and Evaluation (IHME), *GBD Results*, University of Washington, Seattle, W. I, 2020, Available online: www.vizhub.healthdata.org/gbd-results/. (Accessed on: 28 July 2024.).
3. Carlson, K.B., Dilley, A., O'Grady, T., Johnson, J.A., Lopman, B., Viscidi, E., A narrative review of norovirus epidemiology, biology, and challenges to vaccine development. *npj Vaccines*, 299, 1, 94, 2024.
4. Zahorsk, J., Hyperemesis hiemis or the winter vomiting disease. *Arch. Pediatr.*, 46, 391–395, 1929.
5. Adler, J.L. and Zickl, R., Winter vomiting disease. *J. Infect. Dis.*, 119, 6, 668–73, 1969.

6. Kapikian, A.Z., Wyatt, R.G., Dolin, R., Thornhill, T.S., Kalica, A.R., Chanock, R.M., Visualization by immune electron microscopy of a27-nm particle associated with acute infectious nonbacterial gastroenteritis. *J. Virol.*, 10, 1075–1081, 1972.
7. US Centers for Disease Prevention and Control. Common Settings of Norovirus Outbreaks. Available online: https://www.cdc.gov/norovirus/outbreaks/ common-settings.html. (Accessed on: 30 July 2024).
8. Mack, D., Gohl, P., Kolbert, M., Schui, D., Küsters, U., Harzer, O., Pum, J., Zöllner, B., Where have the enteric viruses gone?-Differential effects on frequent causes of infectious diarrhoea by SARS-CoV-2 pandemic lockdown measures. *Infect. Prev. Pract.*, 3, 4, 100184, 2021.
9. Ahn, S.Y., Park, J.Y., Lim, I.S., Chae, S.A., Yun, S.W., Lee, N.M., Kim, S.Y., Choi, B.S., Yi, D.Y., Changes in the occurrence of gastrointestinal infections after COVID-19 in Korea. *J. Korean Med. Sci.*, 36, 24, e180, 2021 Jun 21.
10. Eigner, U., Verstraeten, T., Weil, J., Decrease in norovirus infections in Germany following COVID-19 containment measures. *J. Infect.*, 82, 6, 276–316, 2021.
11. Ando, T., Monroe, S.S., Gentsch, J.R., Jin, Q.I., Lewis, D.C., Glass, R.I., Detection and differentiation of antigenically distinct small round-structured viruses (Norwalk-like viruses) by reverse transcription-PCR and southern hybridization. *J. Clin. Microbiol.*, 33, 1, 64–71, 1995.
12. Zheng, D.P., Ando, T., Fankhauser, R.L., Beard, R.S., Glass, R.I., Monroe, S.S., Norovirus classification and proposed strain nomenclature. *Virology*, 346, 2, 312–23, 2006.
13. Kroneman, A., Vega, E., Vennema, H., Vinjé, J., White, P.A., Hansman, G., Green, K., Martella, V., Katayama, K., Koopmans, M., Proposal for a unified norovirus nomenclature and genotyping. *Arch. Virol.*, 158, 2059–68, 2013.
14. Chhabra, P., de Graaf, M., Parra, G.I., Chan, M.C., Green, K., Martella, V., Wang, Q., White, P.A., Katayama, K., Vennema, H., Koopmans, M.P., Updated classification of norovirus genogroups and genotypes. *J. Gen. Virol.*, 100, 10, 1393–406, 2019.
15. Fonager, J., Stegger, M., Rasmussen, L.D., Poulsen, M.W., Rønn, J., Andersen, P.S., Fischer, T.K., A universal primer-independent next-generation sequencing approach for investigations of norovirus outbreaks and novel variants. *Sci. Rep.*, 7, 1, 813, 2017.
16. Winder, N., Gohar, S., Muthana, M., Norovirus: an overview of virology and preventative measures. *Viruses*, 14, 12, 2811, 2022.
17. Rani, M., Rajyalakshmi, S., Pakalapaty, S., Kammilli, N., Norovirus structure and classification. *Norovirus*, 1, 31, 2021 Sep 1.
18. Graziano, V.R., Wei, J., Wilen, C.B., Norovirus attachment and entry. *Viruses*, 11, 6, 495, 2019 May 30.
19. Pogan, R., Dülfer, J., Uetrecht, C., Norovirus assembly and stability. *Curr. Opin. Virol.*, 31, 59–65, 2018.

20. Yunus, M.A. and Yunus, M.A., Molecular Mechanisms for Norovirus Genome Replication. *Norovirus*, 23, 278–91, 2021.
21. Kobayashi, M., Matsushima, Y., Motoya, T., Sakon, N., Shigemoto, N., Okamoto-Nakagawa, R., Nishimura, K., Yamashita, Y., Kuroda, M., Saruki, N., Ryo, A., Molecular evolution of the capsid gene in human norovirus genogroup II. *Sci. Rep.*, 6, 1, 29400, 2016.
22. Sherman, M.B., Williams, A.N., Smith, H.Q., Pettitt, B.M., Wobus, C.E., Smith, T.J., Structural studies on the shapeshifting murine norovirus. *Viruses*, 13, 11, 2162, 2021.
23. Lindesmith, L.C., Brewer-Jensen, P.D., Mallory, M.L., Jensen, K., Yount, B.L., Costantini, V., Collins, M.H., Edwards, C.E., Sheahan, T.P., Vinjé, J., Baric, R.S., Virus–host interactions between nonsecretors and human norovirus. *Cell. Mol. Gastroenterol. Hepatol.*, 10, 2, 245–67, 2020.
24. Nordgren, J. and Svensson, L., Genetic susceptibility to human norovirus infection: an update. *Viruses*, 11, 3, 226, 2019.
25. Arrouzet, C.J., Ellis, K., Kambhampati, A., Chen, Y., Steele, M., Lopman, B., Population-level human secretor status is associated with genogroup 2 type 4 norovirus predominance. *J. Infect. Dis.*, 221, 11, 1855–63, 2020.
26. Wasielewski, V.V., Itani, T.M., Zakharova, Y.A., Semenov, A.V., Current trends and new approaches for human norovirus replication in cell culture: a literature review. *Arch. Virol.*, 169, 3, 71, 2024.
27. Cheng, C.C., Ke, G.M., Chu, P.Y., Ke, L.Y., Elucidating the Implications of Norovirus N-and O-Glycosylation, O-GlcNAcylation, and Phosphorylation. *Viruses*, 15, 798, 2023.
28. Shimoike, T., Hayashi, T., Oka, T., Muramatsu, M., The predicted stem-loop structure in the 3′-end of the human norovirus antigenomic sequence is required for its genomic RNA synthesis by its RdRp. *J. Biol. Chem.*, 297, 4, 101225, 2021.
29. Han, K.R., Alhatlani, B.Y., Cho, S., Lee, J.H., Hosmillo, M., Goodfellow, I.G., Kim, K.H., Yang, J.M., Identification of amino acids within norovirus polymerase involved in RNA binding and viral replication. *J. Gen. Virol.*, 98, 6, 1311–5, 2017.
30. Giancotti, G., Nannetti, G., Padalino, G., Landini, M., Santos-Ferreira, N., Van Dycke, J., Naccarato, V., Patel, U., Silvestri, R., Neyts, J., Gozalbo-Rovira, R., Structural investigations on novel non-nucleoside inhibitors of human norovirus polymerase. *Viruses*, 15, 1, 74, 2022.
31. Vera-Garcia, M.E., Martinez-Acuña, N., Lozano-Sepulveda, S.A., Arellanos-Soto, D., Galan-Huerta, K.A., Merino-Mascorro, J.A., Ramos-Jimenez, J., Heredia, N., Rivas-Estilla, A.M., Unraveling the mechanisms involved in human norovirus pathogenesis: An update in norovirus pathogenesis. *Med. Universitaria*, 23, 4, 142–7, 2021.
32. Campillay-Véliz, C.P., Carvajal, J.J., Avellaneda, A.M., Escobar, D., Covián, C., Kalergis, A.M., Lay, M.K., Human norovirus proteins: implications in

the replicative cycle, pathogenesis, and the host immune response. *Front. Immunol.*, 16, 11, 961, 2020.

33. Omatola, C.A. and Olaniran, A.O., Epidemiological significance of the occurrence and persistence of rotaviruses in water and sewage: a critical review and proposal for routine microbiological monitoring. *Environ. Sci. Processes Impacts*, 24, 3, 380–99, 2022.
34. Lian, Y., Wu, S., Luo, L., Lv, B., Liao, Q., Li, Z., Rainey, J.J., Hall, A.J., Ran, L., Epidemiology of norovirus outbreaks reported to the Public Health Emergency Event Surveillance System, China, 2014–2017. *Viruses*, 11, 342, 2019.
35. Centers for Disease Control and Prevention. Norovirus Outbreak Linked to Raw Oysters from British Columbia—Norovirus CDC. 2022, Available online: https://www.cdc.gov/norovirus/outbreaks/index.html (Accessed on 30 July 2024).
36. Hardstaff, J.L., Clough, H.E., Lutje, V., McIntyre, K.M., Harris, J.P., Garner, P., O'Brien, S.J., Foodborne and food-handler norovirus outbreaks: a systematic review. *Foodborne Pathog. Dis.*, 15, 10, 589–97, 2018.
37. Hall, A.J., Wikswo, M.E., Pringle, K., Gould, L.H., Parashar, U.D., Vital signs: foodborne norovirus outbreaks—United States, 2009–2012. *Morb. Mortal Wkly Rep.*, 63, 22, 491–5, 2014.
38. Torén, K., Schiöler, L., Nenonen, N.P., Hannoun, C., Roth, A., Andersson, L.M., Westin, J., Bergström, T., Risk factors for norovirus infection in healthcare workers during nosocomial outbreaks: a cross-sectional study. *Antimicrob. Resist. Infect. Control*, 10, 1–9, 2021.
39. Maciel, A.L., de Souza Braga, R.B., Madalosso, G., Padoveze, M.C., Nosocomial outbreaks: A review of governmental reporting systems. *Am. J. Infect. Control*, 50, 2, 185–92, 2022.
40. Adams, C., Peterson, S.R., Hall, A.J., Parashar, U., Lopman, B.A., Associations of infection control measures and norovirus outbreak outcomes in health-care settings: A systematic review and meta-analysis. *Expert Rev. Anti. Infect. Ther.*, 20, 2, 279–90, 2022.
41. Załęski, A., Banasiuk, M., Karpierz, K., Kuchar, E., Podsiadły, E., The clinical course of gastroenteritis due to nosocomial and community-acquired norovirus infections in immunocompromised and immunocompetent children-single center experience. *Epidemiological Rev.*, 74, 1, 23-31, 2020.
42. Ludwig-Begall, L.F., Mauroy, A., Thiry, E., Noroviruses—the state of the art, nearly fifty years after their initial discovery. *Viruses*, 13, 8, 1541, 2021.
43. Summa, M., Henttonen, H., Maunula, L., Human noroviruses in the faeces of wild birds and rodents—New potential transmission routes. *Zoonoses Public Health*, 65, 5, 512–8, 2018.
44. Davis, A., Cortez, V., Grodzki, M., Dallas, R., Ferrolino, J., Freiden, P., Maron, G., Hakim, H., Hayden, R.T., Tang, L., Huys, A., Infectious norovirus is chronically shed by immunocompromised pediatric hosts. *Viruses*, 12, 6, 619, 2020.

45. Parra, G.I., Emergence of norovirus strains: A tale of two genes. *Virus Evol.*, 5, 2, vez048, 2019.
46. Central disease control and Prevention, CaliciNet Surveillance Network. Sept 1, 2023 – June 30, 2024. Available on: https://www.cdc.gov/norovirus/php/reporting/calicinet-data.html (Accessed on: 30 July 2024).
47. Al-Hello, H., Allen, D., Ambert-Balay, K., Botteldoorn, N., Brytting, M., Buesa, J., Cabrerizo, M., Chan, M., Cloak, F., Di, B., Guix, S., Molecular surveillance of norovirus, 2005–16: an epidemiological analysis of data collected from the NoroNet network.
48. Wielick, C., Ludwig-Begall, L.F., Dams, L., Razafimahefa, R.M., Demeuldre, P.F., Napp, A., Laperre, J., Jolois, O., Farnir, F., Haubruge, E., Thiry, E., The use of germicidal ultraviolet light, vaporised hydrogen peroxide and dry heat to decontaminate face masks and filtering respirators contaminated with an infectious norovirus. *Infect. Prev. Pract.*, 3, 1, 100111, 2021.
49. Ludwig-Begall, L.F., Wielick, C., Jolois, O., Dams, L., Razafimahefa, R.M., Nauwynck, H., Demeuldre, P.F., Napp, A., Laperre, J., Thiry, E., Haubruge, E., "Don, doff, discard" to "don, doff, decontaminate"—FFR and mask integrity and inactivation of a SARS-CoV-2 surrogate and a norovirus following multiple vaporised hydrogen peroxide-, ultraviolet germicidal irradiation-, and dry heat decontaminations. *PLoS One*, *16*, 5, e0251872, 2021.
50. Zhang, P., Hao, C., Di, X., Chuizhao, X., Jinsong, L., Guisen, Z., Hui, L., Zhaojun, D., Global prevalence of norovirus gastroenteritis after emergence of the GII. 4 Sydney 2012 variant: a systematic review and meta-analysis. *Front. Public Health*, 12, 1373322, 2024.
51. Omatola, C.A., Mshelbwala, P.P., Okolo, M.L., Onoja, A.B., Abraham, J.O., Adaji, D.M., Samson, S.O., Okeme, T.O., Aminu, R.F., Akor, M.E., Ayeni, G., Noroviruses: Evolutionary Dynamics, Epidemiology, Pathogenesis, and Vaccine Advances—A Comprehensive Review. *Vaccines*, 12, 6, 590, 20242024.
52. Tegegne, D., Gelaw, A., Alemayehu, D.H., Seyoum, T., Leta, D., Ferede, G., Mulu, A., Gelaw, B., Genetic diversity and distribution of noroviruses among all age groups of patients with diarrhea in Amhara National Regional State, Ethiopia. *PLoS One*, 19, 5, e0303887, 2024.
53. Shen, W., Sheng, Y., Weng, J., Li, G., Wang, D., Qiu, D., Lu, Y., Lin, H., Molecular epidemiology of norovirus associated with acute gastroenteritis in Taizhou, China: a retrospective study. *J. Infect. Public Health*, 13, 1, 34–9, 2020.
54. Niendorf, S., Jacobsen, S., Faber, M., Eis-Hübinger, A.M., Hofmann, J., Zimmermann, O., Höhne, M., Bock, C., Steep rise in norovirus cases and emergence of a new recombinant strain GII. P16-GII. 2, Germany, winter 2016. *Eurosurveillance*, 22, 4, 30447, 2017.
55. Tohma, K., Lepore, C.J., Ford-Siltz, L.A., Parra, G.I., Phylogenetic analyses suggest that factors other than the capsid protein play a role in the epidemic potential of GII. 2 norovirus. *MSphere*, 2, 3, 10–128, 2017.

56. Inns, T., Harris, J., Vivancos, R., Iturriza-Gomara, M., O'Brien, S., Community-based surveillance of norovirus disease: a systematic review. *BMC Infect. Diseases*, 17, 1–0, 2017.
57. Zuo, Y., Xue, L., Gao, J., Liao, Y., Liang, Y., Jiang, Y., Cai, W., Qin, Z., Yang, J., Zhang, J., Wang, J., Evolutionary Mechanism of Immunological Cross-Reactivity Between Different GII. 17 Variants. *Front Micro.*, 12, 653719, 2021.
58. Lindesmith, L.C., Beltramello, M., Donaldson, E.F., Corti, D., Swanstrom, J., Debbink, K., Lanzavecchia, A., Baric, R.S., Immunogenetic mechanisms driving norovirus GII. 4 antigenic variation. *PLoS Pathog.*, 8, 5, e1002705, 2012.
59. Bull, R.A. and White, P.A., Mechanisms of GII. 4 norovirus evolution. *Trends Microbiol.*, 19, 5, 233–40, 2011.
60. Zhou, N., Huang, Y., Zhou, L., Li, M., Jin, H., Molecular evolution of RNA-dependent RNA polymerase region in norovirus genogroup I. *Viruses*, 15, 1, 166, 2023.
61. Chen, L.N., Wang, S.J., Wang, S.M., Fu, X.L., Zheng, W.J., Hao, Z.Y., Zhou, H.S., Zhang, X.J., Zhao, Y.L., Qiu, C., von Seidlein, L., Molecular epidemiology analysis of symptomatic and asymptomatic norovirus infections in Chinese infants. *Virol. J.*, 20, 1, 60, 2023.
62. Zhou, H.L., Chen, L.N., Wang, S.M., Tan, M., Qiu, C., Qiu, T.Y., Wang, X.Y., Prevalence and Evolution of Noroviruses between 1966 and 2019, Implications for Vaccine Design. *Pathogens*, 10, 8, 1012, 2021.
63. Tohma, K., Saito, M., Pajuelo, M.J., Mayta, H., Zimic, M., Lepore, C.J., Ford-Siltz, L.A., Gilman, R.H., Parra, G.I., Viral intra-host evolution in immunocompetent children contributes to human norovirus diversification at the global scale. *Emerg. Microbes Infect.*, 10, 1, 1717–30, 2021.
64. Nonthabenjawan, N., Boonyos, P., Phattanawiboon, B., Towayunanta, W., Chuntrakool, K., Ngaopravet, K., Ruchusatsawat, K., Uppapong, B., Sangkitporn, S., Mekada, E., Matsuura, Y., Identification of GII. 14 [P7] norovirus and its genomic mutations from a case of long-term infection in a post-symptomatic individual. *Infect. Genet. Evol.*, 86, 104612, 2020.
65. Hu, M., Zhang, X., Li, J., Chen, L., He, X., Sui, T., Fucosyltransferase 2: a genetic risk factor for intestinal diseases. *Front Micro.*, 13, 940196, 2022.
66. Hong, X., Xue, L., Liao, Y., Wu, A., Jiang, Y., Kou, X., Association of fucosyltransferase 2 gene with norovirus infection: A systematic review and meta-analysis. *Infect. Genet. Evol.*, 96, 105091, 2021.
67. Haga, K., Ettayebi, K., Tenge, V.R., Karandikar, U.C., Lewis, M.A., Lin, S.C., Neill, F.H., Ayyar, B.V., Zeng, X.L., Larson, G., Ramani, S., Genetic manipulation of human intestinal enteroids demonstrates the necessity of a functional fucosyltransferase 2 gene for secretor-dependent human norovirus infection. *MBio.*, 11, 2, 10–128, 2020.
68. Gao, J., Zhang, Z., Xue, L., Li, Y., Cheng, T., Meng, L., Li, Y., Cai, W., Hong, X., Zhang, J., Wang, J., GII. 17 [P17] and GII. 8 [P8] noroviruses showed

different RdRp activities associated with their epidemic characteristics. *J. Med. Virol.*, 95, 1, e28216, 2023.
69. Matsushima, Y., Mizukoshi, F., Sakon, N., Doan, Y.H., Ueki, Y., Ogawa, Y., Motoya, T., Tsukagoshi, H., Nakamura, N., Shigemoto, N., Yoshitomi, H., Evolutionary analysis of the VP1 and RNA-dependent RNA polymerase regions of human norovirus GII. P17-GII. 17 in 2013–2017. *Front Micro.*, 10, 2189, 2019.
70. Hasing, M.E., Hazes, B., Lee, B.E., Preiksaitis, J.K., Pang, X.L., A next generation sequencing-based method to study the intra-host genetic diversity of norovirus in patients with acute and chronic infection. *BMC Genom.*, 17, 1–1, 2016 Dec.
71. Cuevas, J.M., Combe, M., Torres-Puente, M., Garijo, R., Guix, S., Buesa, J., Rodríguez-Díaz, J., Sanjuán, R., Human norovirus hyper-mutation revealed by ultra-deep sequencing. *Infect. Genet. Evol.*, 41, 233–9, 2016.
72. Borg, C., Jahun, A.S., Thorne, L., Sorgeloos, F., Bailey, D., Goodfellow, I.G., Murine norovirus virulence factor 1 (VF1) protein contributes to viral fitness during persistent infection. *J. Gen. Virol.*, 102, 9, 001651, 2021.
73. Walker, F.C., Hassan, E., Peterson, S.T., Rodgers, R., Schriefer, L.A., Thompson, C.E., Li, Y., Kalugotla, G., Blum-Johnston, C., Lawrence, D., McCune, B.T., Norovirus evolution in immunodeficient mice reveals potentiated pathogenicity via a single nucleotide change in the viral capsid. *PLoS Pathog.*, 17, 3, e1009402, 2021.
74. Bull, R.A., Hansman, G.S., Clancy, L.E., Tanaka, M.M., Rawlinson, W.D., White, P.A., Norovirus recombination in ORF1/ORF2 overlap. *Emerg. Infect. Diseases*, 11, 7, 1079, 2005.
75. Vakulenko, Y.A., Orlov, A.V., Lukashev, A.N., Patterns and temporal dynamics of natural recombination in noroviruses. *Viruses*, 15, 2, 372, 2023.
76. Xu, J., Liu, T., Wang, H., Shen, H., Intergenotype recombination among new norovirus GII. 4 variants in the complete genome. *Intervirology*, 60, 4, 138–43, 2018.
77. Nayak, M.K., Balasubramanian, G., Sahoo, G.C., Bhattacharya, R., Vinje, J., Kobayashi, N., Sarkar, M.C., Bhattacharya, M.K., Krishnan, T., Detection of a novel intergenogroup recombinant Norovirus from Kolkata, India. *Virology*, 377, 1, 117–23, 2008.
78. White, P.A., Evolution of norovirus. *Clin. Microbiol. Infect.*, 20, 8, 741–5, 2014 Aug. 1.
79. Ludwig-Begall, L.F., Mauroy, A., Thiry, E., Norovirus recombinants: recurrent in the field, recalcitrant in the lab–a scoping review of recombination and recombinant types of noroviruses. *J. Gen. Virol.*, 99, 8, 970–88, 2018.
80. De Graaf, M., van Beek, J., Koopmans, M.P., Human norovirus transmission and evolution in a changing world. *Nat. Rev. Microbiol.*, 14, 7, 421–33, 2016.
81. Eden, J.S., Tanaka, M.M., Boni, M.F., Rawlinson, W.D., White, P.A., Recombination within the pandemic norovirus GII. 4 lineage. *J. Virol.*, 87, 11, 6270–82, 2013.

82. Arowolo, K.O., Ayolabi, C.I., Adeleye, I.A., Lapinski, B.A., Santos, J.S., Raboni, S.M., Genetic Diversity of Norovirus in Children with Acute Gastroenteritis in Southwest Nigeria, 2015–2017. *Viruses*, 15, 3, 644, 2023.
83. Badjo, A.O., Niendorf, S., Jacobsen, S., Zongo, A., Mas Marques, A., Vietor, A.C., Kabore, N.F., Poda, A., Some, S.A., Ouattara, A., Ouangraoua, S., Genetic diversity of enteric viruses responsible of gastroenteritis in urban and rural Burkina Faso. *PLoS Negl. Trop. Dis.*, 18, 7, e0012228, 2024.
84. Sarmento, S.K., de Andrade, J.D., Miagostovich, M.P., Fumian, T.M., Virological and Epidemiological Features of Norovirus Infections in Brazil, 2017–2018. *Viruses*, 13, 9, 1724, 2021.
85. Tinker, R.J., da Costa, A.C., Tahmasebi, R., Milagres, F.A., dos Santos Morais, V., Pandey, R.P., Jose-Abrego, A., Brustulin, R., Rodrigues Teles, M.D., Cunha, M.S., Araujo, E.L., Norovirus strains in patients with acute gastroenteritis in rural and low-income urban areas in northern Brazil. *Arch. Virol.*, 166, 905–13, 2021.
86. Costa, J.N.D., Siqueira, J.A.M., Teixeira, D.M., Lobo, P.D.S., Guerra, S.D.F.D.S., Souza, I.M., Cardoso, B.T.M., Farias, L.S.S., Resque, H.R., Gabbay, Y.B., Silva, L.D.D., Epidemiological and molecular surveillance of norovirus in the Brazilian Amazon: description of recombinant genotypes and improvement of evolutionary analysis. *Rev. do Instituto de Med. Trop. De São Paulo, 66*, e22, 2024.
87. Bykov, R., Itani, T., Starikova, P., Skryabina, S., Kilyachina, A., Koltunov, S., Romanov, S., Semenov, A., Genetic Diversity and Phylogenetic Relationship of Human Norovirus Sequences Derived from Municipalities within the Sverdlovsk Region of Russia. *Viruses*, 16, 7, 1001, 2024.
88. Lartey, B.L., Quaye, O., Damanka, S.A., Agbemabiese, C.A., Armachie, J., Dennis, F.E., Enweronu-Laryea, C., Armah, G.E., Understanding pediatric norovirus epidemiology: a decade of study among Ghanaian Children. *Viruses*, 12, 11, 1321, 2020.
89. Chen, L., Xu, D., Wu, X., Liu, G., Ji, L., An increasing prevalence of non-GII. 4 norovirus genotypes in acute gastroenteritis outbreaks in Huzhou, China, 2014-2018. *Arch. Virol.*, 165, 5, 1121–8, 2020.
90. Chilaúle, J.J., Munlela, B., Mans, J., Mabasa, V.V., Marques, S., Bauhofer, A.F., Jane, G., Anapakala, E., Oliveira, F., Cossa-Moiane, I., Guimarães, E., Norovirus Genetic Diversity in Children under Five Years Old with Acute Diarrhea in Mozambique (2014–2015). *Viruses*, 14, 9, 2001, 2022.
91. Ennuschat, N., Härtel, S., Pietsch, C., Liebert, U.G., Norovirus epidemiology and genetic diversity in Leipzig, Germany during 2013–2017. *Viruses*, 13, 10, 1961, 2021.
92. Niendorf, S., Faber, M., Tröger, A., Hackler, J., Jacobsen, S., Diversity of noroviruses throughout outbreaks in Germany 2018. *Viruses*, 12, 10, 1157, 2020.
93. Khamrin, P., Kumthip, K., Yodmeeklin, A., Okitsu, S., Motomura, K., Sato, S., Ushijima, H., Maneekarn, N., Genetic recombination and genotype diversity

of norovirus GI in children with acute gastroenteritis in Thailand, 2015-2021. *J. Infect. Public Health*, 17, 3, 379–85, 2024.

94. Dey, S.K., Sharif, N., Billah, B., Siddique, T.T., Islam, T., Parvez, A.K., Talukder, A.A., Phan, T., Ushijima, H., Molecular epidemiology and genetic diversity of norovirus infection in children with acute gastroenteritis in Bangladesh, 2014–2019. *J. Med. Virol.*, 93, 6, 3564–71, 2021.
95. Ji, L., Hu, G., Xu, D., Wu, X., Fu, Y., Chen, L., Molecular epidemiology and changes in genotype diversity of norovirus infections in acute gastroenteritis patients in Huzhou, China, 2018. *J. Med. Virol.*, 92, 12, 3173–8, 2020.
96. Li, J., Zhang, L., Zou, W., Yang, Z., Zhan, J., Cheng, J., Epidemiology and genetic diversity of norovirus GII genogroups among children in Hubei, China, 2017–2019. *Virol. Sin.*, 38, 3, 351–62, 2023.
97. Degiuseppe, J.I., Roitman, K.L., Rivero, K.A., Stupka, J.A., Norovirus passive surveillance as an alternative strategy for genetic diversity assessment in developing countries. *J. Infect. Public Health*, 14, 8, 990–3, 2021.
98. Utsumi, T., Lusida, M.I., Dinana, Z., Wahyuni, R.M., Soegijanto, S., Athiyyah, A.F., Sudarmo, S.M., Ranuh, R.G., Darma, A., Yamani, L.N., Doan, Y.H., Molecular epidemiology and genetic diversity of norovirus infection in children hospitalized with acute gastroenteritis in East Java, Indonesia in 2015–2019. *Infect. Genet. Evol.*, 88, 104703, 2021.
99. Kebe, O., Fernandez-Garcia, M.D., Zinsou, B.E., Diop, A., Fall, A., Ndiaye, N., Vinjé, J., Ndiaye, K., Prevalence and genetic characterization of noroviruses in children with acute gastroenteritis in Senegal, 2007–2010. *J. Med. Virol.*, 94, 6, 2640–4, 2022.
100. Hungerford, D., Jere, K.C., Bar-Zeev, N., Harris, J.P., Cunliffe, N.A., Iturriza-Gómara, M., Epidemiology and genotype diversity of norovirus infections among children aged< 5 years following rotavirus vaccine introduction in Blantyre, Malawi. *J. Clin. Virol.*, 123, 104248, 2020.
101. Lo, M., Mitra, S., De, P., Banerjee, A., Deb, A.K., Miyoshi, S.I., Manna, A., Ghosh, S.K., Okamoto, K., Dutta, S., Chawla-Sarkar, M., Genetic characterization and evolutionary analysis of norovirus genotypes circulating among children in eastern India during 2018-2019. *Arch. Virol.*, 166, 11, 2989–98, 2021.
102. Kırdar, S., Başara, T., Ömürlü, IK., Prevalence and Genetic Diversity of Norovirus in Acute Gastroenteritis Cases in the Southwest Province of Turkey. *Balkan Med. J.*, 39, 2, 153, 2022.
103. Burke, R.M., Mattison, C.P., Pindyck, T., Dahl, R.M., Rudd, J., Bi, D., Curns, A.T., Parashar, U., Hall, A.J., Burden of norovirus in the United States, as estimated based on administrative data: updates for medically attended illness and mortality, 2001–2015. *Clin. Infect. Dis.*, 73, 1, e1–8, 2021.
104. Liao, Y., Hong, X., Wu, A., Jiang, Y., Liang, Y., Gao, J., Xue, L., Kou, X., Global prevalence of norovirus in cases of acute gastroenteritis from 1997 to 2021: an updated systematic review and meta-analysis. *Microb. Pathog.*, 161, 105259, 2021.

105. Zheng, G.L., Zhu, Z.X., Cui, J.L., Yu, J.M., Evolutionary analyses of emerging GII. 2 [P16] and GII. 4 Sydney [P16] noroviruses. *Virus Evol.*, 8, 1, 30, 2022.
106. Ge, L., Chen, X., Liu, J., Zheng, L., Chen, C., Luo, S., Guo, P., Kong, J., Song, Y., Huo, Y., Genomic and biological characterization of a pandemic norovirus variant GII. 4 Sydney 2012. *Virus Genes*, 56, 174–81, 2020.

7

Bridging the Gap: One Health Approach in Norovirus

Sasikanth Kothamasu[1], Naga Rani Kagithala[2], Dheeraj Sharma[2], Shriyansh Srivastava[3], Aroop Mohanty[4] and Rakesh Sahu[5]*

[1]Department of Pharmaceutics, SIMS College of Pharmacy, Mangaldasnagar, Guntur, India
[2]Department of Pharmacology, School of Medical and Allied Sciences, Galgotias University, Greater Noida, Uttar Pradesh, India
[3]Department of Pharmacology, Delhi Pharmaceutical Sciences and Research University, New Delhi, India
[4]Department of Microbiology, AIIMS Gorakhpur, Uttar Pradesh, India
[5]Department of Physiology and Cell Biology, Faculty of Medicine, Ben Gurion University of the Negev, Beersheba, Israel

Abstract
Norovirus is one of the causes of acute gastroenteritis worldwide and it is involved in about 685 million cases yearly. This highly contagious virus can be transmitted through contaminated food, water, objects, and person-to-person contact. Its capacity to live in different environments and resist common cleaning agents makes it very difficult to control. Norovirus is imposing a huge public health stress by serious outbreaks that are quite often observed in the healthcare domain, schools, and food service industries. The groups of people most at risk of developing severe complications from this disease are the ones that consist of young children, the elderly, and immunocompromised individuals. The One Health method that integrates human, animal, and environmental health surveillance systems is of a kind of comprehensive model for managing the multifaceted nature of norovirus transmission. This method also appreciates the role of the environment as a source of infection, as water and food systems serve as vectors for its spread. Surveillance programs that integrate environmental monitoring, like analyzing wastewater, with human health data facilitate the early identification of outbreaks and thus make response strategies more effective. Despite the non-existence of a licensed

**Corresponding author*: rksh9341abc@gmail.com

Sachin Kumar, Shriyansh Srivastava, Rishabha Malviya and Ranjit Sah (eds.) Emerging Infectious Disease: Unravelling Norovirus and Pandemic Dynamics, (187–220) © 2026 Scrivener Publishing LLC

norovirus vaccine, there has been significant development in vaccine production, especially in the manufacture of recombinant virus-like particles (VLP) vaccines. The chapter addresses, among other topics, the epidemiology of norovirus, the means of transmission, and the interplay of cooperation between the different scientific domains in the case of the One Health approach. Furthermore, it will touch upon ways of preventing the virus which include the proper sanitation of surroundings, hygiene, and water treatment methods, as well as the ongoing efforts in vaccine and antiviral production. The text demonstrates the significance of international action by displaying case studies and samples of successful One Health projects.

***Keywords*:** Norovirus, One Health approach, zoonotic disease, norovirus surveillance, hygiene practice

7.1 Overview of Norovirus: Structure and Epidemiology

Norovirus is a nonenveloped and single-stranded RNA virus, which belongs to the Caliciviridae family. Its genome has around 7.5 kb and has three open reading frames (ORFs). ORF1 has non-structural proteins, and the virus cannot replicate without them, ORF2 has the main viral capsid protein VP1, and ORF3 has a minor protein, which is the VP2 protein, and it stabilizes the capsid [1]. The virus has various genotypes, and it is not easy to find one that is similar to the other. It can be divided into seven genogroups, which are GI to GVII. Furthermore, the number of genotypes is over 30. One of the most virulent strains of norovirus is the GII.4, being the global cause of thousands of cases [2]. This new variant that has emerged is the one that has the potential to bypass the immune system, at the same time allowing breaking the vaccine barrier. The problem of the high rate of mutation of viral genes is that the virus is able to sidestep immunity from previous infections, and this may lead to repeated episodes of the outbreaks [3]. Norovirus spreads mainly through the fecal–oral route, which occurs by people directly passing it on to each other, through the consumption of dirty food or water, and *via* the contacting of contaminated surfaces. The virus is also known to be very stable in an environment; it persists on surfaces for weeks and it is not killed in low temperatures or when heated. This durability has led to a number of epidemics, particularly in places where people are closely associated, such as schools, hospitals, and ships [4, 5]. Approximately, 685 million cases of acute gastroenteritis are caused by norovirus every year, and 200,000 deaths result mainly from low income and middle-income countries, where cleanliness levels are low [3].

Its high infectivity and environmental resilience rank it as one of the most common cause of outbreaks in environments such as healthcare facilities, schools, and military barracks [4, 5]. Norovirus, a virus with a high degree of contagiousness, seems to be less dependent on a large infective dose, needing as little as 18 viral particles for infection. Its rapid dispersal during outbreaks is the issue of infectivity to a low dose of virus which is one of the major factors. The virus gets in through various intermediaries such as direct contact, contaminated food, and environmental surfaces, making its spread very hard to control especially in places like long-term care facilities and daycare centers, where vulnerable populations are at a high risk [3]. In addition, asymptomatic carriers who do not show any symptoms of the infection can still spread the virus to other people, thereby increasing the transmission rate [6, 7]. Norovirus is a source of a particular strain on the healthcare industry and is a heavy blow to the economy, with norovirus causing around $2 billion in direct medical costs and lost productivity in the United States alone [8]. Likewise, the economic effects on the food industry are significant with the fact that infected products can result in the recall and the closure of several businesses. Illnesses transmitted through food, usually food handlers or raw produce, remain a regular occurrence at food chains, school cafeterias, and food processors [9, 10]. When trying to manage the norovirus issue in the hospitals and long-term care facilities, the hospitals must undertake the necessary infection control measures which will in turn lead to operation cost increases. The problem of patient overcrowding on the one side and the shortage of labor and equipment on another lifts the already precarious situation and makes matters worse [11, 12]. Norovirus impacts particularly those who are most vulnerable, such as young children, seniors, and immunocompromised individuals, the most [3, 6]. In these groups, viral infection can cause serious dehydration that can require hospitalization and, in severe cases, can even cause death. Kids younger than 5 years and older adults in long-term care facilities, as a consequence of the weaker immune responses of the former and the very close living spaces in the latter, are the most susceptible group in addition to contracting the virus easily. In weakened patients, the virus can be the cause of prolonged disease, as it may be hard for them to get rid of the infection [13, 14]. Apart from the human factor at a personal level, the medical attempts to control the norovirus are a big issue for the budget especially in the healthcare centers. The medical costs increase due to the need to isolate the patient, the increased load placed on the cleaning service, and the demand for extra personnel during outbreaks [15, 16]. Loss of income and tarnished reputation known as norovirus outbreaks can result in facility closures contributing to the problem in the long-term care facilities. Moreover, during massive outbreaks that put stress on the public health sector, especially in areas with scarce resources, this could lead to service disruptions and delays in routine care [3].

7.2 Understanding the One Health Approach

7.2.1 Definition and Principles of One Health

The One Health approach is the cooperation between several sectors including human health, animal health, environmental sciences, and other concepts aimed at the treatment of complex public health problems, such as norovirus [17, 18]. According to the core principle, human health is closely connected with animal health and the environment, recognizing the significant impact of zoonotic diseases and environmentally transmitted pathogens. It is vital to mention the people like the veterinarians, epidemiologists, environmental scientists, and public health officials who have to collaborate to construct an entire plan to eliminate the factors that lead to the spread of the epidemic diseases [19, 20]. The incorporation of human, animal, and environmental health is essential in the eradication and control of infectious diseases like norovirus where water contamination and food transmission are the major contributors. One Health method allows for streamlined actions in observing and attending potential sources of infections to embark on prevention [18]. The insight from the data comes from the environmental monitor (like water quality monitoring), the hospital's admission records, and cancer research (e.g., study in which zoonotic aspects are addressed) that will enable the public health officials to reveal the best way to go about their task [21].

7.2.2 Historical Context and Evolution of the One Health Concept

The notion of One Health is within early positioning of the interconnections among animal and human health. In the 19th century, Rudolf Virchow, a physician from Germany, discussed the significance of veterinary sciences in controlling zoonotic diseases [22]. Initially, these ideas morphed into a more organized whole that embraces human and animal health, in particular. Conditions that can spread from animals to humans like zoonoses were the kind of problems being managed this way [20]. By the 21st century, One Health embraced areas beyond animal–human disease interactions such as the environment and its relation to the infectious disease norovirus. This change was brought about as a response to worldwide problems like change in climate, deforestation, and urbanization, which changed the pattern of disease transmission [23]. Progressive projects with One Health now include collaborations across sectors, such as environmental agencies for the protection of food safety regulatory officials, to investigate the multifaceted nature of the ways that norovirus, for instance, gets transmitted through food, water, and animals [18, 24].

7.2.3 Relevance of One Health to Norovirus Control

7.2.3.1 Environmental Transmission Pathways

Norovirus is highly contagious and spreads through the contaminated water and the food that makes it a perfect candidate for One Health approach. The condition of environmental contamination, which is often related to poor sanitation practices, can result in quick spread in communities. One of the primary culprits, a waste treatment plant where the sewage is not properly treated or is non-existing, is usually spending developing countries and overloading water systems leading to waterborne outbreaks. One Health strategies such as administering water quality and food safety through the perfect environmental tracking are chief in the prevention of transmission pathways [21, 25]. Safety surveillance is one of the fundamental principles of One Health approach, especially in the control of diseases like norovirus that have many pathways of transmission. Surveillance that is both public and private is a mode of systematically collecting data about human, animal, and environmental health. For example, with norovirus, water systems, shellfish farms, and food processing plants should become part of regular monitoring to filter out contaminated food and water. Collective surveillance gives households the chance to detect the disease quickly and to serve it with the least trouble [26, 27]. As an example, the One Health approach of shellfish farm surveillance in The Netherlands is the best illustration. In this area, the virus' effects are high on the list of problems related to contamination because of unevaluated sewage water. A joint healthcare scheme/strategy that consists of environment surveillance, food safety standards, and local health department regulations has helped minimize the frequent potentially fatal foodborne illnesses through the spotting and elimination of the sources of infection, which in most cases manifest during production processes, before the products are sent to consumers [28, 29]. Figure 7.1 Illustrates the major transmission pathways of Norovirus, including person-to-person spread, foodborne or waterborne routes, and contaminated surfaces. The visual emphasizes how poor

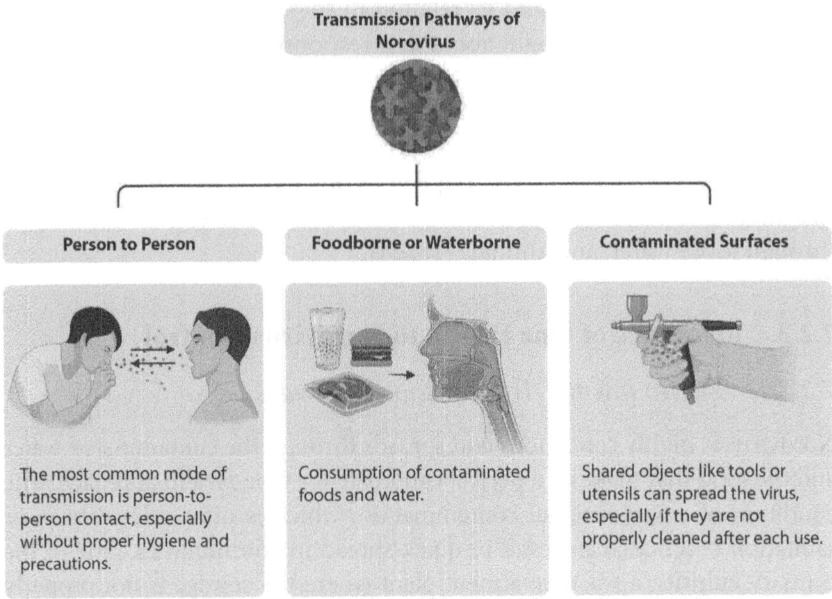

Figure 7.1 Transmission pathways of norovirus.

Table 7.1 Norovirus transmission pathways and key contributing factors.

S. no.	Transmission pathway	Key contributing factors	Examples	References
1.	Human-to-human	Direct physical contact, aerosolized vomit particles, and poor hand hygiene	Spread through contact in hospitals, nursing homes, schools	[32]
2.	Foodborne transmission	Contamination during food preparation or from infected food handlers	Contaminated shellfish, salads, and ready-to-eat foods	[40, 41]

(*Continued*)

Table 7.1 Norovirus transmission pathways and key contributing factors (*Continued*).

S. no.	Transmission pathway	Key contributing factors	Examples	References
3.	Waterborne transmission	Contaminated drinking water or recreational water (e.g., pools)	Outbreaks linked to untreated sewage in drinking water or improperly chlorinated pools	[44, 45]
4.	Environmental Surface contamination	Virus persisting on surfaces, poor disinfection practices	High-contact surfaces in hospitals, public places, or restaurants	[36]
5.	Airborne transmission	Aerosolized particles from vomit can infect individuals nearby.	Vomiting incidents in enclosed spaces like schools or cruise ships	[30]
6.	Asymptomatic carriers	Viral shedding by asymptomatic individuals can continue for days/ weeks.	Undiagnosed carriers spreading the virus in food service or healthcare	[34]
7.	Zoonotic potential	Evidence of norovirus-like viruses in animal reservoirs, especially livestock	Potential zoonotic transmission through contact with animals like pigs or *via* contaminated meat	[38]

hygiene, unsafe food handling, and environmental contamination contribute to outbreaks, particularly in healthcare, school, and food service settings. Table 7.1 outlines the key transmission pathways of Norovirus, detailing contributing factors such as direct contact, contaminated food or water, surface stability, and emerging zoonotic concerns. This structured overview highlights the diverse and complex routes through which Norovirus spreads, reinforcing the need for integrated prevention strategies.

7.3 Transmission Pathways of Norovirus

7.3.1 Human-to-Human Transmission

Norovirus is primarily transmitted *via* the fecal–oral route, specifically through ingestion (direct or indirect), where contaminating excrements or vomit of infected persons enter others' bodies [30]. The most common mode of virus transmission is a person-to-person one, particularly through body fluids in crowded places such as nursing homes, hospitals, schools, and cruise ships [31]. In such situations, the virus is rapidly transmitted by having close physical contact and this makes it hard to contain the infections right after breakout. Hand-to-mouth contact, especially when hygiene measures are not properly followed, still remains a noteworthy virus spreader [32]. Asymptomatic carriers of norovirus act as means of transmission in its developing process. These people accrete the virus through their feces or vomit without giving any symptoms [33]. This prevents the instantaneous detection of the cases; thus, outbreaks are unavoidable. Silent virus transmission can occur in high-risk environmental health facilities or meal service venues where individuals may inadvertently affect surfaces or food products [34]. Moreover, people can keep on with the virus-shedding process for some weeks even after they stop suffering from the acute symptoms, which in turn extends the transmission risk.

7.3.2 Environmental Contamination and Spread

A major factor that accounts for norovirus' high transmissibility is its capability to survive in the environment. The virus can be carried on surfaces such as countertops, door handles, and utensils for up to 2 weeks, especially in places with high humidity and low temperatures [35]. This environmental strength stretches the virus' ability to infect people long after the original cause of contamination has already been removed. Contaminated surfaces in food service environments, public restrooms and clinics, among others, are the most common vehicles for the transportation of the virus. There is also another important aspect that facilitates the virus spreading which is inadequate cleaning and disinfection [36]. Norovirus also can get easily transferred from contaminated surfaces to food preparation areas, utensils, and kitchen equipment. However, these surfaces are less common in food service settings where food handlers not following hygiene rules are the major sources of contamination, which could be, e.g., salads, sandwiches, or fruits [35]. Cross-contamination between contaminated surfaces and food products can happen in household settings too and this underlines the importance of rigorous sanitation measures in both public and private environments. Common items like makeshift kitchen utensils or dinner sets can spread the virus more if they are not cleaned properly after each use [37].

7.3.3 Zoonotic Potential and Animal Reservoirs

Recent research has observed not only norovirus-like viruses in a multitude of animal species, with livestock seemingly being the most affected ones. Along with pigs and cattle, which are known as livestock species, deer and rodents as well as other wild animal species too are found to be infected. These animal strains, which do not signal a direct human infection source, are considered potential vector animals that may cause these viruses to change structurally to cross species barriers if conditions permit [38]. The zoonotic potential of these viruses while not yet fully understood is still a concern for the possibility that animals are becoming of norovirus fulminant transmission. Zoonotic ways of norovirus infections are still under scrutiny, with the possibility of transmission from animals to humans being promoted in the theory but unproven [39]. The occurrence of noroviruses in livestock and wildlife populations as mentioned above indicates a situation where it is just in specific environmental conditions that humans come down with the infection. Paths of diseases, such as through contaminated water or through contact between species, are undoubtedly less for anybody to be safe from the infection. However, the study of animals having a great influence on the creation of outbreaks among human beings would need complete research [38].

7.3.4 Role of Food and Water in Norovirus Outbreaks

Norovirus transmitted *via* food stays mainly in products and shellfish that have been contaminated. Shellfish of this kind like oysters are capable of capturing the norovirus particles when they are taken from contaminated water sources, hence the worrisome factor to be ignored in consuming them raw or undercooked [40, 41]. Likewise, the fresh uncooked vegetables, which have been freshly irrigated with contaminated water, are the primary cause of the foodborne norovirus outbreak. Frequently, this occurs in large-scale food production systems, where a single contamination event—or in extreme cases—can cause the spread of infection across multiple regions [25, 42, 43]. Waterborne transmission forms due to the contamination of drinking water and recreational water sources with norovirus happens most often by the lack of adequate water treatment or the presence of untreated sewage [44, 45]. The developing countries with low-quality sewage systems mostly get measles through the water supplies, resulting in a waterborne infection. They also have had some outbreaks originating from the sports water, like the swimming pools or the lakes, and surrounded by inadequate treatment of the water with lime chlorine, which is not enough to inactivate the virus [46, 47].

7.4 Epidemiology of Norovirus: A Global Perspective

7.4.1 Geographic Distribution and Prevalence

Norovirus is a worldwide pathogen that is reported to be present in all continents. In cases of developed countries, there are recurrent outbreaks in settings such as hospitals, schools, and cruise ships, but the disease burden in low and middle-income countries is very high especially where waterborne transmission is common due to poor sanitation infrastructure [46]. In areas like Sub-Saharan Africa and South Asia, the lack of clean water and proper sanitation remains the main contributor to the virus' spread, and thus the death rate is higher for the little ones, generally aged under 5 years [48]. In developed nations, the primary cause of norovirus is through large and localized outbreaks, which most of the time is linked to food-based transmission. Healthcare settings, cruise ships, and food service environments often fall under this category [49, 50]. On the other hand, in third-world countries, waterborne transmission is the major cause of the epidemic due to lack of access to proper sanitation and drinking water. Furthermore, infections in underdeveloped countries are often more harmful because of the lack of healthcare facilities and rehydration treatments which are urgently needed in controlling the dehydration related to the norovirus diarrhea [51–53].

7.4.2 Risk Factors for Infection

Certain occupational groups, such as healthcare workers, food handlers, and daycare menders, have a very high likelihood of catching and transmitting norovirus [54]. Norovirus infects patients and may thus become a health concern for them in case healthcare workers tend not to apply the infection control techniques [55, 56]. The reason why employees of the catering industry are responsible for outbreaks of the disease is their omission to observe the safety of hygiene rules, especially in the course of preparing foodstuffs [57]. Another place that also constitutes a substantial threat to outbreaks is mainly designated to the places that are occupationally crowded and enclosed such as schools, hospitals, and nursing homes. This is owing to the environs where they are close and the possibility of the virus to be transmitted rapidly among the individuals [58]. Nursing houses and hospitals are two areas where the population of patients, especially the old and the ones who have low immunity, is at risk of outbreaks, which in many cases tend to be due to life-threatening diseases [55]. As schooling and play and care centers are also in danger, however, the risks are greater in these places due to the fact that young children, who might be negligent of hygiene practices, may transmit the virus easily to peers and faculty members as well [59].

7.4.3 Seasonal and Regional Variations in Norovirus Incidence

Norovirus outbreaks follow periodic trends that reflect strong seasonality; the incidences peak in the frost, mainly in the mid-latitude regions [60]. Thus, the illness is typically called "winter vomiting disease" as it is frequently observed with the occurrence of the cold weather [61]. The exact reason for such occurrence is yet not identified; however, this could be due to increasing home indoor gatherings that can promote interperson transmission, which is the case particularly in the winter months. Moreover, environmental factors like lower temperatures may help in increasing the stability and life of the virus on surfaces as well as in the water [62, 63]. Norovirus spread differs from one location to the next due to disparity in environmental factors such as the construction of roads, bridges, and other sanitation infrastructure. In areas that are closer to the equator, on the other hand, the temperatures are generally warmer all year and certain events, such as flights, may also be affected by diseases, which are transmitted through water and air [60, 64]. On the other hand, in temperate zones, the number of norovirus infections is larger during colder seasons, and this is mostly because of the observation that the virus tends to be more stable at lower temperature settings and the high probability of indoor contacts, leading to higher reports of norovirus infections [60]. Figure 7.2 presents the One Health surveillance model for Norovirus, depicting the integration of data from human, animal, and environmental health sectors. This model highlights key intersections, such as wastewater monitoring, veterinary-public health collaboration,

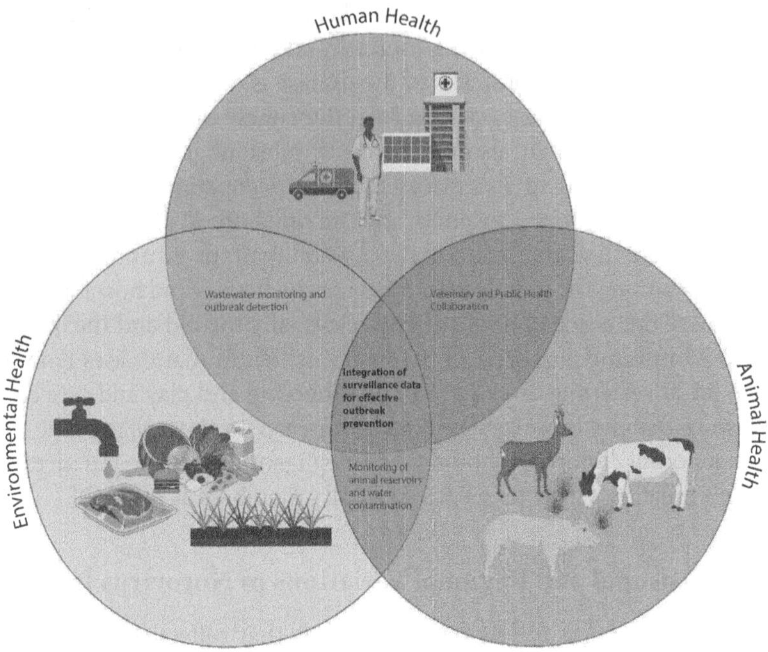

Figure 7.2 The One Health surveillance model for norovirus.

and the monitoring of food and water contamination to support effective outbreak detection and prevention.

7.5 The One Health Approach to Norovirus Surveillance

7.5.1 Integrating Human, Animal, and Environmental Surveillance Systems

Human health surveillance is a basic component to the early detection and management of outbreaks caused by norovirus. Surveillance systems include the tracking of cases of gastroenteritis at healthcare facilities, public health laboratories, and through mandatory reporting systems. Early detection of norovirus cases allows for swift containment measures, such as isolating a person who is infected and enhancing sanitation protocols [65]. Real-time data from hospital admissions, emergency room visits, and laboratory confirmations are collected in some countries and integrated into national databases to monitor and observe the outbreak trends [66]. Environmental monitoring is very important since it can help in the detection of norovirus in water systems, food supplies, and public settings. Waterborne surveillance includes the checking of the norovirus RNA in sewage systems, drinking water supplies, and recreational water sources [67]. By using these kinds of surveillance, we can detect any early signs of community outbreaks before the clinical cases emerge. The supervision of food production, initially in industries such as shellfish farming and large-scale food processing, is as important as preventing foodborne transmission [68]. Norovirus-like viruses in animals are potential zoonotic viruses that can cause infections in humans; thus, animal health surveillance needs to be practiced. Surveillance in livestock, particularly pigs and cattle, can offer relevant information about the probable viral reservoirs and recognize the risks for zoonotic transmission [69]. Seamless utilization of animal health data and human surveillance and environmental monitoring can help trace norovirus transmission routes and steer preventive interventions.

7.5.2 Advances in Diagnostic Techniques for Norovirus Detection

Recent technical developments for polymerase chain reaction (PCR), such as real-time PCR and digital PCR, have significantly changed the possibilities of norovirus detection in clinical and environmental samples. Therefore, these molecular diagnostics, which are very sensitive, detect even small amounts of viral RNA from stool, vomit, and environmental samples [70]. Real-time PCR, the method of choice in hospitals and public health labs for norovirus confirmations during outbreaks, shows by digital PCR even better precision, particularly for use in environmental monitoring [71]. Metagenomic approaches have become more and more prevalent in the study of norovirus diversity and the transmission of the virus. By high-throughput sequencing technologies, the diverse strains of norovirus in a population or environment, even in the absence of PCR primers, can be discovered [72, 73]. This is especially important in underlining the tremendous genetic variety that norovirus presents, implying that it is obligatory for vaccine development to scrutinize the whole spectrum of strains and detect new types that may cause future outbreaks [74, 75].

7.5.3 Case Studies: Successful Implementation of One Health Surveillance

In The Netherlands, they have implemented a robust One Health surveillance system to oversee shellfish farms for shellfish imported, as contaminated shellfish are a major source of norovirus infections. Environmental monitoring coupled with food safety regulations and routine testing have served a great deal in arresting the large number of outbreaks by early identification of contaminations [28]. In the USA, monitoring efforts similar to the ones mentioned above focused on foodborne outbreaks were put on the production of contaminated fruits and vegetables with the usage of both human health records and environmental testing to detect and then prevent the outbreaks [76]. By this method, they can save it from becoming unsuccessful and use it effectively to find out how many people will be affected, and for that, they can use their estimates from a regional sewage system where they can find out about existing viral shedding and predict outbreaks even before the clinical cases appear [77, 78]. Countries such as Japan and the United Kingdom have more than once won the bid by reviewing their virus outbreaks through this approach as they easily anticipate the rise of the viral infections through their early warning system in viral gastroenteritis [79, 80].

7.6 Prevention and Control Strategies

7.6.1 Hygiene and Sanitation Practices

One of the most important measures to prevent the spread of norovirus is correct hand hygiene, especially in medical and public locations where infections are common [81]. Washing hands with soap and water is necessary because norovirus is not impaired by alcohol-based hand sanitizers as it is non-enveloped virus [82]. In hospitals, the adherence to hand hygiene principles should be observed by the personnel and the visitors, particularly when dealing with a risk group such as the elderly or immunocompromised individuals. School-based educational activities with the main focus on proper hand hygiene are an effective means of reducing the spread of infections in such places as schools and long-term care institutions [58]. The resistance of norovirus to the environment necessitates the thorough cleanup of surfaces to prevent outbreaks. The majority of household disinfectants, including those with alcohol as an ingredient, are useless against norovirus. In this regard, chlorine-based sanitizers, with a minimum concentration of 1,000 parts per million (ppm), are advised for cleaning contaminated surfaces in healthcare and food service environments [81, 83]. In the event of an outbreak, disinfection should be doubled for high-touch surfaces like doorknobs, countertops, and medical equipment in daycare facilities, limiting the chances of cross-transmission. Proper execution of training the staff in the use of disinfectants is one of the ways to achieve hygiene standards [84].

7.6.2 Vaccination Development and Challenges

At present, there are no vaccines that have been licensed to treat norovirus, but a few candidates with encouraging results are being tested in clinical trials. Most of the vaccines that utilize virus-like particles (VLPs) have been composed of proteins that replicate the structure of the norovirus capsid and do not contain infectious genetic material [85, 86]. What these vaccines are made of is a VLP which is designed to produce a specific immune response in the body without causing the disease. On the other hand, Takeda Pharmaceuticals and Moderna are both in the final phases of clinical trials for the two vaccine candidates and both have demonstrated promise in their effectiveness against multiple strains of norovirus [55]. A universal norovirus vaccine that can be developed successfully is a daunting task because of the virus' genetic diversity and rapid mutation [55]. There are over 30 genotypes and, within genogroups, there are common variations of strains, which is a real trial to vaccine developers, as it is difficult to elaborate a solution for cross-protection of the two major common forms of norovirus (GI and GII), one of which, most of the time, causes human infections. In addition to this, the fact that there is no measurement available, which reflects the illness severity in humans, makes it difficult to produce the medicines and vaccines [86, 87].

7.6.3 Environmental Management and Water Safety

Ensuring the safety of water supplies in the prevention of waterborne norovirus outbreaks is a critical issue. It is important that water treatment facilities employ a variety of barrier methods, such as filtration and disinfection, to deactivate viruses that end up in drinking water [88, 89]. Chlorination and UV light are common methods used to eliminate norovirus from the water system, but research indicates that the use of higher doses of disinfectants will be vital in some environments to totally inactivate the virus [90, 91]. On the other hand, in situations when the water is highly contaminated, the use of water disinfection systems is likely to be found inevitable. Making sure no viruses are present in the sewage system is also important to avoid contamination in water for drinking and recreational activities [92]. Furthermore, farming is the main source of foodborne norovirus, wherein the use of contaminated water for irrigation is the most frequent cause [93]. In addition, farmers need to verify that the water sources they use for giving water to their crops and washing the vegetables are free from pathogens during the irrigation process. Also, governments and food safety authorities need to establish guidelines regarding water testing in agricultural places, especially in regions with a high risk of pollution [94]. Moreover, GAP is an effective tool in the standardization of farming methods which, in turn, can reduce viral pollution in the production and processing of food [95].

7.6.4 Role of Policy and Legislation in Norovirus Prevention

Food safety regulations are the mainstay of foodborne norovirus outbreaks' prevention. Governments of different countries are to generate specific and comprehensive regulations for food processing, handling, and storage; spot industries such as shellfish farming and the large-scale production of food that have many contaminants most of the time have to concentrate more on such acts [25]. To make the abovementioned dreams come true, food producers must define the way of containing products that are already contaminated and develop appropriate guidelines. In some countries, food handlers who are infected by norovirus are made to be isolated both at home for a given recovery time and at work [96]. Aimed at avoiding foodborne norovirus outbreaks, the World Health Organization (WHO) and the Food and Agriculture Organization (FAO) elaborate upon the policies and plans in the global community. The main focus of the organizations is on water quality inspection, food safety examinations, and public health education so as to bring the rate of viral contamination down [97]. WHO and FAO as well suggest the broader international cooperation about norovirus outbreaks by way of data sharing and improving the strategies of the incident response. It will help to raise the quality of water treatment and sanitation in poor- and middle-income countries [98].

7.7 Interdisciplinary Collaboration in One Health

7.7.1 Building Effective Partnerships Between Health Sectors

Effective norovirus prevention and control strongly depend on the collaboration and effective communication among healthcare providers, environmental scientists, and food safety authorities [99, 100]. In the case of these joint work, the breakout causes can be defined. It is also possible to rapidly take proper preventive and corrective measures and accurately assess the risks caused in food production systems. For instance, public health officials must work hand in hand with the food inspectors to look into outbreaks due to the consumption of contaminated food, etc., while environmental scientists are in charge of monitoring water sources and showing signs of contamination [101]. The interdisciplinary work between the various specialists should be mentioned as a key component [102]. Data sharing between health sectors is the mainstay of the coherent response to norovirus outbreaks. The combined surveillance of both human health *via* electronic systems and the environment *via* video cameras and food safety *via* GPS allows for early detection and efficient intervention [101, 103]. For instance, the real-time data containing information about viral loads in the sewage system that may give a hint about the potential outbreaks can be released to the relevant public health agencies. Cooperation between the different sectors makes it possible to ensure that each of the concerned parties is well informed and takes only the necessary action [78, 104].

7.7.2 Case Studies of Multidisciplinary Efforts in Controlling Norovirus

The collaboration of environmental scientists, epidemiologists, and medical professionals has been critical in norovirus management in Japan, especially those in which contaminated water bodies are implicated [105]. Environmental surveillance systems of the country give early alerts to viral particles in sewage and river water, which serve as the basis for a potential outbreak alarm. Also in the USA, health officials, public health sectors, and environmental monitoring bodies were very successful in the tracking and management of foodborne norovirus outbreaks, particularly in the food production factories [101]. One of the main points of the interdisciplinary level outbreak responses is ensuring the quick exchange of information through real-time data communication and sharing among different sectors. For instance, when a massive norovirus outbreak hit UK, cooperations between the food safety sector and healthcare professionals allowed for the quick identification of the oysters that were contaminated. This knowledge was used to rapidly mobilize the intervention and containment efforts [96, 106, 107]. In addition, the collaborative effort of the different departments of the health sector is also seen as a call for the adoption of uniform protocols that can be easily harnessed instantaneously and used in outbreak-environmental transmission, such that the general health status of the population can be intentionally or unintentionally maintained throughout [108]. Table 7.2 summarizes successful One Health implementation examples from countries like the Netherlands, Japan, and the United States. These case studies illustrate the value of multisectoral collaboration

Table 7.2 Case studies of successful One Health approach in norovirus control.

Country	Focus area	One Health approach implemented	Outcome	References
Netherlands	Shellfish-related outbreaks	Environmental monitoring of shellfish farms combined with food safety regulations	Significant reduction in shellfish-associated norovirus cases	[117]
United States	Foodborne outbreaks	Multisectoral collaboration between food safety officials and public health departments	Early identification and containment of contaminated produce	[76]
Japan	Waterborne outbreaks	Sewage monitoring for viral RNA and collaboration with healthcare and environmental scientists	Early warnings helped prevent large-scale waterborne outbreaks.	[79]
United Kingdom	Healthcare settings	Surveillance combining hospital admissions data and environmental monitoring	Faster containment of hospital-acquired infections	[80]

in detecting and managing Norovirus outbreaks across environmental, food, and healthcare contexts.

7.7.3 Challenges and Opportunities in Collaborative Approaches

Though interdisciplinary collaboration has many benefits, yet the implementation of effective One Health strategies against norovirus still has a number of problems. One of the most important hindrance to data sharing and coordination in the overlapping areas is the absence of standardized protocols, which generally results in delayed outbreak response [23].

Many countries have disunited health surveillance systems that use different databases for human, animal, and environmental health, which causes the difficulty of accessing complete information during an outbreak [109, 110]. Furthermore, the financial and logistic limits frequently strip away the ability for interdisciplinary collaboration [111, 112]. Interdisciplinary cooperation in norovirus control can be done in several ways. The first is the creation of connected surveillance systems enabling real-time information sharing across the different sectors which is the key component for early detection and rapid response [102, 113]. The governments and international organizations must also allocate resources to cross-sector training programs to improve coordination and communication between public health professionals, environmental scientists, and food safety officials [114, 115]. In addition, the strength of global partnerships is also a key aspect that must be kept in the picture, and finding ways of cooperation between nations is the only way to do this. Renovation and development of One Health programs allowing them to reach the global level and innovation in preventive measures against infectious diseases, such as norovirus, can only become true through the partnership and cooperation of nations among other things [116].

7.8 Case Studies: One Health Success Stories

7.8.1 Outbreak Response in Different Regions

Managing shellfish-related norovirus outbreaks has been a preeminent health issue in The Netherlands. By monitoring of shellfish farms and testing for viral contamination, the country has very substantially decreased the risk of norovirus transmission through contaminated shellfish [117]. This program encompasses the collaboration of ecological researchers, public health officials, and food safety authorities to find out the contamination proactively and to prevent the affected products from reaching the consumers [101]. Additionally, EPA and CDC in the USA have jointly initiated certain waterborne norovirus surveillance programs that are aimed at detecting viral contamination in both recreation and drinking water [118]. The assays are done by extracting the norovirus RNA using molecular diagnostics like PCR in water samples. The molecular diagnostic technique identifies the virus in water samples; thus, it provides the warning of a possible outbreak, sometimes even before an outbreak is visible. The watershed of norovirus has never been reported in the regions using this integrated surveillance system [119].

7.8.2 Lessons Learned from Norovirus Control Efforts

One lesson that we can learn from the successful control of norovirus is the value of early detection and rapid intervention. In foodborne and waterborne outbreaks, it is necessary to detect the source of contamination immediately and to take control measures aimed at removing the contaminated food from the source or issuing boil-water advisories to prevent the infection from spreading [120]. Surveillance systems that combine information received from human health, environmental monitoring, and food safety inspections are the best examples of the early detection of outbreaks [121]. Moreover, the cooperation of the community and involving population in the public education are also a part of successful norovirus control efforts. Public health programs that familiarize individuals with the significance of hand hygiene, food safety, and environmental sanitation are an important tool in the prevention of norovirus transmission [122]. The collaboration of local communities in the fight against outbreaks, especially in rural areas and marginalized communities, is very important, and it has to be assured so that people learn about the risks and still kind of keep it fun for them [123].

7.8.3 Scaling Up One Health Strategies Globally

Scaling up One Health initiatives in low- and middle-income countries is critical for reducing the global burden of norovirus [124]. Most of the developing countries are increasingly faced with the problem of water sanitation, hygiene, and food safety; thus, they become most vulnerable to norovirus outbreaks [81]. In these spheres, international organizations are attempting to replicate the successful One Health programs, which have been working elsewhere, into these areas by improving water quality, food safety, and, most essentially, the public health infrastructure [125]. Efficient global surveillance and response to norovirus outbreaks will not be possible without the support of international partnerships. The WHO and FAO have laid out strategies especially in areas with limited local capacities for response [126]. Solidifying these partnerships and making sure that the countries have the funding and professional knowledge to run a One Health model will be very important in dealing with the worldwide problem of reducing the occurrences of and the severity of norovirus [127].

7.9 Future Directions in Norovirus Research and Control

7.9.1 Emerging Trends in Norovirus Epidemiology

Global warming, as well as urbanization, is the main reason for altering the future epidemiology of norovirus. As global temperatures rise, the number of extreme weather events, for example, floods and hurricanes, also increases. This poses risks to the water sanitation systems as they may be the reason for more widespread outbreaks of waterborne diseases [128]. Urbanization, in particular, in developing countries, is also related to overcrowding and inadequacy in sanitation infrastructure, both of which contribute to the viral spread of norovirus [129, 130]. All of these conditions along with the virus that can survive in various environmental conditions imply that you might see more cases of norovirus outbreaks in the next few years. The globalization of the food supply chain as well as the increasing number of international flights are altering the pattern of foodborne and waterborne norovirus outbreaks [131, 132]. Contaminated food products, such as shellfish or fresh produce, that come out of a country can be spread to various countries, which may result in the emergence of pandemics all over the world [28]. This demonstrates the necessity of adopting stricter regulations on food safety and of an increase in international collaboration in the supervision and control of the movement of goods that might be tainted [133].

7.9.2 Innovations in Vaccination and Treatment

Although several vaccine candidates are in development, many difficulties are still left over to obtain the anti-norovirus, which is fundamental among them [87]. One encouraging way is to create a vaccine that focuses on the most common genotypes (GI and GII) that are perceived to be the reason for the infection among humans. Nonetheless, to have lasting immunity and cross-protection, the researcher has to overcome the issue of the fact that the virus strains constantly change [134, 135]. Moreover, the absence of a well-established animal model for human norovirus hampers the ability to assess the vaccine's efficiency and safety [136, 137]. In concert with vaccine creation, antiviral studies for norovirus have been done as well. Several antiviral candidates have exhibited positive outcomes in preclinical tests, like inhibitors of protease, which target viral replication [138, 139].

These remedies, which are brought to the market, could be a means of treatment for disadvantaged groups, which might be the people with weak immune systems who are at risk of grave outcomes from norovirus infection [140].

7.9.3 The Role of Technology in Enhancing One Health Approaches

Artificial intelligence (AI) and machine learning (ML) technologies are being more and more applied to prevent and control norovirus outbreaks. The tools are analyzing a very large scale of information from environmental surveillance, healthcare facilities, and public health records in order to see the patterns and give a prediction of future outbreaks. AI models regularize data from the most varied sources, such as wastewater surveillance and viral genome sequencing, that provide early warning messages by satellite where the mud indicates the possibility of an outbreak to be in that geographical area, thus guiding public health interventions to be made in the most vulnerable population and the targeted areas for which the funding should be allocated [102]. Geographic information systems (GIS) technology and applications for mobile health are the main forces, which allow for the implementation of such systems in real time to monitor outbreaks of norovirus [141, 142]. GIS mapping is the technology that is used by the people in public health authority to show outbreaks so that they will know which areas are more dangerous, while mobile apps are serving as the main vehicle for the rapid reporting of common symptoms or cases by the public [141]. These technologies make it possible to participate in quick and efficient interventions to the norovirus outbreaks in different places of the globe where they occur, particularly in remote or underserved areas.

7.10 Conclusion

7.10.1 Summarizing the Impact of the One Health Approach on Norovirus Control

The One Health approach has been found to be most effective in managing the varied pathways of norovirus transmission, particularly the prevention of foodborne and waterborne outbreaks. They bring together the human health, environmental monitoring, and safety systems for food, resulting in a lower response time of outbreaks and more response steps taken in a coordinated manner. The good implementation of One Health surveillance in The Netherlands and Japan is an important evidence of the success of One Health as a means for addressing the high level of burden of norovirus all over the world. Out of its achievements, challenges still exist in the full integration of One Health strategies for norovirus control. Many countries, mainly low- and medium-income ones, do not have the infrastructure and resources required for setting up integrated surveillance systems. Furthermore, the lack of standard protocols for data sharing between the sectors and the issue of developing a universal norovirus vaccine present two non-negligible obstacles to the comprehensive control of the virus. To succeed in this task, it is essential to increase the investment in the public health system and ensure worldwide collaboration.

7.10.2 Recommendations for Future Action and Research

To further lower the worldwide burden of norovirus, it is important to have strong global collaboration and capacity building. The World Health Organization and the Food and Agriculture Organization must play a dominant role by providing necessary data, training programs, and technical assistance to the countries that are less equipped in terms of public health. Additionally, the involvement of One Health in such initiatives should be increased to include more specific areas with high norovirus incidence in order to improve the global surveillance and the overall response to this infectious threat. In addition, the authorities need to allocate money for the modernization of water cleaning infrastructure to keep the waterborne route of the virus in check. It will be next to impossible to develop a multi-valent norovirus vaccine that is effective against all groups and newly emerging strains unless research focuses on this aspect in the future. The researchers should also prioritize figuring out the right animal model that can mimic human norovirus infections properly so vaccines and antivirals can be devised. The development of molecular diagnostics and metagenomic sequencing should be handled in a similar way as each is important for fixing a whole range of breaking out diseases demonstrating viral evolution. More efforts also should be made to increase information among the public on hygiene and sanitary practices which would help in reducing the spread of norovirus.

AI disclosure statement: The author acknowledges the use of AI-based tools for minor language editing or grammatical correction. However, all intellectual contributions, analyses, and conclusions presented in this chapter are entirely the author's own.

References

1. Vera-Garcia, M.E., Martinez-Acuña, N., Lozano-Sepulveda, S.A., Arellanos-Soto, D., Galan-Huerta, K.A., Merino-Mascorro, J.A., *et al.*, Unraveling the mechanisms involved in human norovirus pathogenesis: An update in norovirus pathogenesis. *Med. Universitaria*, 23, 4, 142–7, 2021.
2. Viviana, B., Matias, S., Daiana, M., Rodney, C., Matias, V., Molecular Characterization of Gastroenteric Viruses in Wastewater from Cities in Uruguay. *Food Environ. Virol.*, 15, 4, 318–30, 2023.
3. Winder, N., Gohar, S., Muthana, M., Norovirus: an overview of virology and preventative measures. *Viruses*, 14, 12, 2811, 2022.
4. Lartey, B.N.L., *Molecular Epidemiology of Noroviruses in Ghanaian Children*, College of Basic and Applied Sciences, University of Ghana, Springer, Cham, 2020. https://doi.org/10.1007/978-3-030-65433-7_8
5. Overbey, K.N., *Bench to health: Integration of 3D enteroids to monitor infectious norovirus and inform microbial risk assessment*, Johns Hopkins University, 2021.
6. Wang, J., Gao, Z., Yang, Z-R., Liu, K., Zhang, H., Global prevalence of asymptomatic norovirus infection in outbreaks: a systematic review and meta-analysis. *BMC Infect. Dis.*, 23, 1, 595, 2023.
7. Steele, M.K., Wikswo, M.E., Hall, A.J., Koelle, K., Handel, A., Levy, K., *et al.*, Characterizing norovirus transmission from outbreak data, United States. *Emerg. Infect. Dis.*, 26, 8, 1818, 2020.
8. Manuel, C.S., Moore, M.D., Jaykus, L.-A., Predicting human norovirus infectivity-Recent advances and continued challenges. *Food Microbiol.*, 76, 337–45, 2018.
9. Owusu-Apenten, R. and Vieira, E., Microbial foodborne disease outbreaks. in: *Elementary Food Science*, Food Science Text Series, pp. 171–96, Springer, Cham, 2022. https://doi.org/10.1007/978-3-030-65433-7_8.
10. Warmate, D. and Onarinde, B.A., Food safety incidents in the red meat industry: A review of foodborne disease outbreaks linked to the consumption of red meat and its products, 1991 to 2021. *Int. J. Food Microbiol.*, 398, 110240, 2023.
11. Harris, J., Infection Management and Control Service. *Infection*, 2018.
12. Body, A., Norovirus Policy. *Assessment*, 12, 10, 2021.
13. Nateghian, A., Moshfegh, F., Shoja, Z., Study of clinical characteristics and clinical complications of norovirus gastroenteritis in admitted children to the hospital. *J. Pharm. Res. Int.*, 30, 1–10, 2019.
14. Tsai, H., Yune, P., Rao, M., Norovirus disease among older adults. *Ther. Adv. Infect. Dis.*, 9, 20499361221136760, 2022.
15. Schulz, C., Wyatt, A., Walker, J., Smoll, N., Field, E., Khandaker, G., Outbreak investigation of norovirus gastroenteritis in a childcare facility in Central Queensland, Australia: a household level case series analysis. *Commun. Dis. Intell.*, 48, 2024.

16. Sandmann, F.G., The true cost of epidemic and outbreak diseases in hospitals: London School of Hygiene & Tropical Medicine, 2018.
17. Ferri, M. and Lloyd-Evans, M., The contribution of veterinary public health to the management of the COVID-19 pandemic from a One Health perspective. *One Health*, 12, 100230, 2021.
18. Qian, J., Wu, Z., Zhu, Y., Liu, C., One Health: A holistic approach for food safety in livestock. *Sci. One Health*, 1, 100015, 2022.
19. Wilcox, B.A. and Steele, J.A., One health and emerging zoonotic diseases: Framework, integration and challenges, in: *Handbook of global health*, pp. 1–49, 2020.
20. Shrestha, K., Acharya, K.P., Shrestha, S., One health: The interface between veterinary and human health. *Int. J. One Health*, 4, 2, 8–14, 2018.
21. O'Brien, E. and Xagoraraki, I., A water-focused one-health approach for early detection and prevention of viral outbreaks. *One Health*, 7, 100094, 2019.
22. Mukhtar, M., Fayyaz, Z., Aftab, M., Nawaz, M., Javed, M., Hussain, B., *et al.*, One health approach to zoonosis: integrating medicine, veterinary science, and environmental science, in: *Zoonosis*, vol. 1, pp. 226–36, Unique Scientific Publishers, Faisalabad, Pakistan, 2023.
23. Singh, S., Sharma, P., Pal, N., Sarma, D.K., Tiwari, R., Kumar, M., Holistic one health surveillance framework: synergizing environmental, animal, and human determinants for enhanced infectious disease management. *ACS Infect. Dis.*, 10, 3, 808–26, 2024.
24. Garcia, S.N., Osburn, B.I., Jay-Russell, M.T., One health for food safety, food security, and sustainable food production. *Front. Sustain. Food Syst.*, 4, 1, 2020.
25. Bosch, A., Gkogka, E., Le Guyader, F.S., Loisy-Hamon, F., Lee, A., Van Lieshout, L., *et al.*, Foodborne viruses: Detection, risk assessment, and control options in food processing. *Int. J. Food Microbiol.*, 285, 110–28, 2018.
26. Liu, D., Zhang, Z., Hewitt, J., Li, X., Hou, P., Wang, D., *et al.*, Surveillance of human norovirus in oysters collected from production area in Shandong Province, China during 2017–2018. *Food Control*, 121, 107649, 2021.
27. Chandran, S. and Gibson, K.E., Improving the Detection and Understanding of Infectious Human Norovirus in Food and Water Matrices: A Review of Methods and Emerging Models. *Viruses*, 16, 5, 776, 2024.
28. Friesema, I.H., Slegers-Fitz-James, I.A., Wit, B., Franz, E., Surveillance and characteristics of food-borne outbreaks in the Netherlands, 2006 to 2019. *Eurosurveillance*, 27, 3, 2100071, 2022.
29. Savini, F., Giacometti, F., Tomasello, F., Pollesel, M., Piva, S., Serraino, A., *et al.*, Assessment of the impact on human health of the presence of norovirus in bivalve molluscs: what data do we miss? *Foods*, 10, 10, 2444, 2021.
30. Magalhães-Guedes, K.T., Foodborne toxinfections caused by virus: Characteristics of the main viruses, prevention, treatment and clinical method of laboratory diagnosis by RT-qPCR, 2019.

31. Lian, Y., Wu, S., Luo, L., Lv, B., Liao, Q., Li, Z., *et al.*, Epidemiology of norovirus outbreaks reported to the public health emergency event surveillance system, China, 2014–2017. *Viruses*, 11, 4, 342, 2019.
32. Kabue Ngandu, J.-P., Molecular characterization of norovirus stains circulating in rural communities of Limpopo Province of South Africa, 2018.
33. Phattanawiboon, B., Nonthabenjawan, N., Boonyos, P., Jetsukontorn, C., Towayunanta, W., Chuntrakool, K., *et al.*, Norovirus transmission mediated by asymptomatic family members in households. *PLoS One*, 15, 7, e0236502, 2020.
34. Wu, Q.-S., Xuan, Z.-L., Liu, J.-Y., Zhao, X.-T., Chen, Y.-F., Wang, C.-X., *et al.*, Norovirus shedding among symptomatic and asymptomatic employees in outbreak settings in Shanghai, China. *BMC Infect. Dis.*, 19, 1–8, 2019.
35. Møretrø, T., Nguyen-The, C., Didier, P., Maître, I., Izsó, T., Kasza, G., *et al.*, Consumer practices and prevalence of Campylobacter, Salmonella and norovirus in kitchens from six European countries. *Int. J. Food Microbiol.*, 347, 109172, 2021.
36. Jin, T., Chen, X., Nishio, M., Zhuang, L., Shiomi, H., Tonosaki, Y., *et al.*, Interventions to prevent surface transmission of an infectious virus based on real human touch behavior: a case study of the norovirus. *Int. J. Infect. Dis.*, 122, 83–92, 2022.
37. Kirchner, M., Goulter, R.M., Bernstein, C., Lavallee, A., Schaffner, D., Chapman, B., *et al.*, The role of hands in cross-contamination of kitchen surfaces during meal preparation. *Am. J. Infect. Control*, 51, 11, A44–A57, 2023.
38. Villabruna, N., Schapendonk, C.M., Aron, G.I., Koopmans, M.P., de Graaf, M., Human noroviruses attach to intestinal tissues of a broad range of animal species. *J. Virol.*, 95, 3, 10–128, 2021. 10.1128/jvi. 01492-20.
39. Villabruna, N., Izquierdo Lara, R.W., Szarvas, J., Koopmans, M.P., Graaf, M., Phylogenetic investigation of norovirus transmission between humans and animals. *Viruses*, 12, 11, 1287, 2020.
40. Gherman, I., Holland, D., Clifford, R., Mahmoudzadeh, N., Uy, F., Adkin, A., *et al.*, Technical report: Review of quantitative risk assessment of foodborne norovirus transmission. *Sci.*, 1, 1, 2020.
41. Tao, J., Chunhui, H., Fanning, S., Nan, L., Jiahui, W., Hongyuan, Z., *et al.*, Norovirus contamination in retail oysters from Beijing and Qingdao, China. *Food Control*, 86, 415–9, 2018.
42. Hida, K., Papafragkou, E., Kulka, M., Testing for human norovirus and recovery of process control in outbreak-associated produce items. *J. Food Prot.*, 81, 1, 105–114, 2018. https://doi.org/10.4315/0362-028X.JFP-17-134
43. Sarno, E., Pezzutto, D., Rossi, M., Liebana, E., Rizzi, V., A review of significant European foodborne outbreaks in the last decade. *J. Food Prot.*, 84, 12, 2059–70, 2021.
44. Özgür, C., Determination of enteric virus removal capacities of wastewater treatment plants with varying processes: Middle East Technical University, 2024.

45. Rexin, D., Rachmadi, A.T., Hewitt, J., Persistence of Infectious Human Norovirus in Estuarine Water. *Food Environ. Virol.*, 16, 1, 58–64, 2024.
46. Ma, J.-X., Wang, X., Pan, Y.-R., Wang, Z.-Y., Guo, X., Liu, J., *et al.*, Data-driven systematic analysis of waterborne viruses and health risks during the wastewater reclamation process. *Environ. Sci. Ecotechnol.*, 19, 100328, 2024.
47. Victoria, M., Guimaraes, F.R., Fumian, T.M., Ferreira, F.F.M., Vieira, C.B., Shubo, T., *et al.*, One year monitoring of norovirus in a sewage treatment plant in Rio de Janeiro, Brazil. *J. Water Health*, 8, 1, 158–65, 2010.
48. Das, R., Haque, M.A., Kotloff, K.L., Nasrin, D., Hossain, M.J., Sur, D., *et al.*, Enteric viral pathogens and child growth among under-five children: findings from South Asia and sub-Saharan Africa. *Sci. Rep.*, 14, 1, 13871, 2024.
49. Leone, C.M., Dharmasena, M., Tang, C., DiCaprio, E., Ma, Y., Araud, E., *et al.*, Prevalence of human noroviruses in commercial food establishment bathrooms. *J. Food Prot.*, 81, 5, 719–28, 2018.
50. Liu-Lastres, B. and Ogbeide, G.-C., Norovirus on Cruise Ships: Should Americans Be Worried? *J. Hosp. Tour. Cases*, 8, 2, 23–33, 2020.
51. Ndou, A., Factors contributing to the development of dehydration among under five years children with diarrhoea in Thulamela B. Clinics, Vhembe District, 2020.
52. Adelodun, B., Ajibade, F.O., Ighalo, J.O., Odey, G., Ibrahim, R.G., Kareem, K.Y., *et al.*, Assessment of socioeconomic inequality based on virus-contaminated water usage in developing countries: a review. *Environ. Res.*, 192, 110309, 2021.
53. Lartey, B.L., Quaye, O., Damanka, S.A., Agbemabiese, C.A., Armachie, J., Dennis, F.E., *et al.*, Understanding pediatric norovirus epidemiology: a decade of study among Ghanaian Children. *Viruses*, 12, 11, 1321, 2020.
54. Munzhedzi, L., Characterization of diarrhoeagenic escherichia coli and human norovirus from daycare centres in the Vhembe Region of the Limpopo Province, 2020.
55. Carlson, K.B., Dilley, A., O'Grady, T., Johnson, J.A., Lopman, B., Viscidi, E., A narrative review of norovirus epidemiology, biology, and challenges to vaccine development. *npj Vaccines*, 9, 1, 94, 2024.
56. Yu, F., Fu, J., Tan, M., Xu, R., Tian, Y., Jia, L., *et al.*, Norovirus outbreaks in hospitals in China: a systematic review. *J. Hosp. Infect.*, 1, 142, 32–8, 2023.
57. Teixeira, A.I.P., [Mini-review] The Global Impact and Management of Foodborne RNA Viruses. *Qeios*, 2024. doi:10.32388/0M7IQN.2.
58. Hsu, S., How Much Effect Does Alcohol Have Against Alcohol-Resistant Norovirus?
59. Jianbin, L., Yanjun, Y., Shaomin, W., Dan, X., Analysis of epidemiological characteristics of public health emergencies in Panyu District, Guangzhou from 2013 to 2022. *Int. Med. Health Guid. News*, 29, 23, 3515, 2023.
60. de Almeida, N.A.A., Pimenta, Y.C., de Oliveira Bonfim, F.F., de Almeida, N.C.A., Leite, J.P.G., Olivares, A.I.O., *et al.*, Association between detection

rate of norovirus GII and climatic factors in the Northwest Amazon region. *Heliyon*, 10, 16, 2024.

61. Morral-Puigmal, C., Martínez-Solanas, E., Villanueva, C.M., Basagaña, X., Weather and gastrointestinal disease in Spain: A retrospective time series regression study. *Environ. Int.*, 121, 649–57, 2018.
62. Beck-Friis, T., Sundell, N., Gustavsson, L., Lindh, M., Westin, J., Andersson, L.-M., Outdoor absolute humidity predicts the start of norovirus GII epidemics. *Microbiol. Spectrum*, 11, 2, e02433–22, 2023.
63. Kauppinen, A., Pathogenic viruses in Finnish waters: occurrence, fate and control: Itä-Suomen yliopisto, 2018.
64. Grembi, J.A., Nguyen, A.T., Riviere, M., Heitmann, G.B., Patil, A., Athni, T.S., *et al.*, Influence of hydrometeorological risk factors on child diarrhea and enteropathogens in rural Bangladesh. *PLOS Neglected Trop. Dis.*, 18, 5, e0012157, 2024.
65. Kim, E. and Kim, B.I., Characteristics and related factors of waterborne and foodborne infectious disease outbreaks before and after the onset of the COVID-19 pandemic (2017–2021) in the Republic of Korea: a descriptive study. *Osong Public Health Res. Perspectives*, 14, 6, 483, 2023.
66. Giri, S., Chhabra, P., Kulkarni, R., Reju, S., Sabapathy, S.K., Selvarajan, S., *et al.*, Hospital-based norovirus surveillance in children< 5 years of age from 2017 to 2019 in India. *J. Med. Virol.*, 96, 1, e29384, 2024.
67. Díaz, S.M., Barrios, M.E., Galli, L., Cammarata, R.V., Torres, C., Fortunato, M.S., *et al.*, Microbiological hazard identification in river waters used for recreational activities. *Environ. Res.*, 247, 118161, 2024.
68. Trudel-Ferland, M., Collard, M.-E., Goulet-Beaulieu, V., Jubinville, E., Hamon, F., Jean, J., Evaluation of a new automated viral RNA extraction platform for hepatitis A virus and human norovirus in testing of berries, lettuce, and oysters. *Int. J. Food Microbiol.*, 416, 110664, 2024.
69. Olaimat, A.N., Taybeh, A.O., Al-Nabulsi, A., Al-Holy, M., MmM, H., Alzyoud, J., *et al.*, Common and Potential Emerging Foodborne Viruses: A Comprehensive Review. *Life*, 14, 2, 190, 2024.
70. Khiewbanyang, S., Kaewpradab, Y., Kanchanaudom, W., Pholprasert, S., Koyta, T., Hoontong, P., *et al.*, An Investigation of a Norovirus Outbreak Linked to Contaminated Vegetables in Mueang District, Chanthaburi Province, Thailand, December 2021–January 2022. *OSIR J.*, 16, 4, 217–23, 2023.
71. Boxman, I.L., Molin, R., Persson, S., Juréus, A., Jansen, C.C., Sosef, N.P., *et al.*, International inter-laboratory study to compare digital PCR with ISO standardized qPCR assays for the detection of norovirus GI and GII in oyster tissue. *AnFood Microbiol.*, 120, 104478, 2024.
72. Strubbia, S., Schaeffer, J., Besnard, A., Wacrenier, C., Le Mennec, C., Garry, P., *et al.*, Metagenomic to evaluate norovirus genomic diversity in oysters: Impact on hexamer selection and targeted capture-based enrichment. *Int. J. Food Microbiol.*, 323, 108588, 2020.

73. Nieuwenhuijse, D.F. and Koopmans, M.P., Metagenomic sequencing for surveillance of food-and waterborne viral diseases. *Front. Microbiol.*, 8, 230, 2017.
74. Hallowell, B.D., Parashar, U.D., Hall, A.J., Epidemiologic challenges in norovirus vaccine development. *Hum. Vaccin. Immunother.*, 15, 6, 1279–83, 2019.
75. Lucero, Y. and Vidal, R., Norovirus vaccines under development. *Vaccine*, 36, 36, 5435–41, 2018.
76. Crowe, S.J., Mahon, B.E., Vieira, A.R., Gould, L.H., Vital signs: multistate foodborne outbreaks—United States, 2010–2014. *MMWR Morb. Mortal Wkly Rep.*, 64, 43, 1221–5, 2015.
77. Chacón Jiménez, L.M., Morales Mora, E., Valiente Álvarez, C.I., Reyes Lizano, L., Barrantes Jiménez, K., Wastewater-based epidemiology of enteric viruses and surveillance of acute gastrointestinal illness outbreaks in a resource-limited region, 2021.
78. Walker, D.I., Witt, J., Rostant, W., Burton, R., Davison, V., Ditchburn, J., *et al.*, Piloting wastewater-based surveillance of norovirus in England. *Water Res.*, 263, 122152, 2024.
79. Kazama, S., Masago, Y., Tohma, K., Souma, N., Imagawa, T., Suzuki, A., *et al.*, Temporal dynamics of norovirus determined through monitoring of municipal wastewater by pyrosequencing and virological surveillance of gastroenteritis cases. *Water Res.*, 92, 244–53, 2016.
80. Campos, C.J. and Lees, D.N., Environmental transmission of human noroviruses in shellfish waters. *Appl. Environ. Microbiol.*, 80, 12, 3552–61, 2014.
81. Barclay, L., Park, G., Vega, E., Hall, A., Parashar, U., Vinjé, J., *et al.*, Infection control for norovirus. *Clin. Microbiol. Infect.*, 20, 8, 731–40, 2014.
82. Nicole, W., Douglas, D., Xueling, S., Persistent virucidal activity in an alcohol-based sanitizer formulation (ProtecTeaV) for potential use against norovirus. *Int. J. Microbiol. Infect. Dis.*, 2, 1, 1–8, 2018.
83. Cook, N., Knight, A., Richards, G.P., Persistence and elimination of human norovirus in food and on food contact surfaces: A critical review. *J. Food Prot.*, 79, 7, 1273–94, 2016.
84. Heijne, J.C., Teunis, P., Morroy, G., Wijkmans, C., Oostveen, S., Duizer, E., *et al.*, Enhanced hygiene measures and norovirus transmission during an outbreak. *Emerg. Infect. Dis.*, 15, 1, 24, 2009.
85. Riddle, M.S. and Walker, R.I., Status of vaccine research and development for norovirus. *Vaccine*, 34, 26, 2895–9, 2016.
86. Baehner, F., Bogaerts, H., Goodwin, R., Vaccines against norovirus: state of the art trials in children and adults. *Clin. Microbiol. Infect.*, 22, S136–S9, 2016.
87. Armah, G., Lopman, B.A., Vinjé, J., O'Ryan, M., Lanata, C.F., Groome, M., *et al.*, Vaccine value profile for norovirus. *Vaccine*, 41, S134–S52, 2023.
88. Ghernaout, D., Water treatment challenges towards viruses removal. *Open Access Lib. J.*, 7, 5, 1–22, 2020.

89. Kunz, J.M., Surveillance of Waterborne Disease Outbreaks Associated with Drinking Water—United States, 2015–2020. *MMWR Surveillance Summaries*, 73, 2024.
90. Campos, C.J.A., Avant, J., Lowther, J., Till, D., Lees, D., Levels of norovirus and E. coli in untreated, biologically treated and UV-disinfected sewage effluent discharged to a shellfish water. *J. Water Resour. Prot.*, 5, 10, 978–82, 2013.
91. Mariita, R.M., Davis, J.H., Randive, R.V., Illuminating human Norovirus: a perspective on disinfection of water and surfaces using UVC, norovirus model organisms, and radiation safety considerations. *Pathogens*, 11, 2, 226, 2022.
92. Vieira, C.B., ACdO, M., FR, G., TM, F., Leite, J.P.G., Gaspar, A.M.C., *et al.*, Detection of enteric viruses in recreational waters of an urban lagoon in the city of Rio de Janeiro, Brazil. *Memórias do Instituto Oswaldo Cruz*, 107, 778–84, 2012.
93. Rodrigues, C., da Silva, A.L.B.R., Dunn, L.L., Factors impacting the prevalence of foodborne pathogens in agricultural water sources in the southeastern United States. *Water*, 12, 1, 51, 2019.
94. Banach, J. and Van Der Fels-Klerx, H., Microbiological reduction strategies of irrigation water for fresh produce. *J. Food Prot.*, 83, 6, 1072–87, 2020.
95. Bozkurt, H., Phan-Thien, K.-Y., van Ogtrop, F., Bell, T., McConchie, R., Outbreaks, occurrence, and control of norovirus and hepatitis a virus contamination in berries: A review. *Crit. Rev. Food Sci. Nutr.*, 61, 1, 116–38, 2021.
96. Rumble, C., Addiman, S., Balasegaram, S., Chima, K., Ready, D., Heard, J., *et al.*, Role of food handlers in Norovirus outbreaks in London and South East England, 2013 to 2015. *J. Food Prot.*, 80, 2, 257–64, 2017.
97. Organization, W.H., Safety and quality of water used in food production and processing: meeting report, 2019.
98. Peres, M.R., Risk-based tools to prevent excreta-borne disease transmission in semi-arid rural communities: University of Brighton, 2017.
99. Okumus, B., Norovirus and Coronavirus Risks in Food Service Settings: A Systematic Review for Future Research. *J. Culinary Sci. Technol.*, 21, 1, 71–98, 2023.
100. Newell, D.G., Koopmans, M., Verhoef, L., Duizer, E., Aidara-Kane, A., Sprong, H., *et al.*, Food-borne diseases—the challenges of 20 years ago still persist while new ones continue to emerge. *Int. J. Food Microbiol.*, 139, S3–S15, 2010.
101. White, A.E., Tillman, A.R., Hedberg, C., Bruce, B.B., Batz, M., Seys, S.A., *et al.*, Foodborne illness outbreaks reported to national surveillance, United States, 2009–2018. *Emerg. Infect. Dis.*, 28, 6, 1117, 2022.
102. Cuadros, D.F., Chen, X., Li, J., Omori, R., Musuka, G., Advancing Public Health Surveillance: Integrating Modeling and GIS in the Wastewater-Based Epidemiology of Viruses, a Narrative Review. *Pathogens*, 13, 8, 685, 2024.

103. Jin, M., Wu, S., Kong, X., Xie, H., Fu, J., He, Y., *et al.*, Norovirus outbreak surveillance, China, 2016–2018. *Emerg. Infect. Dis.*, 26, 3, 437, 2020.
104. Ammerman, M.L., Mullapudi, S., Gilbert, J., Figueroa, K., de Paula Nogueira Cruz, F., Bakker, K.M., *et al.*, Norovirus GII wastewater monitoring for epidemiological surveillance. *PLOS Water*, 3, 1, e0000198, 2024.
105. Machida, T., Yanahara, H., Tomita, S., Asakura, C., KyokoSetoguchi, S.H., *et al.*, Coping behaviors and collaboration among staff of multiple occupation classifications during norovirus. *GSTF J. Nurs. Health Care (JNHC)*, 4, 1, 2345–7198, 2020.
106. Diplock, K., Foodborne illness outbreaks: Roles and responsibilities, 2023.
107. Li, S., Jiang, Q., Zhang, P., Factors influencing the health behavior during public health emergency: A case study on norovirus outbreak in a university. *Data Inf. Manage.*, 5, 1, 27–39, 2021.
108. Calo, W.A., Francis, E., Kong, L., Hogentogler, R., Heilbrunn, E., Fisher, A., *et al.*, Implementing infection control and quality of Life best Practices in nursing homes with Project ECHO: protocol for a patient-centered randomized controlled trial. *JMIR Res. Protoc.*, 11, 5, e34480, 2022.
109. Chen, J., He, J., Bergquist, R., Challenges and response to pandemics as seen in a One Health perspective. *Sci. One Health*, 1, 100010, 2022.
110. Burkom, H., Loschen, W., Wojcik, R., Holtry, R., Punjabi, M., Siwek, M., *et al.*, Electronic surveillance system for the early notification of community-based epidemics (ESSENCE): overview, components, and public health applications. *JMIR Public Health Surveillance*, 7, 6, e26303, 2021.
111. Khorram-Manesh, A., Burkle Jr., F.M., Goniewicz, K., Pandemics: past, present, and future: multitasking challenges in need of cross-disciplinary, transdisciplinary, and multidisciplinary collaborative solutions. *Osong Public Health Res. Perspect.*, 15, 4, 267–85, 2024.
112. Yazdani, M. and Haghani, M., Logistics of patient evacuation in response to disease Outbreaks: Critical considerations for transportation planning. *Transp. Res. Interdiscip. Perspect.*, 22, 100975, 2023.
113. Donia, A., S-u, H., Zhang, X., Al-Madboly, L., Bokhari, H., COVID-19 Crisis Creates Opportunity towards Global Monitoring & Surveillance. *Pathogens*, 10, 256, 2021.
114. Alves, F., Artursson, K., Bloch, J., Brisabois, A., Imberechts, H., Jokelainen, P., *et al.*, A multi-country One Health foodborne outbreak simulation exercise: cross-sectoral cooperation, data sharing and communication. *Front. Public Health*, 11, 1121522, 2023.
115. Buschhardt, T., Günther, T., Skjerdal, T., Torpdahl, M., Gethmann, J., Filippitzi, M.-E., *et al.*, A one health glossary to support communication and information exchange between the human health, animal health and food safety sectors. *One Health*, 13, 100263, 2021.
116. Agarwal, K., Srivastava, S., Singh, V., Rohilla, R., Zaman, K., Rukadikar, A., *et al.*, One health concept and its applications in clinical practice: a comprehensive review. *Evidence*, 2, 1, 1-0, 2024.

117. Dirks, R.A., Jansen, C.C., Hägele, G., Zwartkruis-Nahuis, A.J., Tijsma, A.S., Boxman, I.L., Quantitative levels of norovirus and hepatitis A virus in bivalve molluscs collected along the food chain in the Netherlands, 2013–2017. *Int. J. Food Microbiol.*, 344, 109089, 2021.
118. Gerdes, M.E., Miko, S., Kunz, J.M., Hannapel, E.J., Hlavsa, M.C., Hughes, M.J., *et al.*, Estimating waterborne infectious disease burden by exposure route, United States, 2014. *Emerging Infect. Dis.*, 29, 7, 1357, 2023.
119. Chen, H. and Hu, Y., Suppl-1, M5: Molecular Diagnostic Methods for Detection and Characterization of Human Noroviruses. *Open Microbiol. J.*, 10, 78, 2016.
120. Han, M.S., Chung, S.M., Kim, E.J., Lee, C.J., Yun, K.W., Choe, P.G., *et al.*, Successful control of norovirus outbreak in a pediatric ward with multi-bed rooms. *Am. J. Infect. Control*, 48, 3, 297–303, 2020.
121. Gomes, B.M., Rebelo, C.B., de Sousa, L.A., Public health, surveillance systems and preventive medicine in an interconnected world. *One Health*, 33–71, 2022. https://doi.org/10.1016/B978-0-12-822794-7.00006-X
122. N.H.N.R., W.M.W.M., N.J.N.I., N.H.N.N., A.R.R., The effect of Food Safety Education on handwashing practices in school canteens' food handlers. *Sains Malaysiana*, 47, 9, 2119–28, 2018.
123. Magwalivha, M., Molecular characterization of human sapoviruses circulation in the rural communities of Limpopo Province, South Africa, 2022.
124. Hosangadi, D., Smith, P.G., Kaslow, D.C., Giersing, B.K., WHO consultation on ETEC and Shigella burden of disease, Geneva, 6–7th April 2017: Meeting report. *Vaccine*, 37, 50, 7381–90, 2019.
125. Nguyen-Viet, H., Bett, B.K., Unger, F., Murphy, M., Muloi, D.M., Dang-Xuan, S., Protecting human health through a One Health approach, 2022.
126. Organization, W.H., Shiga Toxin-producing Escherichia Coli (STEC) and Food: Attribution, Characterization and Monitoring: World Health Organization, 2019.
127. Saleem, F., Faraz, A., Irtaza, A., Ishaq, H., Ilyas, M., Khalid, R., *et al.*, Policies to control zoonotic disease transmission in Pakistan, in: *Zoonosis*, vol. 1, pp. 348–60, Unique Scientific Publishers, Faisalabad, Pakistan, 2023.
128. Rupasinghe, R., Chomel, B.B., Martínez-López, B., Climate change and zoonoses: A review of the current status, knowledge gaps, and future trends. *Acta Trop.*, 226, 106225, 2022.
129. Tariq, A. and Mushtaq, A., Untreated wastewater reasons and causes: a review of most affected areas and cities. *Int. J. Chem. Biochem. Sci.*, 23, 1, 121–43, 2023.
130. Shokri, A., Sabzevari, S., Hashemi, S.A., Impacts of flood on health of Iranian population: Infectious diseases with an emphasis on parasitic infections. *Parasite Epidemiol. Control*, 9, e00144, 2020.

131. Hassan, A., Khan, M.K.I., Fordos, S., Hasan, A., Khalid, S., Naeem, M.Z., (Eds.), Emerging Foodborne Pathogens: Challenges and Strategies for Ensuring Food Safety. *Biol. Life Sci. Forum*, 31, 1, 32, 2023. https://doi.org/10.3390/ECM2023-16596
132. Latif, M., Danish, W., Waheed, M., Sattar, M., Ali, M., Sudheer, N., *et al.*, Relationship between zoonotic diseases and food safety. *Int. J. Agric. Biosci. Zoonosis*, 1, 338–47, 2023.
133. Eruaga, M.A., Enhancing global food safety standards through international collaboration and policy harmonization, 2024.
134. Chachu, K.A., LoBue, A.D., Strong, D.W., Baric, R.S., Virgin, H.W., Immune mechanisms responsible for vaccination against and clearance of mucosal and lymphatic norovirus infection. *PLoS Pathog.*, 4, 12, e1000236, 2008.
135. Debbink, K., Mechanisms of GII. 4 norovirus antigenic variation and evolution: The University of North Carolina at Chapel Hill, 2014.
136. Tan, M., Norovirus vaccines: current clinical development and challenges. *Pathogens*, 10, 12, 1641, 2021.
137. Tan, M. and Jiang, X., Vaccine against norovirus. *Hum. Vaccin. Immunother.*, 10, 6, 1449–56, 2014.
138. Netzler, N.E., Enosi Tuipulotu, D., White, P.A., Norovirus antivirals: Where are we now? *Med. Res. Rev.*, 39, 3, 860–86, 2019.
139. Campillay-Véliz, C.P., Carvajal, J.J., Avellaneda, A.M., Escobar, D., Covián, C., Kalergis, A.M., *et al.*, Human norovirus proteins: implications in the replicative cycle, pathogenesis, and the host immune response. *Front. Immunol.*, 11, 961, 2020.
140. van Beek, J., van der Eijk, A., Fraaij, P., Caliskan, K., Cransberg, K., Dalinghaus, M., *et al.*, Chronic norovirus infection among solid organ recipients in a tertiary care hospital, the Netherlands, 2006–2014. *Clin. Microbiol. Infect.*, 23, 4, 265. e9–.e13, 2017.
141. Fallatah, D.I. and Adekola, H.A., Digital epidemiology: harnessing big data for early detection and monitoring of viral outbreaks. *Infect. Prev. Pract.*, 6, 3, 100382, 2024. https://doi.org/10.1016/j.infpip.2024.100382
142. Li, X., Sapp, A.C., Singh, N., Matthias, L., Bailey, C., *et al.*, Detecting foodborne disease outbreaks in Florida through consumer complaints. *J. Food Prot.*, 83, 11, 1877–88, 2020.

8

Recent Insights into the Dynamics of Norovirus Outbreak

Sejal Porwal[1], Sonali Sundram[1], Rishabha Malviya[1*] and Shriyansh Srivastava[2]

[1]Department of Pharmacy, School of Medical and Allied Sciences, Galgotias University, Greater Noida, India
[2]Department of Pharmacology, Delhi Pharmaceutical Sciences and Research University, New Delhi, India

Abstract

Norovirus is a highly infectious RNA virus that is a major cause of gastroenteritis epidemics worldwide. Its spread occurs *via* dirty water, food, or direct contact, with clinical manifestations ranging from diarrhea and vomiting to extreme dehydration, particularly impacting children and the elderly. The genetic diversity and ability of viruses to evade immunity contribute to persistent outbreaks and challenges in vaccine development. This study explores the latest insights into norovirus epidemiology, transmission mechanisms, environmental persistence, and strategies for prevention and management. Norovirus outbreaks are driven by low infectious doses and environmental resilience of the virus, enabling rapid transmissions in communal settings like institutes and cruise ships. Efforts to develop vaccines are hindered by the genetic variability of the virus, the absence of robust culture systems, and the limited understanding of immune protection correlates. Public health interventions have shown promise, including improved hygiene, safe food handling, and innovative vaccine strategies like VLP-based formulations. Clinical trials demonstrate potential vaccine efficacy, yet multivalent formulations addressing emerging genotypes are essential. Addressing norovirus outbreaks requires a multifaceted approach integrating vaccination development, public health measures, and ongoing studies into viral evolution and immunity.

***Keywords*:** Transmission, gastroenteritis, epidemiology, outbreak, RNA virus, foodborne disease, vaccine-like particles

**Corresponding author*: rishabhamalviya19@gmail.com; rishabha.malviya@galgotiasuniversity.edu.in

Sachin Kumar, Shriyansh Srivastava, Rishabha Malviya and Ranjit Sah (eds.) Emerging Infectious Disease: Unravelling Norovirus and Pandemic Dynamics, (221–248) © 2026 Scrivener Publishing LLC

8.1 Introduction

Norovirus (NoV) is a non-enveloped, positive-sense single-stranded RNA ($ssRNA^{(+)}$) virus that fits the family *Caliciviridae* and has been divided into 10 genogroups [1]. The norovirus was detected for the first time in 1968 when it was isolated from the excreta of patients suffering from acute gastroenteritis in Norwalk, Ohio, USA [2]. Ever since, stool samples from gastroenteritis patients throughout the world have regularly shown this virus. Globally, noroviruses continue to be important viral reasons for foodborne as well as waterborne gastroenteritis outbreaks in people [3, 4]. Ingestion of dirty food or drink, direct interaction with contaminated environmental reservoirs, or contact with sick individuals are the main ways that viruses are spread through the fecal-oral pathway [5]. In low-to-middle-income countries, children under 5 years old bear a lot of the diarrheal problem, with gastroenteritis linked to norovirus accounting for over 212,489 diarrheal fatalities annually [6].

Symptoms emerge 12–48 h after coming into touch with infected food or an infected individual and include cramps, nausea, diarrhea, as well as stomach ache. In healthy people, norovirus sickness resolves on its own, but in sensitive groups, such as young or undernourished pediatric and geriatric or people with compromised immune, it can be deadly [7]. Most norovirus outbreaks happen in semi-enclosed locations like childcare centers, schools, colleges, nursing homes, and cruise ships. Norovirus is exceedingly contagious. Winter vomiting disease is another moniker for norovirus infections because, in temperate climates, outbreaks of the virus generally occur during the coldest months of the year. Tropical places have less of this seasonality [8].

The genetically heterogeneous groups of newly forming RNA viruses known as noroviruses are still evolving in people as a result of both genome recombination and point mutations, as shown in Figure 8.1 [9]. Notably, alterations in the amino acid content of pathological proteins connected to mutations in the genome have produced a longer duration of virus shedding, the avoidance of protection from prior infections, and a reduction in virus clearance in patients diseased with heterologous strains [10–12]. Worldwide, eight genogroups with 60 distinct P-types, as well as 49 capsid genotypes have been related to infections in humans as well as animals [13], although it is conceivable to detect numerous viruses co-circulating. Generally, only one virus produces full-blown epidemics and moves to neighboring nations [7]. While new mutations develop generally every 2 to 5 years, genotype GII.4 is the most frequent worldwide [14].

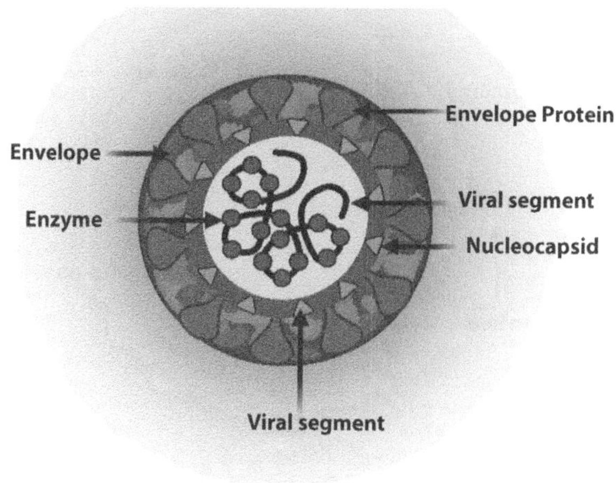

Figure 8.1 Diagram illustrating the basic structure of norovirus.

VP1 epitope domain point mutation accumulation generally leads to a change in antigenic features, which is connected to changes escaping herd immunity [15]. This study analyzes the newest insights into the dynamics of norovirus epidemics, concentrating on epidemiology, transmission mechanisms, environmental persistence, and measures for prevention and management.

8.2 Symptoms Related to Norovirus

The major symptoms of a NoV infection are diarrhea and emesis [16, 17]. Occasionally, these symptoms are accompanied by nausea, cramping, abdominal discomfort, anorexia, malaise, and low-grade pyrexia, as shown in Figure 8.2 [18–20]. Common side effects might include convulsions, necrotizing enterocolitis, benign seizures, benign seizures, and aggravation of inflammatory bowel disease [21–25]. Norovirus may also be linked to respiratory disorders, according to the discovery of respiratory secretions from kids who had respiratory symptoms [26]. Furthermore, based on clinical signs, mild meningoencephalitis has been linked to norovirus infection in a few cases [27].

The primary symptoms of a norovirus infection varied from person to person, with immunocompromised people, the elderly, and babies showing a greater incidence [28]. According to studies, symptoms are more severe in young children, and protective immunity to prevent reinfection

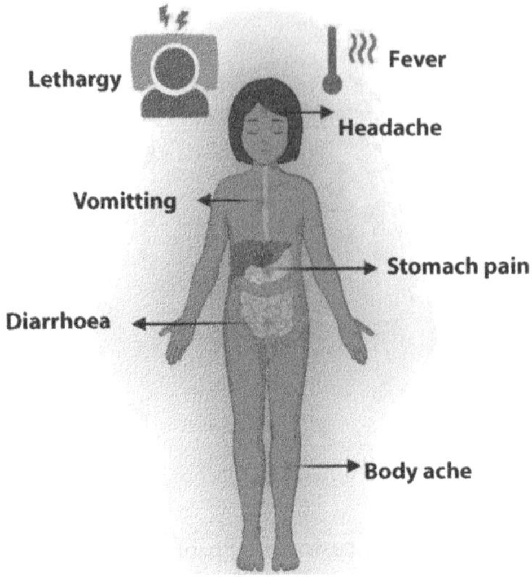

Figure 8.2 Illustration depicting norovirus symptoms.

is usually not established before the age of two years [29]. Watery diarrhea is one of the main symptoms of diarrhea in kids under 1 year old [30]. In extreme situations, this condition can result in electrolyte imbalance, seizures, dehydration, and even death, as mentioned in Table 8.1.

Table 8.1 Overview of clinical manifestations associated with norovirus infections.

Category	Common symptoms	Rare symptoms	Complications	References
Gastrointestinal	Nausea, vomiting, watery diarrhea	Abdominal cramps, anorexia	Severe dehydration, electrolyte imbalance	[40]
Systemic	Fatigue, low-grade fever	Chills, malaise	Exacerbation of inflammatory bowel disease	[41]
Neurological	Headache	Benign seizures, febrile seizures	Mild meningoencephalitis, convulsions	[42]
Respiratory		Respiratory symptoms (rare in children)	Potential respiratory infections (rarely observed)	[43, 44]

Furthermore, individuals who have viral gastroenteritis with convulsions also have temporary vomiting that lasts for around 2 days when they do not have a fever [31]. The incubation period of infection for individuals with immunocompetent norovirus gastroenteritis is usually brief, lasting between 24 and 48 h [32]. The symptoms of this illness include diarrhea, cramping in the abdomen, nausea, and vomiting, all of which usually go away after 72 h.

Additionally, the different virus strains would show different clinical signs. A previous study indicates that the GI genotype is only seen in a small percentage of human norovirus infections, but the GII genotype is found worldwide [33, 34]. In light of seasonal fluctuations, GI genotypes are more common in the spring, whereas GI genotypes are more common in the fall [35–37]. Severe diarrhea and vomiting are more common in cases of norovirus GII infection than in cases of other genotypes [38]. In addition, a higher percentage of individuals with genotype GI infection had symptoms like increased bowel sounds, diarrhea that lasted fewer than 3 days, and nausea [39].

It will be easier to improve the accuracy of the diagnosis and develop evidence-based treatment and prevention strategies if dominant strains and their relationship to clinical symptoms are understood.

8.3 Epidemiology of Norovirus Outbreak

The genus Norovirus and family *Caliciviridae* contain a hereditarily varied group of nonenveloped, positive-sense RNA viruses known as noroviruses [9]. The 10 distinct genogroups of the rapidly evolving RNA virus (GI–GX) are further subdivided into 60 unique P-types and 49 genotypes according to variations in the amino acid gene sequences of the most common viral protein shell (VP1) and the nucleotide sequences of the RNA-dependent RNA polymerase, respectively. The isolates that infected humans have genotypes nine, 27, and two and are members of genogroups GI, GII, and GIV [13]. The most prevalent norovirus genotype observed in humans globally is genotype II, which contains several genotypes. Moreover, 51%–79% of the norovirus disease burden is produced by the extremely varied and developing GII.4 genotype of the GII viruses [45]. Notably, strains of NoV are unstable both genetically as well as antigenically, and by epochal evolution, new variants develop virtually every 2 to 3 years [46].

The incidence of diverse NoV strains, including GII.4 as well as non-GII.4 genotypes in cases of acute gastroenteritis, has grown significantly during the previous 5 to 7 years. Nevertheless, non-GII.4 viruses have briefly supplanted GII.4 viruses in certain nations. For instance, the GII. P16/GII.2 genotype produced a dramatic spike in acute norovirus gastroenteritis cases in China in 2017, but the GII.3 genotype predominated over GII.4 viruses in Ethiopia in 2009 [47]. A surge in sporadic cases of acute gastroenteritis in Europe as well as Asia during the 2016–2017 epidemic season was also caused by the GII.2[P16] variant, indicating that it may spread internationally [48]. In China in 2014/2015, a new strain of GII.17 swiftly supplanted the previously dominant GII.4 viruses, most likely as a result of the virus' fast development and genetic dynamics [49]. The initial pandemic happened in June and July of 2021 in the coastal region of Alappuzha in Kerala. Public health facilities alerted the district surveillance unit of higher AGE in kids. The second pandemic happened in October 2021 in Kerala's hilly region of Wayanad. A substantial surge in AGE on a university campus was reported to the DSU of Wayanad district by the local primary health clinic [50].

8.4 Norovirus Genome and Structure

The positive-sense, single-stranded RNA genomes of NoV are about 7.3 to 7.5 kb long [51]. Viral particles include a subgenomic RNA that is roughly equivalent to the last 2.3 kb of the genome. Infected cells also produce this RNA in higher quantities than the viral genomic RNA [52]. The viral protein VPg binds to the 50 ends of NoV genomic as well as subgenomic RNA, whereas the 30 ends are polyadenylated [53–55]. The highly conserved non-coding RNA stem-loop structure located upstream of the starting site for subgenomic RNA at the overlap of ORFs 1 and 2 serves as the principal promoter for NoV subgenomic RNA synthesis [51, 56–58]. Short untranslated regions (UTRs) in NoV genomes have evolutionarily conserved RNA secondary structures that are recurrent across the genome and extend into the coding regions. These structures associate with host and viral elements to execute functional roles in viral replication, translation, and pathogenicity [59, 60].

The NoV genome is divided into a trio ORFs [61]. The 50 proximal ORF1 encodes a substantial polyprotein that is co- and post-translationally cleaved into six nonstructural proteins (NS) that participate in polyprotein processing (NS6), genome linkage (NS5 and VPg), genome replication (NS7 and RdRp), and the formation of the replication complex [62, 63]. Major viral proteins, as well as minor viral proteins, the structural elements of the virion, are encoded by ORF2 along with ORF3, which are both translated from subgenomic RNA. Exclusive to MNVs, ORF4 is translated mostly from subgenomic RNA, overlays ORF2, and encodes the virulence factor 1, which controls innate immunity and apoptosis [61, 64].

8.5 Virion Morphology

Norovirus capsids have T = 3 icosahedral symmetry and are normally 27–30 nm in diameter. The threefold symmetry axis and the fivefold symmetry axis of calicivirus capsids are cup-shaped depressions, respectively [65]. The monomeric primary structural protein VP1 is present in 180 copies in capsids, producing 90 dimeric capsomers. Every VP1 has a projecting domain (P), a shell domain (S), and a short N-terminal arm with an unknown function [66]. Viral RNA is surrounded by a continuous surface developed by the capsid N-terminal S domain. The C-terminal section of VP1 is represented by the P domain, which is joined to the S domain by a flexible hinge. It is believed to control the viral particle size and to stabilize the icosahedral capsid [67]. The P domain is alienated into two subdomains: the highly variable distal P2 and the proximal P1 stalk close to the base of the arches. The exposed P2 subdomain, which is found at the arches' points, interacts with neutralizing antibodies and has both a potential receptor binding site for HuNoVs and an established host receptor binding site for MNVs [68, 69]. Based on bile salts and buffer pH, cryo-electron microscopy shows distinct P domain conformations for MNV and some HuNoV strains (GII.10). A "floating" P domain floats over the shell in physiological buffers. Nevertheless, the P domain spins and compresses into the shell surface when bile salts are present, perhaps improving stability and receptor affinity [70]. Smith and Smith's analysis in 2019 describes the flexibility of the *Caliciviridae P* domain in response to environmental signals, whereas Tenge *et al.* in 2021 review the impact of bile salts as well as relations with viral capsids [71, 72].

8.6 Transmission Mechanism

The fecal–oral channel is the predominant mechanism of the human-to-human spread of NoV, especially for the outbreak strain GII.4, but vomiting episodes might aid proliferation [73]. NoV can capably persist in the atmosphere and tolerate cold temperatures, hot temperatures (60°C), and disinfection by alcohol or chlorine [74]. This facilitates uncleanness of food, drink, and infected surfaces and eventually causes epidemics, particularly in close quarters like nursery schools, military camps, ships, and others [75, 76]. The low infectious dose of NoV makes it exceedingly contagious; a single particle has a 50% chance of infecting someone. However, there is a dose–response link, suggesting that people who are exposed to more viruses have a larger infectious rate [77, 78].

NoV shedding in feces lasts for an average of 4 weeks, peaking in the first 24 to 48 h following the initiation of symptoms [79]. For months after the infection, virus shedding in feces could continue in immunocompromised hosts. Human NoV is disseminated by direct human-to-human touch or by consuming contaminated diet or water [80]. They are exceedingly durable and resistant to numerous virucidal cure and decontamination approaches [81, 82]. Environmental waters can be contaminated by infectious viruses through the direct release of inappropriately treated sewage from large, insignificant private wastewater treatment plants or urban runoff, particularly during periods of flooding or precipitation, which has been connected to a high prevalence of HuNoVs in coastal waters as shown in Figure 8.3 [83].

Noroviruses that are suspended or precipitated have been identified up to 10 km from their site of release and have been proven to hold their infectiousness for weeks to months; 14% of HuNoV outbreaks are considered

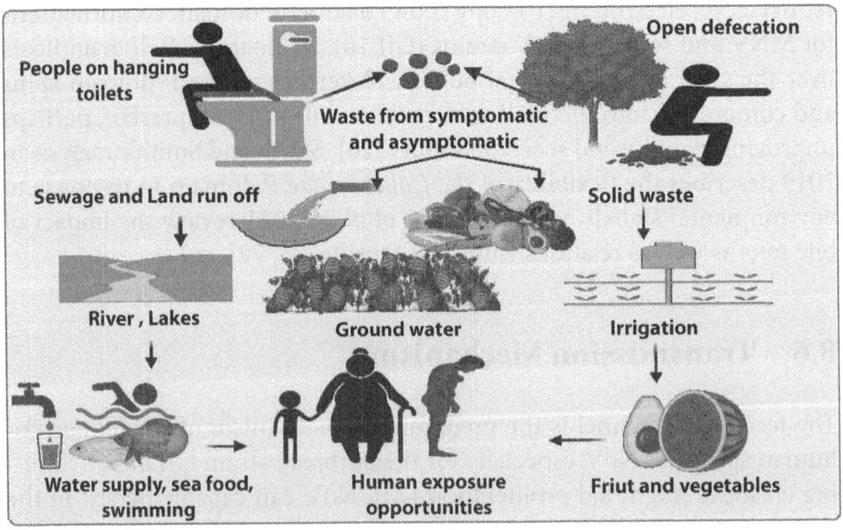

Figure 8.3 Diagram illustrating norovirus emission pathway.

to be foodborne [84]. Either diseased food handlers or direct contact with excrement at the source of contaminated foods was involved in the outbreaks [85]. The most popular food carriers continue to be raw or undercooked shellfish (bivalve molluscs), fresh soft fruits and vegetables, and ready-to-eat products (like salads and sandwiches) that merely need to be handled and do not need to be cooked further [86]. Cockles, mussels, clams, scallops, and oysters are examples of bivalve molluscs that adopt filter feeding to gather noroviruses. Depuration treatments, which aim to eliminate these bioaccumulated pathogen charges, are useless when there is a norovirus infection.

Noroviruses are filtered and concentrated by nonspecific interactions, although they also adhere to molluscan gastrointestinal carbohydrate structures in a manner that is dependent on genogroup and strain, which is increasingly considered to be the underlying source of this phenomenon [87]. Bivalve molluscs are recognized as significant vectors for the introduction of novel recombinant strains into the human population [88]. Similar to this, bivalve molluscs have been tentatively connected to the assumed technique of introducing both human as well as diverse animal noroviruses into a single host by filtering human and animal feces, potentially functioning as interfaces of shared species exposure [88].

In the setting of group meals at picnics, schools, restaurants, festivals, cruise ships, and military facilities, norovirus outbreaks are regularly observed [89]. Closed living areas and poor personal cleanliness are permissive to the transmission of viruses from a common source. Concern has grown over the contamination of healthcare staff, especially rookie users, with chronic respiratory diseases that might pollute masks during PPE donning and doffing (especially in light of the ongoing coronavirus pandemic) [90].

8.6.1 Pattern of Viral Invasion into Host

The gastrointestinal tract macrophages and dendritic cells are the major entrance routes for noroviruses into the body [91]. These viruses prefer to infect the gastrointestinal tract in humans; enterocytes may acquire waste from the intestinal lumen and transport it to prospective immune cells, and NoV can utilize M cells across the intestinal epithelial barrier [91]. Because it can be difficult to effectively cultivate norovirus in human cells, the majority of studies exploring this mechanism have mainly utilized mice or other caliciviruses.

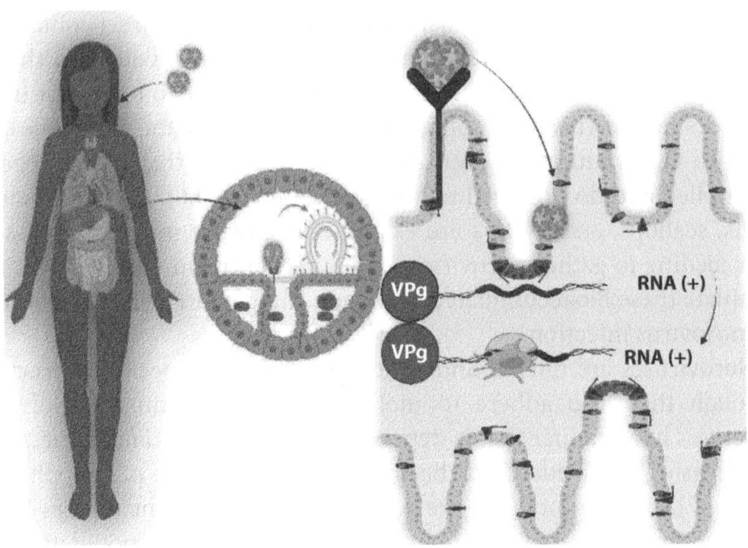

Figure 8.4 Diagram illustrating the pathophysiology of norovirus infection.

The five critical phases of norovirus invasion are attachment, receptor interaction, endocytosis, uncoating, and translation [92]. It has been observed that bile acids, sialic acids, and cations help in the attachment of murine norovirus to the HBG structure on the cell membrane. The norovirus subsequently binds to CD300lf [93]. An essential aspect of the binding process is ceramide. Gerondopoulos *et al.* indicate that norovirus largely requires dynein and cholesterol for endocytosis, in contrast to the majority of viruses that depend on pH value [94]. Norovirus RNA, which lacks the 5′ cap structure, needs the VPg protein to connect to initiation factors and initiate the translation process, as shown in Figure 8.4 [95].

8.7 Administration of Pharmaceuticals and Challenges in Vaccination

Preventing NoV infections by vaccination is of crucial significance as NoV is the leading cause of AGEs worldwide. It is connected to severe morbidity and death, and there is no distinct anti-viral therapy for it [96]. However, there have been several hurdles along the process of manufacturing a NoV vaccine. These include the lack of a suitable cell line for viral replication, the genetic diversity of NoVs as a result of the sporadic synthesis of new mutants and recombinants, the lack of suitable and approachable animal models, and the inadequate comprehension of correlates of protection.

One of the most investigated norovirus vaccine candidates is Takeda Pharmaceuticals International AG (TAK-214), which is a bivalent vaccine based on adjuvanted VLP [97]. There are two sorts of VLPs in it: one from the GI.1 strain of the Norwalk virus (NV) and another from a consensus of GII.3 GII.4 mutations to yield the four sequences. To improve its potential protective immunity, both GI and GII norovirus antigens are added [98].

Numerous phase 1 clinical trials have demonstrated that both antigen components in various formulations elicit significant immune responses in healthy persons, coupled with great safety and tolerability. The perseverance of blood antibodies to the two VLP components of the vaccination was established in a follow-up test done 1 year later using memory probe immunization [99, 100].

The TAK-214 vaccine initial in-human phase 2b field efficacy investigation involved more than 4,600 volunteers and was carried out over two wintertime seasons in a US Navy training facility. The results indicated a 61.8% immunization effectiveness against moderate to severe AGE brought on by noroviruses of diverse genotypes, such as GI.1, G1.7. With a 57.4% vaccination efficiency against GII.2 noroviruses, the TAK-214 vaccine is known to generate a broad immune response and give cross-genotype protection [101].

A bivalent vaccine based on recombinant VP1 (VXA NVV-104) is being established by Vaxart Pharmaceutical Inc. to prevent norovirus infection. Through the use of recombinant adenovirus-based vectors that carry genes for noroviral VP1, the vaccine stimulates mucosal immunity in the recipients' intestines by prompting them to generate VP1 antibodies. To produce double-stranded RNA for greater immunogenicity, the AD vector moreover contains particular RNA-encoding sequences [102].

The goal of the tablet vaccine is to prevent norovirus infection by activating the intestinal mucosal immune response *via* oral delivery. The oral vaccination produced mucosal-primed NV-specific circulating antibody-secreting cells, IgA^+ memory B cells with gut-homing receptor ($\alpha 4\beta 7$), and fecal IgA, according to the phase 1 clinical study. It was also well tolerated. These findings hint at significant mucosal and systemic immune responses that may help defend against norovirus infection. Vaxart has initiated a new stage 1b NoV dose-ranging which is experimental in older individuals by utilizing a two-dose vaccination schedule to assess the vaccine care, tolerability, immunogenicity, and efficacy [103].

The protection and immunogenicity of Hansenulapolymorpha, a bivalent VLP-based immunization produced by China National Vaccination and Serum Institute (NVSI), are being examined. Two recombinant VLPs representing the GI.1 and GII.4 genotypes made up the vaccine, which was delivered to 510 healthy participants vacillating in age from 6 months to 59 years.

The development of norovirus vaccines confronts various difficulties, including the necessity for multivalent immunizations episodic apprises to guard against a variety of existing and newly developing epidemiologically relevant genotypes. These problems underscore the necessity for vaccination effectiveness in light of the diversity and continual development of norovirus strains. The absence of a generally accepted indicator of guard against the virus, known variation in seroresponse, and the uncertainty around the amount of time that vaccination or infection gives long-term immunity are noteworthy barriers to the expansion of a NoV vaccine [104]. However, two vaccines are now undergoing clinical trials: a recombinant adenovirus vector vaccine that expresses GII.4 VP1 with a univalent or bivalent dose and a bivalent GII.4 VLP [105]. Additional vaccines are either in the preclinical stage of study or have been licensed for testing in clinical trials.

According to their mechanisms, antiviral medicines are split into three broad classes. HBGAs are critical in boosting viral adherence and entry, while the first category aims to block viruses from sticking to host cell tops. Caramel carbohydrates have shown the capacity to prevent viral activity, much like citric acid salt [106]. Targeting the catalytic degradation of replication-related proteins by viral proteases, the second category focused on inhibitors of virus proteases. Within this category, rufintrivir is a remarkable antiviral medicine that is largely used to treat human rhinovirus, although it also demonstrates efficiency against other picornaviruses, coronaviruses, and caliciviruses [107].

The most representative cases of RdRp inhibitors lie within the third group, which comprises nucleoside analogs. By preventing nucleotide binding, these chemicals hinder the transmission of viruses. The only approved medicine for treating norovirus infections at the time is purine nucleoside CM521 [108]. Black raspberry seed, algal polysaccharide, and green tea are examples of natural plant-based products whose antiviral activities have been researched [109, 110]. It has also been proven that synthetic medications such as dasabuvir (DSB), which was initially intended to treat hepatitis C, suppress norovirus proliferation in the body [111, 112].

8.8 Immunity to Norovirus Infection

Protective immunity against human noroviruses is a multifaceted poorly unstated topic since there is no appropriate cell culture system and humans are constantly bare to antigenically distinct virus strains throughout time [113]. Human exposure to norovirus is frequent, and by maturity, anti-norovirus antibody prevalence is almost 80% worldwide, according to sero-prevalence studies. However, people continuously exhibit a high amount of norovirus susceptibility in both natural and artificial tests [114].

Many norovirus-infected volunteers were susceptible to recurrence by either of the homotypic and heterologous strains, according to early human challenge experiments. Except for those who had been showing to the homologous strain often between 3 and 6 months, some people's preexisting antibodies did not offer total protection against infection. Some trial participants did not generate long-term protective antibodies following an experimental trial with a Hu NoV GI strain despite one study finding that an early elevation in mucosal IgA was linked to resistance to infection [115].

The discovery of innate inherited resistance has advanced our consideration of adaptive protection in the current years [116]. Blood antibody levels determined by ELISA have been shown to be an unreliable indication of illness susceptibility or resistance to norovirus sickness [117]. Nevertheless, in a number of human infection investigations, larger levels of the obstructive antibodies have been related to a reduced danger of disease in people, which makes HBGA-blocking assays detect pre-existing serum antibodies that impede NoV binding to HBGAs more relevant. Wang *et al.* recently devised the capable surrogate neutralization test for the cellular assessment of the norovirus vaccination. A study on HBGA blockade has produced comparative indicators between the capacity to evade blockade antibody responses and the genetic evolution of human norovirus genotypes [118]. There is further evidence linking a decreased risk of norovirus illness to higher titers of virus-specific memory B lymphocytes and IgA from the salivary mucosa before the viral trial [119].

Various clinical trial data related to norovirus are mentioned in Table 8.2.

Table 8.2 List of clinical-trial-based details for norovirus infections.

Sn. no.	Trail ID	Sponsor name	Title	Status	Ref.
1	NCT05508178	Icon Genetics GmbH	Recombinant, Bivalent, Norovirus Vaccine Study	Completed	[120, 121]
2	NCT02661490	Takeda	Safety and Immunogenicity of Norovirus GI.1/GII.4 Bivalent Virus-Like Particle Vaccine in an Elderly Population	Completed	[122]
3	NCT02153112	Takeda	Safety and Immunogenicity of Norovirus GI.1/ GII.4 Bivalent Virus-Like Particle (VLP) Vaccine in Children	Completed	[123]
4	NCT00806962	LigoCyte Pharmaceuticals, Inc.	Phase 1 Norwalk Vaccine Study	Completed	[124, 125]
5	NCT05212168	Vaxart	Norovirus Challenge Study (G1-1 Challenge)	Completed	[126]
6	NCT04174560	National Institute of Allergy and Infectious Diseases (NIAID)	Phase I Norovirus Challenge Model	Completed	[127]
7	NCT04854746	Vaxart	Ph 1b: Safety and Immunogenicity of Ad5 Based Oral Norovirus Vaccine (VXA-NVV-104)	Completed	[128]
8	NCT02337842	National Institute of Allergy and Infectious Diseases (NIAID)	Phase I Study to Determine the Optimal Human Challenge Dose for a Norovirus GII.4 Challenge Stock (CIN-1)	Completed	[129]
9	NCT01435811	Children Hospital Medical Center, Cincinnati	Dose Range Evaluation of Norovirus Challenge Pool (GII.4, CIN-1) (CIN-002)	Completed	[130]

(*Continued*)

Table 8.2 List of clinical-trial-based details for norovirus infections. (*Continued*)

Sn. no.	Trail ID	Sponsor name	Title	Status	Ref.
10	NCT04875676	Vaxart	Immunogenicity and Safety Study of Adenovirus Type 5 (AD5) Based Oral Norovirus Vaccines (VXA-NVV-105)	Completed	[131]
11	NCT05213728	Vaxart	A Phase 1, Open-Label, Safety and Immunogenicity Study of an Oral Multi-dose Administration Regimen With an Adenoviral-Vector Based Tablet Norovirus Vaccine (VXA-G1.1-NN) Administered to Healthy Adult Volunteers (VXA-G1-1-NN)	Completed	[132]

8.9 Public Health Measures to Avoid Being Infected with Norovirus

The complexity of the human ordinary immune response to norovirus, the lack of animal models for vaccine testing, the hereditary assortment within the viral norovirus genome, and the nonappearance of a dependable viral *in vitro* culture system are some of the obstacles facing norovirus research and development. Consequently, there is currently no authorized norovirus vaccine [133, 134].

The good news is that several studies show that zebrafish may be employed as a novel HuNov model, giving valuable data for norovirus vaccine development [135]. Currently, however, the principal treatment for acute gastroenteritis caused by noroviruses is oral rehydration treatment. This comprised both a rehydration phase that entailed taking oral rehydration electrolytes and water as well as a maintenance stage therapy that needed ingesting electrolytes, water, and the necessary meals [136].

Ondansetron has been proven in trials to help ease the symptoms of vomiting in children. In the interim, the community healthcare program needs to be enhanced even further in order to appropriately thwart viral infections.

It is also vital to take actions to enhance everyday living and lower the chance of catching a virus. These measures' key aspects include frequent hand washing, paying specific attention to sanitizing surfaces in the surroundings, and avoiding direct contact with infectious sources [137].

8.10 Conclusion

Managing norovirus outbreaks necessitates an integrated approach that includes public health initiatives, vaccine development, and continuous study into viral behavior and immunity. Norovirus remains a chronic public health concern due to its low infectious dosage, considerable genetic diversity, and environmental resilience. Current preventative efforts, such as frequent handwashing, thorough surface disinfection, and better food and water safety, are still crucial to minimizing transmission, especially in high-risk communal settings like schools and hospitals. Vaccines are a potential method for norovirus control. In clinical studies, virus-like particle (VLP)-based vaccinations and recombinant adenovirus vector vaccines produced positive results in generating immune responses and giving partial cross-genotype protection. However, genetic variety and the periodic appearance of new strains of the virus highlight the importance of multivalent vaccinations that target many genotypes. Addressing these problems requires regular vaccination updates and a greater knowledge of long-term immunity. Furthermore, the absence of standardized immunological correlates and appropriate animal models for vaccination testing hampers the route to widespread vaccine use.

Antiviral medicines are another area of possibility, with developments in compounds that target replication pathways or viral attachment to host cells. Furthermore, investigating natural chemicals and repurposing current antiviral drugs might result in cost-effective remedies. To minimize the worldwide norovirus burden, particularly among vulnerable groups including children, the elderly, and immunocompromised people, further investment in vaccine development, new therapeutics, and public health interventions is required. Addressing these diverse difficulties together will be crucial for reducing norovirus and limiting its worldwide economic and health repercussions.

AI disclosure statement: The author acknowledges the use of AI-based tools for minor language editing or grammatical correction. However, all intellectual contributions, analyses, and conclusions presented in this chapter are entirely the author's own.

References

1. Xu, D., Li, J., Han, L., Chen, D., Bao, W., Li, L., *et al.*, Epidemics and diversity of norovirus variants with acute gastroenteritis outbreak in Hongshan District, Wuhan City, China, 2021-2023. *J. Infect. Public Health*, 17, 9, 102499, 2024.
2. Kulkarni, R., Insights into the Biology of Norovirus: The So-Called "Perfect" Human Pathogen. *Emerg. Hum. Viral Dis.*, II [Internet], 367–409, 2024 [cited 2025 Jan 24]. Available from: https://link.springer.com/chapter/10.1007/978-981-97-4480-0_13.
3. Afework, D.T., Shumie, M.K., Endalew, G.F., Adugna, A.G., Tarekegn, B.G., Pooled prevalence and genetic diversity of norovirus in Africa: a systematic review and meta-analysis. *Virol. J.*, 19, 115, 2022.
4. Mans, J., Armah, G.E., Steele, A.D., Taylor, M.B., Norovirus epidemiology in Africa: A review. *PLoS One*, 11, e0146280, 2016.
5. Ji, L., Hu, G., Xu, D., Wu, X., Fu, Y., Chen, L., Molecular epidemiology and changes in genotype diversity of norovirus infections in acute gastroenteritis patients in Huzhou, China, 2018. *J. Med. Virol.*, 92, 12, 3173–3178, 2020.
6. Bartsch, S.M., Lopman, B.A., Ozawa, S., Hall, A.J., Lee, B.Y., Global economic burden of norovirus gastroenteritis. *PLoS One*, 11, 4, e0151219, 2016.
7. Parra, G.I., Emergence of norovirus strains: A tale of two genes. *Virus Evol.*, 5, 2, vez048, 2019.
8. Rohayem, J., Norovirus seasonality and the potential impact of climate change. *Clin. Microbiol. Infect.*, 15, 524–527, 2009.
9. Vinjé, J., Estes, M.K., Esteves, P., Green, K.Y., Katayama, K., Knowles, N.J., *et al.*, ICTV Virus Taxonomy Profile: Caliciviridae. *J. Gen. Virol.*, 100, 11, 1469–1470, 2019.
10. Sajewski, E.T., Vinjé, J., Glass, R.I., Lopman, B.A., Noroviruses, Sapoviruses, and Astroviruses, in: *Viral Infections of Humans*, 2022.
11. Lindesmith, L.C., Brewer-Jensen, P.D., Mallory, M.L., Yount, B., Collins, M.H., Debbink, K., *et al.*, Human Norovirus Epitope D Plasticity Allows Escape from Antibody Immunity without Loss of Capacity for Binding Cellular Ligands. *J. Virol.*, 93, 2, 10–1128, 2019.
12. Gustavsson, L., Nordén, R., Westin, J., Lindh, M., Andersson, L.M., Slow clearance of norovirus following infection with emerging variants of genotype GII.4 strains. *J. Clin. Microbiol.*, 55, 5, 1533–1539, 2017.
13. Chhabra, P., de Graaf, M., Parra, G.I., Chan, M.C.W., Green, K., Martella, V., *et al.*, Updated classification of norovirus genogroups and genotypes. *J. Gen. Virol.*, 100, 10, 1393–1406, 2019.
14. Flynn, T.G., Olortegui, M.P., Kosek, M.N., Viral gastroenteritis. *Lancet*, 403, 862–876, 2024.
15. Zhang, Q., Zhu, S., Zhang, X., Su, L., Ni, J., Zhang, Y., *et al.*, Recent insights into reverse genetics of norovirus. *Virus Res.*, 325, 199046, 2023.

16. Lin, C.Y., Chiu, N.C., Lee, H.C., Chuang, C.K., Lin, S.P., Yeung, C.Y., The Emerging Importance of Norovirus as the Etiology of Pediatric Gastroenteritis in Taipei. *J. Microbiol. Immunol. Infect.*, 43, 2, 105–110, 2010.
17. Alsedà, M., Godoy, P., Bach, P., Soldevila, N., Cornejo, T., Corominas, L., *et al.*, Two successive outbreaks of acute gastroenteritis due to norovirus GII.6 in a holiday camp house. *Sci. Rep.*, 13, 1, 15558, 2023.
18. Sun, W., Pang, Z., He, Y., Chen, Y., Pan, J., Gao, J., *et al.*, Improper handling of vomitus as a risk factor in the human norovirus outbreak in a kindergarten in Wuyi County, Zhejiang Province, China. *Epidemiol. Infect.*, 150, e111, 2022.
19. Hellysaz, A., Neijd, M., Vesikari, T., Svensson, L., Hagbom, M., Viral Gastroenteritis: Sickness Symptoms and Behavioral Responses. *mBio.*, 14, 2, e03567–22, 2023.
20. Fang, Y., Zhang, Y., Wang, H., Shi, O., Wang, W., Hou, M., *et al.*, Molecular epidemiology of norovirus infections in children with acute gastroenteritis in 2017–2019 in Tianjin, China. *J. Med. Virol.*, 94, 2, 616–624, 2022.
21. Khan, R.R., Lawson, A.D., Minnich, L.L., Martin, K., Nasir, A., Emmett, M.K., *et al.*, Gastrointestinal norovirus infection associated with exacerbation of inflammatory bowel disease. *J. Pediatr. Gastroenterol. Nutr.*, 48, 3, 328–333, 2009.
22. Chen, S.Y., Tsai, C.N., Lai, M.W., Chen, C.Y., Lin, K., Lin, T.Y., *et al.*, Norovirus infection as a cause of diarrhea-associated benign infantile seizures. *Clin. Infect. Dis.*, 48, 7, 849–855, 2009.
23. Chan, C.M.V., Chan, C.W.D., Ma, C.K., Chan, H.B., Norovirus is the cause of benign convulsion associated with gastro-enteritis. *J. Paediatr. Child Health*, 47, 6, 373–377, 2011.
24. Shi, K., Jiang, D., Yang, J., Li, Y., Chen, W., Li, P., Clinical characteristics and follow-up of children with norovirus-associated benign convulsions with mild gastroenteritis. *Epilepsia Open*, 8, 3, 1049–1053, 2023.
25. Turcios-Ruiz, R.M., Axelrod, P., St. John, K., Bullitt, E., Donahue, J., Robinson, N., *et al.*, Outbreak of Necrotizing Enterocolitis Caused by Norovirus in a Neonatal Intensive Care Unit. *J. Pediatr.*, 153, 3, 339–344, 2008.
26. Esposito, S., Daleno, C., Scala, A., Senatore, L., Ascolese, B., Principi, N., Detection of norovirus in respiratory secretions in children with respiratory tract infection. *Pediatr. Infect. Dis. J.*, 33, 3, 314–316, 2014.
27. Kimura, E., Goto, H., Migita, A., Harada, S., Yamashita, S., Hirano, T., *et al.*, An adult norovirus-related encephalitis/encephalopathy with mild clinical manifestation. *BMJ Case Rep.*, 1, bcr0320102784, 2010.
28. Lopman, B.A., Reacher, M.H., Vipond, I.B., Sarangi, J., Brown, D.W.G., Clinical manifestation of norovirus gastroenteritis in health care settings. *Clin. Infect. Diseases*, 39, 3, 318–324, 2004.
29. Sai, L., Sun, J., Shao, L., Chen, S., Liu, H., Ma, L., Epidemiology and clinical features of rotavirus and norovirus infection among children in Ji'nan, China. *Virol. J.*, 10, 302, 2013.

30. Shang, L.H., Xiong, L.J., Liu, L.R., Deng, X.Z., Xie, X.L., Epidemiological and clinical features of calicivirus-associated diarrhea in hospitalized children in Chengdu, China from 2012 to 2014. *Chin. J. Contemp. Pediatr.*, 18, 10, 1005–1008, 2016.
31. Chen, Y.F.E., Wang, C.Y., Chiu, C.H., Kong, S.S., Chang, Y.J., Chen, S.Y., Molecular epidemiology and clinical characteristics of norovirus gastroenteritis with seizures in children in Taiwan, 2006-2015. *Med. (United States)*, 98, 40, e17269, 2019.
32. Lively, J.Y., Johnson, S.D., Wikswo, M., Gu, W., Leon, J., Hall, A.J., Clinical and epidemiologic profiles for identifying norovirus in acute gastroenteritis outbreak investigations. *Open Forum Infect. Dis.*, 5, 4, ofy049, 2018.
33. Sun, Z., Niu, P., Jin, M., Zhang, R., Gao, Q., Wang, H., *et al.*, Genotyping and traceability analysis of norovirus in Yantai between 2017 and 2019. *J. Med. Virol.*, 95, 11, e29220, 2023.
34. Shigemura, K., Fardah Athiyyah, A., Kitagawa, K., Agustina, N., Darma, A., Ranuh, R., *et al.*, Clinical manifestation of norovirus infection in children aged less than five years old admitted with acute diarrhea in Surabaya, Indonesia: A cross-sectional study. *F1000Res*, 8, 1–28, 2020.
35. Chiu, S.C., Hu, S.C., Liao, L.M., Chen, Y.H., Lin, J.H., Norovirus Genogroup II Epidemics and the Potential Effect of Climate Change on Norovirus Transmission in Taiwan. *Viruses*, 14, 3, 641, 2022.
36. Chan, M.C.W., Kwok, K., Zhang, L.Y., Mohammad, K.N., Lee, N., Lui, G.C.Y., *et al.*, Bimodal seasonality and alternating predominance of norovirus GII.4 and Non-GII.4, Hong Kong, China, 2014-2017. *Emerg. Infect. Dis.*, 24, 4, 767, 767, 2018.
37. Cao, R.R., Ma, X.Z., Li, W.Y., Wang, B.N., Yang, Y., Wang, H.R., *et al.*, Epidemiology of norovirus gastroenteritis in hospitalized children under five years old in western China, 2015–2019. *J. Microbiol. Immunol. Infect.*, 54, 5, 918–925, 2021.
38. Melhem, N.M., Zaraket, H., Kreidieh, K., Ali, Z., Hammadi, M., Ghanem, S., *et al.*, Clinical and epidemiological characteristics of norovirus gastroenteritis among hospitalized children in Lebanon. *World J. Gastroenterol.*, 22, 48, 10557, 2016.
39. Xue, Y., Pan, H., Hu, J., Wu, H., Li, J., Xiao, W., *et al.*, Epidemiology of norovirus infections among diarrhoea outpatients in a diarrhoea surveillance system in Shanghai, China: A cross-sectional study. *BMC Infect. Dis.*, 15, 1, 183, 2015.
40. Del Castillo, S. and Catton, K., Dietary methods to treat acute gastroenteritis, in: *Nutrition in Infancy*, 2013.
41. Singh, A. and Nandy, A., Lifestyle diseases in adolescents: Diseases, disorders, and preventive measures. *Lifestyle Dis. Adolesc. Diseases Disorders Preventative Meas.*, 1, 1–207, 2024 Jul 23.

42. Bartolini, L., Mardari, R., Toldo, I., Calderone, M., Battistella, P.A., Laverda, A.M., *et al.*, Norovirus gastroenteritis and seizures: An atypical case with neuroradiological abnormalities. *Neuropediatrics*, 42, 4, 167–169, 2011.
43. Khokhlova, N.I., Kapustin, D.V., Krasnova, E.I., Izvekova, I.Y., Norovirus infection (systematic review). *Jurnal Infektologii*, 10, 5–14, 2018.
44. Van Asten, L., Van Den Wijngaard, C., Van Pelt, W., Van De Kassteele, J., Meijer, A., Van Der Hoek, W., *et al.*, Mortality attributable to 9 common infections: Significant effect of influenza A, respiratory syncytial virus, influenza B, norovirus, and parainfluenza in elderly persons. *J. Infect. Dis.*, 206, 5, 628–639, 2012.
45. Bucardo, F., Reyes, Y., Svensson, L., Nordgren, J., Predominance of norovirus and sapovirus in nicaragua after implementation of universal rotavirus vaccination. *PLoS One*, 9, 5, e98201, 2014.
46. Desdouits, M., de Graaf, M., Strubbia, S., Oude Munnink, B.B., Kroneman, A., Le Guyader, F.S., *et al.*, Novel opportunities for NGS-based one health surveillance of foodborne viruses. *One Health Outlook*, 2, 1, 14, 2020.
47. Shen, W., Sheng, Y., Weng, J., Li, G., Wang, D., Qiu, D., *et al.*, Molecular epidemiology of norovirus associated with acute gastroenteritis in Taizhou, China: A retrospective study. *J. Infect. Public Health*, 13, 1, 34–39, 2020.
48. Niendorf, S., Jacobsen, S., Faber, M., Eis-Hübinger, A.M., Hofmann, J., Zimmermann, O., *et al.*, Steep rise in norovirus cases and emergence of a new recombinant strain GII.P16-GII.2, Germany, winter 2016. *Eurosurveillance*, 22, 4, 30447, 2017.
49. Fu, J., Ai, J., Jin, M., Jiang, C., Zhang, J., Shi, C., *et al.*, Emergence of a new GII.17 norovirus variant in patients with acute gastroenteritis in jiangsu, China, september 2014 to march 2015. *Eurosurveillance*, 20, 24, 21157, 2015.
50. Rajeevan, A., Sakthivel, M., Menon, N., K.C.S., Sudersanan, H., Nagarajan, R., *et al.*, Norovirus outbreaks due to contaminated drinking water and probable person-to-person transmission, Kerala, India, 2021. *J. Infect. Public Health*, 17, 12, 102568, 2024.
51. Thorne, L.G. and Goodfellow, I.G., Norovirus gene expression and replication. *J. Gen. Virol.*, 95, 278–291, 2014.
52. Asanaka, M., Atmar, R.L., Ruvolo, V., Crawford, S.E., Neill, F.H., Estes, M.K., Replication and packaging of Norwalk virus RNA in cultured mammalian cells. *Proc. Natl. Acad. Sci. U.S.A.*, 102, 29, 10327–10332, 2005.
53. Lambden, P.R., Caul, E.O., Ashley, C.R., Clarke, I.N., Sequence and genome organization of a human small round-structured (Norwalk-like) virus. *Sci. (1979)*, 259, 5094, 516–519, 1993.
54. Lee, J.H., Park, B.S., Han, K.R., Biering, S.B., Kim, S.J., Choi, J., *et al.*, Insight into the interaction between RNA polymerase and VPg for murine norovirus replication. *Front. Microbiol.*, 9, JUL, 1466, 2018.
55. Goodfellow, I., The genome-linked protein VPg of vertebrate viruses - A multifaceted protein. *Curr. Opin. Virol.*, 1, 355–362, 2011.

56. Bull, R.A., Hansman, G.S., Clancy, L.E., Tanaka, M.M., Rawlinson, W.D., White, P.A., Norovirus recombination in ORF1/ORF2 overlap. *Emerg. Infect. Dis.*, 11, 7, 1079, 2005.
57. Lin, X., Thorne, L., Jin, Z., Hammad, L.A., Li, S., Deval, J., *et al.*, Subgenomic promoter recognition by the norovirus RNA-dependent RNA polymerases. *Nucleic Acids Res.*, 43, 1, 446–460, 2015.
58. Yunus, M.A., Lin, X., Bailey, D., Karakasiliotis, I., Chaudhry, Y., Vashist, S., *et al.*, The Murine Norovirus Core Subgenomic RNA Promoter Consists of a Stable Stem-Loop That Can Direct Accurate Initiation of RNA Synthesis. *J. Virol.*, 89, 2, 1218–1229, 2015.
59. Simmonds, P., Karakasiliotis, I., Bailey, D., Chaudhry, Y., Evans, D.J., Goodfellow, I.G., Bioinformatic and functional analysis of RNA secondary structure elements among different genera of human and animal caliciviruses. *Nucleic Acids Res.*, 36, 8, 2530–2546, 2008.
60. Bertolotti-Ciarlet, A., Crawford, S.E., Hutson, A.M., Estes, M.K., The 3′ End of Norwalk Virus mRNA Contains Determinants That Regulate the Expression and Stability of the Viral Capsid Protein VP1: a Novel Function for the VP2 Protein. *J. Virol.*, 77, 21, 11603–11615, 2003.
61. McFadden, N., Bailey, D., Carrara, G., Benson, A., Chaudhry, Y., Shortland, A., *et al.*, Norovirus regulation of the innate immune response and apoptosis occurs via the product of the alternative open reading frame 4. *PLoS Pathog.*, 7, 12, e1002413, 2011.
62. Sosnovtsev, S.V., Belliot, G., Chang, K.O., Prikhodko, V.G., Thackray, L.B., Wobus, C.E., *et al.*, Cleavage Map and Proteolytic Processing of the Murine Norovirus Nonstructural Polyprotein in Infected Cells. *J. Virol.*, 80, 16, 7816 7831, 2006.
63. Thorne, L., Bailey, D., Goodfellow, I., High-Resolution Functional Profiling of the Norovirus Genome. *J. Virol.*, 86, 21, 11441–11456, 2012.
64. Roth, A.N. and Karst, S.M., Norovirus mechanisms of immune antagonism. *Curr. Opin. Virol.*, 16, 24–30, 2016.
65. Prasad, B.V., Rothnagel, R., Jiang, X., Estes, M.K., Three-dimensional structure of baculovirus-expressed Norwalk virus capsids. *J. Virol.*, 68, 8, 5117–5125, 1994.
66. Prasad, B.V.V., Hardy, M.E., Dokland, T., Bella, J., Rossmann, M.G., Estes, M.K., X-ray crystallographic structure of the Norwalk virus capsid. *Sci. (1979)*, 286, 5438, 287–290, 1999.
67. Bertolotti-Ciarlet, A., White, L.J., Chen, R., Prasad, B.V.V., Estes, M.K., Structural Requirements for the Assembly of Norwalk Virus-Like Particles. *J. Virol.*, 76, 8, 4044–4055, 2002.
68. Hutson, A.M., Atmar, R.L., Estes, M.K., Norovirus disease: Changing epidemiology and host susceptibility factors. *Trends Microbiol.*, 12, 279–287, 2004.
69. Graziano, V.R., Walker, F.C., Kennedy, E.A., Wei, J., Ettayebi, K., Strine, M.S., *et al.*, CD300lf is the primary physiologic receptor of murine norovirus but not human norovirus. *PLoS Pathog.*, 16, 4, e1008242, 2020.

70. Sherman, M.B., Williams, A.N., Smith, H.Q., Nelson, C., Wilen, C.B., Fremont, D.H., *et al.*, Bile Salts Alter the Mouse Norovirus Capsid Conformation: Possible Implications for Cell Attachment and Immune Evasion. *J. Virol.*, 93, 19, 10–1128, 2019.
71. Tenge, V.R., Murakami, K., Salmen, W., Lin, S.C., Crawford, S.E., Neill, F.H., *et al.*, Bile goes viral. *Viruses*, 13, 998, 2021.
72. Smith, H.Q. and Smith, T.J., The dynamic capsid structures of the noroviruses. *Viruses*, 11, 235, 2019.
73. Izquierdo-Lara, R.W., Villabruna, N., Hesselink, D.A., Schapendonk, C.M.E., Ribó Pons, S., Nieuwenhuijse, D., *et al.*, Patterns of the within-host evolution of human norovirus in immunocompromised individuals and implications for treatment. *EBioMedicine*, 109, 105391, 2024 Nov 1.
74. Stuempfig, N.D., Seroy, J., Labat-Butler, J.R., Viral Gastroenteritis (Nursing). *StatPearls*, 1, 1–10, 2021.
75. Ahmed, K., Dony, J.J.F., Mori, D., Haw, L.Y., Giloi, N., Jeffree, M.S., *et al.*, An outbreak of gastroenteritis by emerging norovirus GII.2[P16] in a kindergarten in Kota Kinabalu, Malaysian Borneo. *Sci. Reports 2020 10:1 [Internet]*, 10, 1, 1–6, 2020 Apr 28 [cited 2025 Jan 24]. Available from: https://www.nature.com/articles/s41598-020-64148-4.
76. Wang, X., Wei, Z., Guo, J., Cai, J., Chang, H., Ge, Y., *et al.*, Norovirus Activity and Genotypes in Sporadic Acute Diarrhea in Children in Shanghai During 2014-2018. *Pediatr. Infect. Dis. J. [Internet]*, 38, 11, 1085–9, 2019 Nov 1 [cited 2025 Jan 24]. Available from: https://pubmed.ncbi.nlm.nih.gov/31626043/.
77. Wulandari, P.S., Juniastuti, RM, W., Amin, M., LN, Y., Matondang, M.Q.Y., *et al.*, Predominance of norovirus GI.4 from children with acute gastroenteritis in Jambi, Indonesia, 2019. *J. Med. Virol. [Internet]*, 92, 12, 3165–72, 2020 Dec 1 [cited 2025 Jan 24]. Available from: https://scholar.unair.ac.id/en/publications/predominance-of-norovirus-gi4-from-children-with-acute-gastroente.
78. Gaythorpe, K.A.M., Trotter, C.L., Lopman, B., Steele, M., Conlan, A.J.K., Norovirus transmission dynamics: a modelling review. *Epidemiol. Infect. [Internet]*, 146, 2, 147–58, 2018 Jan 1 [cited 2025 Jan 24]. Available from: https://pubmed.ncbi.nlm.nih.gov/29268812/.
79. Lucero, Y., Matson, D.O., Ashkenazi, S., George, S., O'ryan, M., Norovirus: Facts and reflections from past, present, and future. *Viruses*, 13, 2399, 2021.
80. Ludwig-Begall, L.F., Mauroy, A., Thiry, E., Noroviruses—the state of the art, nearly fifty years after their initial discovery. *Viruses*, 13, 1541, 2021.
81. Ludwig-Begall, L.F., Wielick, C., Jolois, O., Dams, L., Razafimahefa, R.M., Nauwynck, H., *et al.*, "Don, doff, discard" to "don, doff, decontaminate"-FFR and mask integrity and inactivation of a SARS-CoV-2 surrogate and a norovirus following multiple vaporised hydrogen peroxide-, ultraviolet germicidal irradiation-, and dry heat decontaminations. *PLoS One*, 16, e0251872, 5 May, 2021.

82. Ludwig-Begall, L.F., Wielick, C., Dams, L., Nauwynck, H., Demeuldre, P.F., Napp, A., *et al.*, The use of germicidal ultraviolet light, vaporized hydrogen peroxide and dry heat to decontaminate face masks and filtering respirators contaminated with a SARS-CoV-2 surrogate virus. *J. Hosp. Infect.*, 106, 3, 577–584, 2020.
83. Hassard, F., Sharp, J.H., Taft, H., LeVay, L., Harris, J.P., McDonald, J.E., *et al.*, Critical Review on the Public Health Impact of Norovirus Contamination in Shellfish and the Environment: A UK Perspective. *Food Environ. Virol.*, 9, 123–141, 2017.
84. Campos, C.J.A., Goblick, G., Lee, R., Wittamore, K., Lees, D.N., Determining the zone of impact of norovirus contamination in shellfish production areas through microbiological monitoring and hydrographic analysis. *Water Res.*, 124, 556–565, 2017.
85. Ehuwa, O., Jaiswal, A.K., Jaiswal, S., Salmonella, food safety and food handling practices. *Foods*, 10, 5, 907, 2021.
86. Martinez-Albores, A., Lopez-Santamarina, A., Rodriguez, J.A., Ibarra, I.S., Del Carmen Mondrag6n, A., Miranda, J.M., *et al.*, Complementary methods to improve the depuration of Bivalves: A review. *Foods*, 9, 2, 129, 2020.
87. Almand, E.A., Moore, M.D., Jaykus, L.A., Norovirus binding to ligands beyond histo-blood group antigens. *Front. Microbiol.*, 8, 2549, 2017.
88. Begall, L.F.L., Mauroy, A., Thiry, E., Norovirus recombinants: Recurrent in the field, recalcitrant in the lab - a scoping review of recombination and recombinant types of noroviruses. *J. Gen. Virol.*, 99, 970–988, 2018.
89. Rha, B., Lopman, B.A., Alcala, A.N., Riddle, M.S., Porter, C.K., Incidence of norovirus-associated medical encounters among active duty United States military personnel and their dependents. *PLoS One*, 11, 4, e0148505, 2016.
90. Tsai, H., Yune, P., Rao, M., Norovirus disease among older adults. *Ther. Adv. Infect. Dis.*, 9, 20499361221136760, 2022.
91. Karst, S.M. and Tibbetts, S.A., Recent advances in understanding norovirus pathogenesis. *J. Med. Virol.*, 88, 1837–1843, 2016.
92. Marsh, M. and Helenius, A., Virus entry: Open sesame. *Cell*, 124, 729–740, 2006.
93. Sherman, M.B., Williams, A.N., Smith, H.Q., Pettitt, B.M., Wobus, C.E., Smith, T.J., Structural studies on the shapeshifting murine norovirus. *Viruses*, 13, 2162, 2021.
94. Shivanna, V., Kim, Y., Chang, K.O., Endosomal acidification and cathepsin L activity is required for calicivirus replication. *Virology*, 464–465, 1, 287–295, 2014.
95. Emmott, E., Sorgeloos, F., Caddy, S.L., Vashist, S., Sosnovtsev, S., Lloyd, R., *et al.*, Norovirus-mediated modification of the translational landscape via virus and host-induced cleavage of translation initiation factors. *Mol. Cell. Proteomics*, 16, 4, S215–S229, 2017.
96. Cates, J.E., Vinjé, J., Parashar, U., Hall, A.J., Recent advances in human norovirus research and implications for candidate vaccines. *Expert Rev. Vaccines*, 19, 539–548, 2020.

97. Baehner, F., Bogaerts, H., Goodwin, R., Vaccines against norovirus: state of the art trials in children and adults. *Clin. Microbiol. Infect.*, 22, S136–S139, 2016.
98. Lopman, B.A., Steele, D., Kirkwood, C.D., Parashar, U.D., The Vast and Varied Global Burden of Norovirus: Prospects for Prevention and Control. *PLoS Med.*, 13, e1001999, 2016.
99. Leroux-Roels, G., Cramer, J.P., Mendelman, P.M., Sherwood, J., Clemens, R., Aerssens, A., *et al.*, Safety and immunogenicity of different formulations of norovirus vaccine candidate in healthy adults: A randomized, controlled, double-blind clinical trial. *J. Infect. Dis.*, 217, 4, 597–607, 2018.
100. Atmar, R.L., Baehner, F., Cramer, J.P., Lloyd, E., Sherwood, J., Borkowski, A., *et al.*, Persistence of Antibodies to 2 Virus-Like Particle Norovirus Vaccine Candidate Formulations in Healthy Adults: 1-Year Follow-up with Memory Probe Vaccination. *J. Infect. Dis.*, 221, 2, 603–614, 2020.
101. Sherwood, J., Mendelman, P.M., Lloyd, E., Liu, M., Boslego, J., Borkowski, A., *et al.*, Efficacy of an intramuscular bivalent norovirus GI.1/GII.4 virus-like particle vaccine candidate in healthy US adults. *Vaccine*, 38, 41, 6442–6449, 2020.
102. Vaxart, Inc., Vaxart Announces First Subject Enrolled in Phase 1b Norovirus Dose-Ranging Trial in Elderly Adults [Internet]. [cited 2025 Jan 27]. Available from: https://investors.vaxart.com/news-releases/news-release-details/vaxart-announces-first-subject-enrolled-phase-1b-norovirus-dose.
103. Kim, L., Liebowitz, D., Lin, K., Kasparek, K., Pasetti, M.F., Garg, S.J., *et al.*, Safety and immunogenicity of an oral tablet norovirus vaccine, a phase I randomized, placebo-controlled trial. *JCI Insight*, 3, 13, e121077, 2018.
104. Hallowell, B.D., Parashar, U.D., Hall, A.J., Epidemiologic challenges in norovirus vaccine development. *Hum. Vaccin. Immunother.*, 15, 6, 1279–1283, 2019.
105. Treanor, J., Sherwood, J., Cramer, J.P., Le Cam Bouveret, N., Lin, S., Baehner, F., *et al.*, A phase 2 study of the bivalent VLP norovirus vaccine candidate in older adults; impact of MPL adjuvant or a second dose. *Vaccine*, 38, 36, 5842–5850, 2020.
106. Ali, E.S., Rajapaksha, H., Lundborg, M., Carr, J.M., Petrovsky, N., Norovirus drug candidates that inhibit viral capsid attachment to human histo-blood group antigens. *Antiviral Res.*, 133, 14–22, 2016.
107. Chen, J.L., Cheng, Z.C., Chen, J., Qian, L., Wang, H., Liu, Y.W., Advances in human norovirus research: Vaccines, genotype distribution and antiviral strategies. *Virus Res.*, 350, 199486, 2024.
108. Santos-Ferreira, N., Van Dycke, J., Neyts, J., Rocha-Pereira, J., Current and future antiviral strategies to tackle gastrointestinal viral infections. *Microorganisms*, 9, 1599, 2021.
109. Kudkyal, V.R., Matsuura, I., Hiramatsu, H., Hayashi, K., Kawahara, T., Phenol Derivatives Obtained from Grape Seed Extract Show Virucidal Activity against Murine Norovirus. *Molecules*, 27, 22, 7739, 2022.

110. Oh, C., Chowdhury, R., Samineni, L., Shisler, J.L., Kumar, M., Nguyen, T.H., Inactivation Mechanism and Efficacy of Grape Seed Extract for Human Norovirus Surrogate. *Appl. Environ. Microbiol.*, 88, 9, e02247-21, 2022.
111. Hayashi, T., Murakami, K., Hirano, J., Fujii, Y., Yamaoka, Y., Ohashi, H., *et al.*, Dasabuvir Inhibits Human Norovirus Infection in Human Intestinal Enteroids. *mSphere*, 6, 6, e00623-21, 2021.
112. Živković, I., Šavikin, K., Živković, J., Zdunić, G., Janković, T., Lazić, D., *et al.*, Antiviral Effects of Pomegranate Peel Extracts on Human Norovirus in Food Models and Simulated Gastrointestinal Fluids. *Plant Foods Hum. Nutr.*, 76, 2, 203–209, 2021.
113. Karst, S.M., Wobus, C.E., Goodfellow, I.G., Green, K.Y., Virgin, H.W., Advances in norovirus biology. *Cell Host Microbe*, 15, 668–680, 2014.
114. Sajewski, E.T., Vinjé, J., Glass, R.I., Lopman, B.A., Noroviruses, Sapoviruses, and Astroviruses, in: *Viral Infections of Humans*, 2023.
115. Omatola, C.A., Mshelbwala, P.P., Okolo, M.L.O., Onoja, A.B., Abraham, J.O., Adaji, D.M., *et al.*, Noroviruses: Evolutionary Dynamics, Epidemiology, Pathogenesis, and Vaccine Advances—A Comprehensive Review. *Vaccines (Basel) [Internet]*, 12, 6, 590, 2024 Jun 1 [cited 2025 Jan 27]. Available from: https://pmc.ncbi.nlm.nih.gov/articles/PMC11209302/.
116. Bányai, K., Estes, M.K., Martella, V., Parashar, U.D., Viral gastroenteritis. *Lancet*, 392, 175–186, 2018.
117. Malm, M., Uusi-Kerttula, H., Vesikari, T., Blazevic, V., High serum levels of norovirus genotype-specific blocking antibodies correlate with protection from infection in children. *J. Infect. Dis.*, 210, 11, 1755–1762, 2014.
118. Wang, X., Wang, S., Zhang, C., Zhou, Y., Xiong, P., Liu, Q., *et al.*, Development of a surrogate neutralization assay for norovirus vaccine evaluation at the cellular level. *Viruses*, 10, 1, 27, 2018.
119. Ludwig-Begall, L.F., Lu, J., Hosmillo, M., De Oliveira-Filho, E.F., Mathijs, E., Goodfellow, I., *et al.*, Replicative fitness recuperation of a recombinant murine norovirus - *In vitro* reciprocity of genetic shift and drift. *J. Gen. Virol.*, 101, 5, 510–522, 2020.
120. Leroux-Roels, I., Maes, C., Joye, J., Jacobs, B., Jarczowski, F., Diessner, A., *et al.*, A randomized, double-blind, placebo-controlled, dose-escalating phase I trial to evaluate safety and immunogenicity of a plant-produced, bivalent, recombinant norovirus-like particle vaccine. *Front. Immunol.*, 13, 1021500, 2022 Oct 7.
121. Study Details | Recombinant, Bivalent, Norovirus Vaccine Study | ClinicalTrials.gov [Internet]. [cited 2025 Jan 25]. Available from: https://clinicaltrials.gov/study/NCT05508178?cond=norovirus&rank=1.
122. Study Details | Safety and Immunogenicity of Norovirus GI.1/GII.4 Bivalent Virus-Like Particle Vaccine in an Elderly Population | ClinicalTrials.gov [Internet]. [cited 2025 Jan 25]. Available from: https://clinicaltrials.gov/study/NCT02661490?cond=norovirus&rank=2.

123. Study Details | Safety and Immunogenicity of Norovirus GI.1/GII.4 Bivalent Virus-Like Particle (VLP) Vaccine in Children | ClinicalTrials.gov [Internet]. [cited 2025 Jan 25]. Available from: https://clinicaltrials.gov/study/NCT02153112?cond=norovirus&rank=3.
124. Study Details | Phase 1 Norwalk Vaccine Study | ClinicalTrials.gov [Internet]. [cited 2025 Jan 25]. Available from: https://clinicaltrials.gov/study/NCT00806962?cond=norovirus&rank=4.
125. El-Kamary, S.S., Pasetti, M.F., Mendelman, P.M., Frey, S.E., Bernstein, D.I., Treanor, J.J., *et al.*, Adjuvanted intranasal norwalk virus-like particle vaccine elicits antibodies and antibody-secreting cells that express homing receptors for mucosal and peripheral lymphoid tissues. *J. Infect. Dis.*, 202, 11, 1649–58, 2010 Dec 1.
126. Study Details | Norovirus Challenge Study | ClinicalTrials.gov [Internet]. [cited 2025 Jan 25]. Available from: https://clinicaltrials.gov/study/NCT05212168?cond=norovirus&rank=5.
127. Study Details | Phase I Norovirus Challenge Model | ClinicalTrials.gov [Internet]. [cited 2025 Jan 25]. Available from: https://clinicaltrials.gov/study/NCT04174560?cond=norovirus&rank=6.
128. Study Details | Ph 1b: Safety & Immunogenicity of Ad5 Based Oral Norovirus Vaccine | ClinicalTrials.gov [Internet]. [cited 2025 Jan 25]. Available from: https://clinicaltrials.gov/study/NCT04854746?cond=norovirus&rank=8.
129. Study Details | Phase I Study to Determine the Optimal Human Challenge Dose for a Norovirus GII.4 Challenge Stock (CIN-1) | ClinicalTrials.gov [Internet]. [cited 2025 Jan 25]. Available from: https://clinicaltrials.gov/study/NCT02337842?cond=norovirus&rank=9.
130. Study Details | Dose Range Evaluation of Norovirus Challenge Pool (GII.4, CIN-1) | ClinicalTrials.gov [Internet]. [cited 2025 Jan 25]. Available from: https://clinicaltrials.gov/study/NCT01435811?cond=norovirus&rank=10.
131. Study Details | Immunogenicity & Safety Study of Adenovirus Type 5 (AD5) Based Oral Norovirus Vaccines | ClinicalTrials.gov [Internet]. [cited 2025 Jan 25]. Available from: https://clinicaltrials.gov/study/NCT04875676?cond=norovirus&page=2&rank=11.
132. Study Details | A Phase 1, Open-label, Safety and Immunogenicity Study of an Oral Multi-dose Administration Regimen With an Adenoviral-vector Based Tablet Norovirus Vaccine (VXA-G1.1-NN) Administered to Healthy Adult Volunteers | ClinicalTrials.gov [Internet]. [cited 2025 Jan 25]. Available from: https://clinicaltrials.gov/study/NCT05213728?cond=norovirus&page=2&rank=15.
133. Ettayebi, K., Crawford, S.E., Murakami, K., Broughman, J.R., Karandikar, U., Tenge, V.R., *et al.*, Replication of human noroviruses in stem cell-derived human enteroids. *Sci. (1979)*, 353, 6306, 1387–1393, 2016.
134. Ford-Siltz L, A. and Tohma K, I., Parra G. Understanding the relationship between norovirus diversity and immunity. *Gut Microbes*, 13, 1, 1900994, 2021.

135. Zhang, M., Fu, M., Hu, Q., Advances in human norovirus vaccine research. *Vaccines*, 9, 732, 2021.
136. Bonvanie, I.J., Weghorst, A.A.H., Holtman, G.A., Russchen, H.A., Fickweiler, F., Verkade, H.J., *et al.*, Oral ondansetron for paediatric gastroenteritis in primary care: A randomised controlled trial. *Br. J. Gen. Pract.*, 71, 711, 1–23, 2021.
137. Office NS, T., Case, C., ID, W., Roberts, L., Chartier, Y., Chartier, O., *et al.*, Membrane Filtration for Counting Bacteria *. *Eur. Neuropsychopharmacol.*, 5, 1–23, May, 2019.

9

Norovirus Excretion Patterns and their Role in Spread

Cynthia Lizzie Lobo[1], Soumya Patil[2], Raushan Kumar Chaudhary[3] and Pukar Khanal[4*]

[1]Department of Pharmaceutics, NGSM Institute of Pharmaceutical Sciences (NGSMIPS), Nitte (Deemed to be University), Mangalore, Karnataka, India
[2]Department of Pharmacy Practice, NGSM Institute of Pharmaceutical Sciences (NGSMIPS), Nitte (Deemed to be University), Mangalore, Karnataka, India
[3]Centre for Integrative Omics Data Science (CIODS), Yenepoya (Deemed to be University), Mangalore, Karnataka, India
[4]Department of Pharmacology, NGSM Institute of Pharmaceutical Sciences (NGSMIPS), Nitte (Deemed to be University), Mangalore, Karnataka, India

Abstract

Norovirus, a highly contagious pathogen, is responsible for causing acute gastroenteritis globally. The key route of norovirus transmission is the fecal–oral route and the virus spreads through person-to-person contact, contaminated water or food, and touching contaminated surfaces. On ingesting norovirus, gastrointestinal symptoms consist of abdominal cramps, vomiting, nausea, diarrhea, and fever. The norovirus is primarily excreted through feces and other routes such as vomitus and saliva. Having a deep understanding of the routes of norovirus excretion is essential to controlling norovirus transmission. Norovirus shedding begins soon after the onset of symptoms and continues even after the symptoms have diminished. This virus also spreads through asymptomatic carriers, who may unintentionally cause transmission as they will be involved in daily activities, while the individuals who are ill will refrain from attending school, work, or any other public gatherings. The symptomatic and asymptomatic individuals shed

*Corresponding author: pukarkhanal58@gmail.com; ORCID: https://orcid.org/0000-0002-8187-2120
Cynthia Lizzie Lobo: ORCID: https://orcid.org/0000-0003-0627-6019
Soumya Patil: ORCID: https://orcid.org/0000-0002-5804-3498
Raushan Kumar Chaudhary: ORCID: https://orcid.org/0000-0002-5862-3636

Sachin Kumar, Shriyansh Srivastava, Rishabha Malviya and Ranjit Sah (eds.) Emerging Infectious Disease: Unravelling Norovirus and Pandemic Dynamics, (249–274) © 2026 Scrivener Publishing LLC

comparable viral loads, where the duration of shedding depends on various factors related to host, viral strain and genotype, and environment. Norovirus excreted in stool, vomitus, and saliva has been quantified using different measurement techniques, with RT-PCR being the most predominant. Norovirus excretion plays a vital role in the transmission of norovirus and causes environmental contamination. Implementation of effective management strategies, including protocols for quarantine and hygiene practices, is necessary to prevent and manage outbreaks. Further discovery focusing on antiviral therapies and especially research on therapeutic interventions that could significantly reduce the viral load, which will ultimately control the spread of norovirus, should be encouraged.

***Keywords*:** Aerosolization, excretion, fecal shedding norovirus, spread, transmission, asymptomatic carriers, vomiting, diarrhea

9.1 Introduction

Norovirus, a single-stranded RNA virus, belongs to the family Caliciviridae and is globally known as the leading cause of acute gastroenteritis [1], which is accountable for many mortalities and mortalities [2]. This pathogen is highly contagious, with the fecal–oral route being the primary transmission route. Norovirus is reported to spread through direct person-to-person contact and contaminated water, food, and surfaces. Following a norovirus infection, there are quick-onset symptoms such as abdominal cramps, vomiting, and diarrhea, ultimately resulting in dehydration. The norovirus infection poses a public health concern as the World Health Organization (WHO) estimated that around 685 million cases of gastroenteritis are due to norovirus infection yearly, with 2,000,000 deaths worldwide [3]. Also, 50,000 child deaths have been observed per year and are mainly seen in developing countries, where there is poor sanitation and hygiene along with inaccessibility of clean water facilities. Another factor for the widespread transmission of norovirus is its ability to withstand its transmission potential in harsh conditions and environmental resistance. There are 10 genogroups of norovirus, GI to GX [4], where genogroup II (GII) is reported to be the most common in outbreaks. Norovirus outbreaks are primarily seen in closed environments such as schools, hospitals, and cruise ships, where the virus can quickly spread among people, as only 18 norovirus particles are sufficient to infect an individual [5, 6].

Understanding the various excretion patterns of norovirus is essential for preventing transmission. Norovirus is highly contagious; its shedding mainly occurs through feces, but it has also been reported to be excreted in vomitus and saliva [7]. Norovirus transmission occurs through the fecal–oral route, which involves direct person-to-person contact, aerosolized norovirus particles from the vomitus, and foodborne, waterborne, or environmental fomites [8]. The noroviral shedding peaks shortly after symptoms are seen, where the shedding duration varies among individuals and lasts for several days to weeks [9]. The most concerning matter here is that even the asymptomatic cases shed almost similar amounts of norovirus, where the virus could spread unknowingly [10]. The prolonged shedding of norovirus makes it difficult to control outbreaks due to extended transmission capability.

This chapter aims to deliver an overview of norovirus excretion patterns and their role in its spread. Studying norovirus excretion patterns is essential to understanding the mechanism of noroviral spreading into the population and the environment. By unraveling the norovirus spread patterns, public awareness could be given to prevent norovirus outbreaks.

9.2 Routes of Norovirus Excretion

Norovirus is excreted through feces or oral mucous (Figure 9.1), which contaminates water, food, and surfaces and is accountable for transmitting this virus. The norovirus enters the gastrointestinal tract through transmission *via* the fecal–oral route [11]. Norovirus infection is a multi-step process involving the attachment of viral particle, receptor involvement, endocytosis-based entry followed by uncoating, and, finally, the viral genome release [12]. It travels to the small intestine, infecting the intestinal cells lining the intestinal wall. Norovirus is believed to enter the host cells through special cells known as M cells, which are part of the gut-associated lymphoid tissue (GALT). Upon entering the host cells, the norovirus replicates rapidly and disrupts normal cellular functions, giving rise to gastrointestinal symptoms such as diarrhea, nausea, vomiting, abdominal cramps, and fever [13].

9.2.1 Fecal Shedding

Post-infection, there is an onset of norovirus shedding in feces, a key feature of norovirus pathogenesis and transmission [10]. Shedding in feces occurs because of norovirus replication in the intestinal epithelium and successive disruption of cellular integrity. Upon infection of the cells lining the gastrointestinal tract, the norovirus captures the cellular machinery to replicate, which will ultimately cause significant stress on the infected cells. The compromised cells from the epithelial lining are eliminated by the stress that

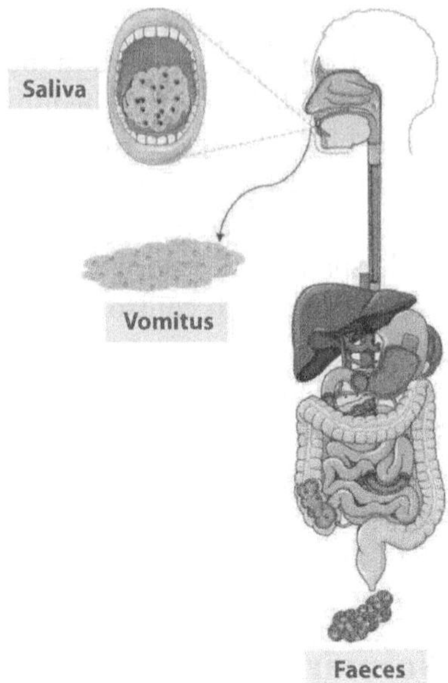

Figure 9.1 Routes of norovirus excretion.

triggers apoptosis, a programmed cell death. Norovirus infection also stimulates an inflammatory response in the host. To clear the virus, the immune system subsequently releases pro-inflammatory cytokines and chemokines, which leads to epithelial cell detachment from the intestinal lining. The detachment often occurs by damaging the tight junctions and adhesion molecules responsible for maintaining the epithelial barrier integrity. Thus, the norovirus epithelial cells are shed into the intestinal lumen, where the cells are fragmented, releasing the norovirus particles.

Subsequently, the feces of an infected individual contain the excreted norovirus, which will be released into the environment and enter into a new host. Since norovirus shedding continues even after the clinical symptoms have stopped, this concern highlights the high transmission ability and environmental robustness of norovirus. This extended shedding of norovirus has a critical role in the spread of norovirus, exclusively in localities with scarce sanitation and hygiene. The norovirus is released into the intestinal lumen due to the combined effect of immune responses that are virus- and host-mediated. Norovirus is shed in stool as viral clusters inside the vesicles, where they remain intact during fecal–oral transmission. Additionally, these clusters are reported to be more virulent than the free norovirus particles [14].

Besides, the norovirus particles released are infectious and can withstand the harsh conditions of the gastrointestinal tract and the environment [15]. In order to restrict the spread of norovirus, there is a need to understand the shedding of the virus while focusing on vigorous sanitation and hygiene practices.

9.2.2 Vomitus

One of the common symptoms of norovirus infection is vomiting, which plays a critical role in the spread of norovirus. During vomiting, the material that is expelled from the stomach through the mouth is known as vomitus, which consists of partially digested food and stomach acid. Additionally, it consists of mucus accumulation due to the activation of immune response by the norovirus infection in the gastrointestinal tract [16]. Remarkably, the norovirus particles are present in the vomitus of an infected individual, contributing to the spread of infection. The occurrence of norovirus in vomitus is a matter of great concern as it can result in aerosolized transmission. The forceful expulsion of the gastric contents during the vomiting event generates tiny droplets of norovirus, which are subsequently released into the air [17]. These small droplets carrying the norovirus particles can travel noteworthy distances from the vomiting site. The norovirus droplets can travel up to 3 m, which poses a risk of environmental contamination and infection. This transmission mode is of concern, particularly in closed environments, including schools, nursing homes, hospitals, and cruise ships, where there is crowding and poor ventilation [18]. The aerosolized transmission of norovirus highlights its importance in the spread of the virus, as a small amount of the virus is sufficient to cause an infection due to its very low infectious dose of around 18 viral particles [19]. Therefore, an individual could get infected even when exposed to contaminated surfaces, inhaling aerosolized norovirus particles, or even standing closer to the infected individual during a vomiting episode. Limiting the spread of norovirus through this route necessitates strict hygiene and infection control procedures. Effective disinfection of surfaces, hand hygiene, and isolation of infected people is necessary. In settings that could lead to outbreaks, mainly in the healthcare facilities, the caregivers and all healthcare workers should use personal protective equipment (PPE) further to diminish the risk of aerosolized transmission of norovirus [20]. It is, furthermore, understanding the excretion pattern of norovirus through vomitus which plays a critical role in outbreak management and prevention. Exploring effective methods to neutralize aerosolized norovirus particles will critically enhance efforts to control the spread of norovirus.

9.2.3 Saliva

Scientists at the National Institutes of Health have revealed that norovirus can grow in the salivary glands of mice and hence can spread through their saliva. This discovery indicates that norovirus can potentially spread through saliva and the clearly evidenced fecal–oral pathway. However, these findings must be confirmed in humans, in which we will explore new paths to understand norovirus circulation within populations [21]. A study on neonatal mouse pups infected with norovirus found that the virus can persist and replicate in the salivary glands, which suggests that saliva serves as a reservoir for the virus and will be present in all oral secretions, which again poses a risk for transmission. Therefore, talking, coughing, sneezing, and sharing food, drinks, and utensils will contribute to the spread of the virus. In a study focusing on neonatal pups, a novel transmission mode involved maternal interaction during nursing. The norovirus-infected pups, while suckling, were found to transmit the virus directly to the mother's mammary glands. This indicates that the saliva contained norovirus and served as a medium for transmission, and transmission occurred through backflow during suckling [22]. Maternal infections due to the presence of norovirus in the mammary glands can influence the lactation process and cause further spread of the virus. These findings suggest that it can sustain circulation in people as norovirus adapts to diverse transmission pathways [23]. Although the implications in humans are not well known, this finding indicates an added mode of norovirus spread that has been overlooked earlier. It also emphasizes the importance of salivary gland examination and related behaviors for potential virus reservoirs [24].

Additional studies are necessary to confirm these findings in humans and further explore the clinical and epidemiological significance of salivary gland investigation in norovirus infections. Such studies will deliver perception about the ability of the virus to persist and spread within the human population. This new approach could also lead to the development of targeted interventions, including vaccines and antiviral therapies, whose primary focus will be to disrupt these new transmission modes. These innovative studies unveil the complicated nature of norovirus malleability and transmission tactics, highlighting the importance of extended research into biology and epidemiology. Establishing these mechanisms in humans will pave the way for cutting-edge approaches to prevent and control norovirus outbreaks.

9.3 Duration of Noroviral Shedding

Norovirus shedding in feces begins shortly after infection, as early as 18 h after exposure to the virus. It was first detected by RT-PCR and lasted for a median of 28 days post-infection. Some cases showed norovirus shedding within 3 to 14 h before the onset of any clinical signs and symptoms [10]. After the onset of symptoms, virus shedding in feces peaked within 2 to 5 days. Studies indicate that the highest norovirus concentrations in feces are detected in experimental settings at around 4 days post-infection. However, the highest virus fecal concentration was detected after their symptoms disappeared. The median peak amount of virus shedding was 95×10^9 genomic copies/g feces, with an even higher amount of viral shedding of $\geq 100 \times 10^6$ genomic copies/g feces in some individuals until at least 14 days [25].

9.3.1 Factors Influencing Noroviral Shedding Patterns

9.3.1.1 Host Factors

The duration of norovirus shedding differs significantly among various age groups and is influenced by the immune system and individual health status. In infants under 6 months of age, infants have extended viral shedding duration in feces, mainly due to their immature immune system, which lacks the total capacity to elicit an immune response to clear the virus from the body. Therefore, norovirus replication and shedding continue longer, increasing the chances of transmission. In children over 1 year, the norovirus shedding duration is comparatively shorter than that of infants [26]. Developing a more robust immune system in this age group aids in a more effective viral clearance, reducing the shedding period. Nevertheless, variability in immune responses may still lead to sporadic differences in the shedding period among children.

However, prolonged viral shedding that lasts from several months to even years may be seen in immunocompromised individuals, where the immune system fails to eliminate the virus from the body [27, 28]. The complications in effectively controlling the norovirus infection are mainly due to the extended shedding, highlighting the crucial need for targeted interventions to prevent chronic shedding in vulnerable individuals.

9.3.1.2 *Viral Factors*

Depending on the norovirus strain and their distinct genotypes, the shedding properties of norovirus vary significantly, demonstrating differences in transmission potential. For instance, compared to other genotypes, the infections caused by GII.4 strain exhibit higher viral loads and more severe symptoms [29, 30]. Infections caused by GII.4 norovirus strain show more intense and extended viral replication, resulting in increased viral shedding in the feces with more severe gastrointestinal symptoms involving recurring vomiting and diarrhea, which ultimately enables effective transmission of norovirus. The GII.4 strain, having unique genetic and structural features, allows the ability to bind to host cells and circumvent immune responses, which are alleged to subsidize majorly to the outbreaks. The variances relating to strain and genotypes highlight the importance of genotype studies in managing norovirus outbreaks and the advanced studies on targeted prevention approaches focusing on vaccines and antiviral treatments [31].

9.3.1.3 *Environmental Factors*

The norovirus shedding characteristics and the environmental factors influence the severity of the infection. Co-infections, particularly with any other gastrointestinal pathogen, may cause prolonged shedding of norovirus and more severe symptoms [32]. In the cohabitation of norovirus with one or more pathogens, there is an intense inflammatory response and increased epithelial damage in the gastrointestinal tract due to the combined infection. This leads to more severe clinical consequences and higher transmission potential of norovirus as the synergistic effect can hamper norovirus clearance and extend the infectious period. On the other hand, a healthy gut microbiota diminishes the impact of norovirus infection. The gut microbiome, consisting of assorted communities of beneficial microorganisms, is responsible for maintaining homeostasis, regulating immune responses, and impeding pathogenic infections. Norovirus replication could be reduced with a balanced microbiota by competing for nutrients and attachment sites present on the gut epithelium while at the same time reducing the inflammation triggered by the viral infection. The composition and functionality of the gut microbiome mainly depend on an individual's dietary choices. The growth of beneficial microbes is boosted by consuming diets rich in fiber, vitamins, and probiotics, which are well known to sustain gut health [33]. Remarkably, probiotic supplements have been associated with improved immune function and reduced inflammation of the gastrointestinal tract, which could drop the severity of norovirus symptoms, shortening the viral shedding period [34]. These findings highlight the remarkable role of dietary habits in mitigating a norovirus infection. A nutritious diet will promote healthy gut microbiomes, ultimately improving the gastrointestinal tract's overall well-being and reducing the norovirus infection severity and duration. Future studies must investigate the gut microbiome's influence on norovirus dynamics and open new opportunities for novel therapeutic and preventive approaches.

9.3.2 Asymptomatic Excretion of Norovirus

Asymptomatic norovirus excretion is of great concern in public health as it can significantly impact the transmission dynamics of norovirus. Symptomatic individuals limit their interactions due to sickness, whereas the asymptomatic ones remain involved in daily activities [34], which leads to the unintentional hushed spread of norovirus within the public [27]. Controlling the norovirus outbreak is an extensive challenge due to this unnoticed transmission. Studies have reported that asymptomatic individuals shed similar norovirus levels compared to symptomatic individuals, with the peak viral load ranging from 105 to 109 genome copies/g feces and a shedding period lasting 8 to 60 days [35]. Prolonged shedding of norovirus increases the chances of norovirus persistence and, consequently, spreading in community and healthcare facilities. The epidemiological impact of asymptomatic carriers is developed by their ability to remain invisible in routine observation and outbreak investigations. In the absence of typical clinical signs, including vomiting or diarrhea, the asymptomatic individuals are less likely to be recognized as the sources of norovirus infection. Subsequently, they can continue interacting with others by sharing food and water and contaminating surfaces, creating numerous possibilities for the virus to spread [36]. Several studies have underscored the role of asymptomatic individuals in spreading norovirus, mainly in daycare centers, nursing homes, and food service spots, where close interaction occurs and resources are often shared. Furthermore, the risk posed by the asymptomatic individuals is intensified due to the high transmissibility of the norovirus along with the low infectious dose. Even a few of the 18 viral particles are sufficient to cause a norovirus infection, which suggests that even a slight contamination by asymptomatic carriers can initiate the spread and lead to an outbreak [19]. This shows the importance of considering asymptomatic carriers in public health approaches.

Even without clinical symptoms, enhancing diagnostic proficiencies to identify the carriers can control the spread of norovirus from the asymptomatic norovirus. Asymptomatic individuals could be identified to prevent their unintentional role in outbreaks by regular screening in high-risk environments such as healthcare facilities and food handling industries [37]. Implementing rigorous hygienic practices, including systematic washing of hands and surface disinfection, is essential to limit transmission from the asymptomatic carriers. Educating the public about health and good hygiene practices will encourage awareness and preventive actions within societies where asymptomatic transmission is likely to occur [38]. Furthermore, a detailed study of the mechanisms involved in asymptomatic infections, such as discrepancies in gut microbiota composition and immune responses, will lead to the development of targeted interventions involving vaccines and antiviral therapies. Asymptomatic excretion of norovirus signifies a silent yet effective driver of its transmission [19]. Addressing this aspect is critical for effective outbreak management and long-term control of norovirus within populations.

9.4 Quantitative Analysis of Noroviral Load

Quantitative analysis of noroviral load excretion refers to measuring the norovirus in feces, vomitus, or saliva samples [39]. Feces are the first choice of sample for the detection of norovirus. However, it is often challenging to collect feces, and the healthcare staff may miss the opportunity to collect the sample in a clinical setting. There are some alternatives to fecal samples, such as rectal swabs [40], vomit samples [41], and mouthwash samples collected after vomiting [42]. The amount of norovirus present is expressed as copies of viral RNA per gram of sample collected. A quantitative analysis of viral load helps to understand the viral shedding pattern, assess the outbreaks' severity, and instigate necessary control measures. The various methods used for the quantification of norovirus are presented in Table 9.1.

9.4.1 Measurement Techniques

9.4.1.1 Reverse Transcriptase–Polymerase Chain Reaction (RT-PCR)

The RT-PCR technique is the gold standard for detecting norovirus in samples such as stool and vomitus. This technique amplifies and quantifies noroviral RNA and accurately measures viral loads. An added advantage of RT-PCR is that it can identify different norovirus genotypes [43, 44]. Firstly, the viral nucleic acid is purified from the experimental sample, followed by transcription of RNA to cDNA using reverse transcriptase and norovirus-specific oligonucleotide primers. Finally, the cDNA is amplified with norovirus-specific oligonucleotide primers by the DNA polymerase [45].

9.4.1.2 *Quantitative PCR (qPCR)*

qPCR, a type of RT-PCR, offers quantitative data on viral loads. Assessment of the duration of viral shedding in patients that lasts from days to weeks, even after the symptoms have resolved, could be assessed by this method [46]. Norovirus was quantified in the hands of mothers with children under 5 years of age in Bagamoyo, Tanzania, using the qPCR technique. The mother's hands were tested as they play a vital role in the spread of norovirus, and the results proved that the mother's hands were a source of the norovirus infection of young children in Tanzania [47].

9.4.1.3 *Immune Electron Microscopy (IEM)*

Electron microscopy and antibodies specific to norovirus have been used in the IEM technique to identify norovirus particles in stool samples [48]. This technique is less favorable than the RT-PCR technique as it has low sensitivity and necessitates a higher concentration of norovirus particles with at least 10^6 particles/mL [49]. Moreover, the IEM technique is employed for research purposes rather than routine diagnostic cases.

9.4.1.4 *Antigen Detection Immunoassays*

These tests are rapid and detect norovirus antigens present in stool samples, and they utilize antibodies for detection. The presence of norovirus in saliva was detected by using the salivary antibody immunoassay, where the individual who tested positive had a fourfold increase in anti-Norwalk IgG [50]. A multiplex salivary immunoassay was employed to study norovirus infections among visitors to a Lake Michigan beach. The saliva was added to Luminex microspheres that were united to recombinant antigens of genogroup I (GI) and II (GII) noroviruses and incubated. The presence of norovirus was indicated by immunoconversion, defined as at least a fourfold increase in anti-norovirus IgG antibody response [51]. Although they provide quick results, their sensitivity is often below 50%, making them less reliable for diagnosing erratic cases [52]. However, they are of great use in outbreak situations where rapid screening is essential.

Table 9.1 Methods for quantification of norovirus.

Sl. no.	Technique	Sample collected	Mean viral load	References
1.	Real-time RT-qPCR	Saliva	3.16 ± 1.08 log10 genome copies/mL	[53]
2.	PMA-qPCR	Air samples from healthcare facilities	1.35 × 101 to 2.35 × 103 genomes/m3	[54]
3.	Quantitative RT-PCR	Hand rinse samples	1.8 ± 3.2 log10 genome copies/two hands	[47]
4.	Immune electron microscopy	Stool	106 viral particles/mL of stool	[49]
5.	Enzyme-linked immunosorbent assay (ELISA)	Stool	104 to 106 viral particles/mL of stool	[55]
6.	Nake-eye biosensor	Stool	2.7 × 103 to 2.7 × 105 genome copies/mL of stool	[56]
7.	Immunochromatography (ICG)	Stool	107 viral particles/mL of stool	[57]

9.4.1.5 Biosensors

Emergent technologies such as optical biosensors and surface-enhanced Raman scattering (SERS) are being explored for norovirus detection [58]. These methods can distinguish the norovirus strains based on their spectral signatures but are still under development and, hence, not widely used in clinical studies.

9.4.2 Noroviral Load Decay Rates

Understanding the viral decay rates is vital for managing norovirus transmission and infection control. Although norovirus shedding can persist for an extended period, studies have shown a steady decline in viral load over time. On day 5 post-infection, a slight decrease in viral load is observed compared to day 0, demonstrating the commencement of viral decay. Despite this decline, norovirus remains measurable in feces for up to 28 days, indicative of its capacity for prolonged shedding [26].

The prolonged existence of the virus in feces, even at diminished levels, poses a substantial amount of risk for sustained environmental contamination and secondary transmission. This perseverance highlights the challenge of eradicating norovirus from affected settings, predominantly in outbreaks. The capability to detect the virus weeks after early infection highlights the importance of constant hygiene practices, consistent environmental disinfection, and observation to alleviate the spread of norovirus in spite of the usual decay in viral load [11].

9.5 Transmission Modes of Norovirus

Norovirus is primarily transmitted through the fecal–oral route, which has several specific modes of transmission (Figure 9.2). Understanding these modes of transmission is very crucial to prevent and control outbreaks.

9.5.1 Direct Person-to-Person Contact

Direct person-to-person contact is the most common and significant norovirus transmission mode, accounting for many outbreaks. This transmission route occurs when an infected individual comes into close contact with a vulnerable individual, enabling the spread of the virus. Everyday activities such as shaking hands, providing care to an infected person, or sharing personal items (e.g., towels and utensils) can transfer viral particles from one person to another. The virus is classically spread through contaminated hands that have come into contact with fecal material or vomitus, which may then be transferred to the uninfected individual's mucous membranes (mouth, eyes, etc.). This transmission mode is chiefly prevalent in environments with frequent close interactions, such as households, healthcare facilities, daycare centers, and schools [59]. Preventing direct

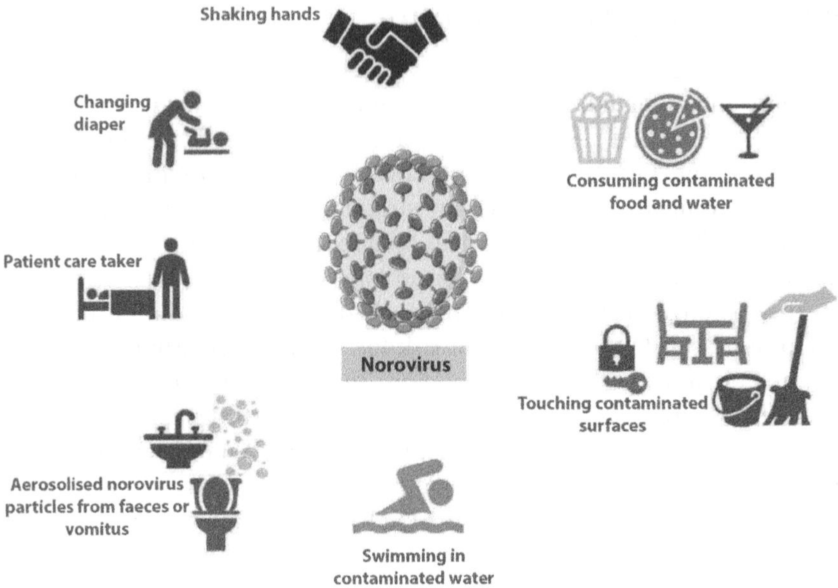

Figure 9.2 Transmission modes of norovirus.

person-to-person transmission requires strict hygiene measures, such as frequent handwashing, disinfecting surfaces, and isolating infected individuals to reduce the risk of outbreak spread.

9.5.2 Aerosolized Norovirus Particles

Aerosolization of norovirus refers to a process in which the virus particles get suspended in a gaseous medium such as air [60]. The disease-causing microbial aerosols in the environment have particle sizes ranging from 0.1 to 20 μm [61]. Aerosol particle size influences the residence time in the air and the likelihood of being inhaled by humans [62]. Norovirus aerosol particles are obtained by diarrhea or vomiting of norovirus-infected individuals. The vomitus of a norovirus-infected person contains norovirus. Due to the expulsion force, the norovirus in the vomitus gets aerosolized during the vomiting episodes, which remains suspended in the air for varying periods and may get deposited on any surface or object after a while [19]. The aerosolized particles smaller than 1 μm in diameter remain suspended in the air for extended periods, which may be inhaled into the upper respiratory tract. The virus could be swallowed upon reaching the pharynx, allowing the virus to reach the GIT and subsequently infect the intestinal cells [19, 54]. Additionally, the site of vomit contaminates the ground surface and other objects around, and eventually, the airflow around these contaminated surfaces and objects disperses the norovirus into the air [63].

Diarrhea is the most prominent symptom of norovirus infection and contains many norovirus particles excreted through feces. Norovirus aerosolization is likely to occur during stressful defecation and when the individual flushes the toilet after defecation [10, 64].

9.5.3 Contaminated Food or Water

Norovirus transmission through contaminated food or water is an important route, often contributive to prevalent outbreaks. The chief mechanism for foodborne transmission is when infected food handlers, who shed the virus in their feces or vomitus, fail to exercise proper hygiene, such as washing hands meticulously after using the washroom. Infected individuals may contaminate food directly through contact or indirectly by handling utensils and surfaces that come into contact with the food [65]. Common foods linked with outbreaks include shellfish, fruits, and vegetables, especially when raw or not properly cooked, as these are mostly inclined to contamination during harvesting or preparation [59, 66]. On the other side, water can also serve as a medium for norovirus transmission, exclusively in cases where it is contaminated with fecal matter or inappropriately treated sewage [67]. Consuming contaminated water or using it for washing food, preparing beverages, or even in recreational surroundings such as swimming pools upsurges the risk of infection. The virus is extremely robust, enduring in water and on surfaces for long periods, making it a significant public health concern. Averting norovirus transmission *via* food and water necessitates stringent food safety practices as well as proper hand hygiene among food handlers and suitable water treatment to prevent contamination [68].

9.5.4 Contaminated Surfaces

Contaminated surfaces are an important way of norovirus transmission. Infected people shed large amounts of viruses in their feces and vomitus, which later contaminate various surfaces such as doorknobs, faucets, countertops, and utensils. The virus can be transmitted when another person touches these contaminated surfaces and afterward touches their mouth, nose, or food [69]. This system of indirect transmission is particularly challenging in high-touch areas, where numerous individuals may come into contact with the same surfaces. This perseverance upsurges the risk of transmission, exclusively in crowded, shared places such as hospitals, schools, nursing homes, and cruise ships, where frequent interactions and communal settings enable the spread of the virus [70, 71]. Effective infection control actions, such as frequent cleaning and disinfection of high-touch surfaces and encouraging good hand hygiene, are crucial to diminish the transmission risk in these atmospheres.

9.6 Role of Excretion Patterns in Norovirus Spread

9.6.1 Contribution to Disease Transmission

The excreted norovirus plays a significant role in disease transmission, whereas asymptomatic individuals play a critical role in transmission. A glimpse of norovirus transmission cases is given in Table 9.2. Among norovirus-infected individuals, up to 30% of infections will be asymptomatic infections, which shed the virus at lower levels than symptomatic infections [17, 72]. This silent spread of the norovirus makes it difficult for outbreak management as these asymptomatic individuals may unknowingly contribute to transmission, particularly in community settings. Also, the prolonged noroviral shedding post-infection, even after the resolution of symptoms, further aggravates this issue. The super-shedders can significantly worsen outbreak management as they shed very high amounts of norovirus. Vomiting causes aerosolization of norovirus particles, where the tiny droplets containing the virus can become airborne, which could be inhaled by others or settle on surfaces. Here the norovirus spreads even in the absence of an infected individual. Additionally, the infection dose is significantly low, with only 18 noroviral particles, which indicates the ease at which norovirus can spread [73].

9.6.2 Environmental Contamination

Environmental contamination significantly influences norovirus transmission, as the excreted norovirus present in feces and vomitus could lead to water, food, and surface contamination. Norovirus can survive in water for prolonged periods, where it can remain infectious even in chlorinated water for 30 to 60 min, which shows that there is a chance for transmission at water parks and swimming pools, where the water can be contaminated with the fecal material of an infected individual [44]. The food could be contaminated during food preparation by an infected food handler. Also, food such as oysters can accumulate norovirus from contaminated water, which is also the cause of noroviral transmission through food. Norovirus can remain on surfaces for up to 2 weeks. The environmental samples

Table 9.2 Norovirus transmission cases.

Sl. no.	Setting	Infection source	Mode of transmission	Outcome	References
1.	Cruise ship	Contaminated food or surfaces	Person-to-person contact or touching contaminated surfaces	All of the passengers and crew members reported symptoms of norovirus infection.	[74, 75]
2.	Nursing home	Contaminated food (Turkey dinner)	Person-to-person contact	The residents and staff of the nursing home were affected, and their household contacts were infected.	[76]
3.	Hotel restaurant	Diner	Person-to-person contact	Norovirus intake was higher in diners in the same table.	[77]
4.	Medical college	Gargle and stool suspension of the patient	Airborne, fecal–oral	Of the 32 volunteers who inhaled gargle aerosol, 17 developed symptoms.	[78]
5.	Kindergarten	The first case vomited in the classroom	Aerosol-vomitus	Children closely exposed to the vomit developed symptoms more than those who were exposed at a distance	[79]
6.	School	The first case vomited at the playground	Aerosol-vomitus	Vomitus in the playground was not disinfected on time, leading to aerosolization of norovirus.	[80]
7.	NA	Raw oysters	Foodborne	Over 400 norovirus cases were reported due to consumption of raw oysters.	[81, 82]
8.	NA	Infant	Direct contact during breastfeeding	The norovirus present in the saliva of the infected infant has transmitted norovirus to the mother.	[22]

collected during outbreaks showed significant viral loads on flush buttons, door handles, and kitchen counters [83].

Furthermore, the resistance of norovirus to common disinfectants makes it challenging to control the spread of infections. Effective disinfection requires significantly higher concentrations of sodium hypochlorite [84]. This resistance makes the disinfection process rigorous, which is not always feasible in busy environments. The cleanup process of feces and vomitus poses a risk for transmission as there are chances for noroviral particle aerosolization. Cleanup should be performed with minimum agitation to avoid the suspension of noroviral particles in the air [85].

9.7 Public Health Implications of Excretion Patterns

The highly infectious nature and the norovirus excretion patterns related to infected individuals possess substantial public health challenges. Effective management strategies, including protocols for quarantine, hygiene practices, and policy recommendations, are necessary to control the spread of norovirus. The individuals who have contracted the virus should quarantine for a sufficient time as viral shedding can continue even after the symptoms have resolved. Infected persons should refrain from attending work, school, or public gatherings. In outbreak situations, separate designated areas should be established for infected individuals to minimize contact with healthy individuals [86].

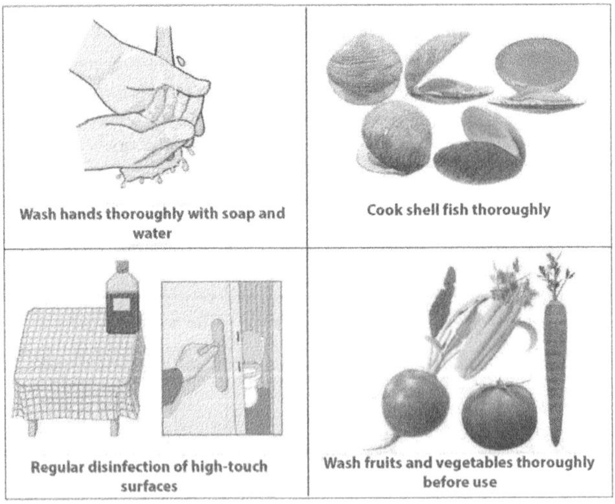

Figure 9.3 Preventive measures toward norovirus infections.

Hygiene and sanitation are critical in preventing the transmission of norovirus. Washing hands frequently with soap and water is recommended, especially after using the washroom and before handling and consuming food. The surfaces should be regularly cleaned and disinfected with effective disinfectants such as sodium hypochlorite, reducing the surface's viral load [59]. Also, the fruits and vegetables should be washed meticulously, and consuming raw shellfish from contamination should be avoided [87]. The preventive measures to be practiced in order to protect against norovirus infection are given in Figure 9.3.

Several proactive policy recommendations could be implemented for norovirus outbreak prevention and control. Public education campaigns can be launched to create awareness of the transmission routes of norovirus, with a focus on hygiene practices and safe food handling. Also, strict regulations could be implemented for harvesting and processing shellfish. Water quality, sanitation, and hygiene infrastructure in rural areas should be improved, and access to clean water should also be provided.

9.8 Research Gaps and Future Directions

Although substantial research has been done on norovirus excretion patterns and its transmission, some research gaps and challenges hinder our understanding of norovirus transmission and its consequences on public health. Addressing these gaps is essential for developing effective interventions and controlling outbreaks. The primary challenge in understanding norovirus excretion pattern is the exertion in sampling and monitoring asymptomatic individuals, who are the key drivers of the outbreak and confounds the identification of potential outbreak sources. Current surveillance systems typically rely only on symptomatic cases for sampling, which fails to capture the wider range of infection. To overcome this, advanced methodologies are required to identify and monitor asymptomatic individuals. Sampling could be done in high-risk environments and using non-invasive sampling techniques, such as environmental sampling or self-collected stool samples, could be implemented in high-risk environments like hospitals and schools, which allows for easy monitoring and identification of potential outbreak sources.

Longitudinal studies are also vital for understanding chronic norovirus shedding, which may persevere in specific individuals, particularly in immunocompromised individuals. These studies would help illuminate the long-term dynamics of viral excretion, providing perceptions into the influence of host factors, including immune response and gut microbiota composition on the perseverance and clearance of the virus. Future research should explore the relationship between gut microbiota and norovirus infection. Understanding the effects of microbiome on viral replication and immune responses could open new avenues for intervention, such as manipulating the microbiota through diet, probiotics, or other therapies to decrease norovirus shedding and transmission. Satisfying these research gaps will significantly improve our understanding of norovirus dynamics and deliver valuable information for controlling outbreaks, eventually reducing the public health burden of this extremely infectious virus.

Abbreviations

RT-PCR	reverse transcriptase–polymerase chain reaction
RNA	ribonucleic acid
GII	genotype II
qPCR	quantitative polymerase chain reaction
GALT	gut-associated lymphoid tissue
IEM	immune electron microscopy
PMA-qPCR	propidium monoazide–quantitative polymerase chain reaction
ICG	immunochromatography
ELISA	enzyme-linked immunosorbent assay

AI disclosure statement: The author acknowledges the use of AI-based tools for minor language editing or grammatical correction. However, all intellectual contributions, analyses, and conclusions presented in this chapter are entirely the author's own.

References

1. Chen, R., Mukhopadhyay, S., Merits, A., Bolling, B., Nasar, F., Coffey, L.L., *et al.*, ICTV Virus Taxonomy Profile. *J. Gen. Virol.*, 99, 6, 761–2, 2018.
2. Mirza, S. and Hall, A., Norovirus | CDC Yellow Book 2024 [Internet]. [cited 2024 Nov 30]. Available from: https://wwwnc.cdc.gov/travel/yellowbook/2024/infections-diseases/norovirus.
3. Centers for Disease Control and Prevention. Norovirus Facts and Stats [Internet]. Cdc. 2024 [cited 2024 Dec 2]. Available from: https://www.cdc.gov/norovirus/data-research/.

4. Chhabra, P., de Graaf, M., Parra, G.I., Chan, M.C.W., Green, K., Martella, V., *et al.*, Updated classification of norovirus genogroups and genotypes. *J. Gen. Virol.*, 100, 10, 1393–406, 2019.
5. Robilotti, E., Deresinski, S., Pinsky, B.A., Norovirus. *Clin. Microbiol. Rev.*, 28, 1, 134–64, 2015.
6. Cui, C., Pan, L., Wang, Y., Xue, C., Zhu, W., Zhu, L., *et al.*, An outbreak of acute GII.17 norovirus gastroenteritis in a long-term care facility in China: The role of nursing assistants. *J. Infect. Public Health [Internet]*, 10, 6, 725–9, 2017. Available from: https://www.sciencedirect.com/science/article/pii/S1876034117300424.
7. Omatola, C.A., Mshelbwala, P.P., Okolo, M.L.O., Onoja, A.B., Abraham, J.O., Adaji, D.M., *et al.*, Noroviruses: Evolutionary Dynamics, Epidemiology, Pathogenesis, and Vaccine Advances—A Comprehensive Review. *Vaccines*, 12, 6, 1–26, 2024.
8. Bányai, K., Estes, M.K., Martella, V., Parashar, U.D., Viral gastroenteritis. *Lancet.*, 392, 10142, 175–86, 2018 Jul.
9. Baric, R., Carolina, N., Kang, C., College, C.M., Lee, B., Global Burden of Norovirus and Prospects for Vaccine Development Table of Contents, 2015.
10. Atmar, R.L., Opekun, A.R., Gilger, M.A., Estes, M.K., Crawford, S.E., Neill, F.H., *et al.*, Norwalk virus shedding after experimental human infection. *Emerg. Infect. Dis*, 14, 10, 1553–7, 2008.
11. Unidos, E. and Unidos, E., Norovirus. Wikipedia [Internet], 1–6, 1968 [cited 2024 Nov 24. Available from: https://en.wikipedia.org/wiki/Norovirus.
12. Marsh, M. and Helenius, A., Virus entry: open sesame. *Cell*, 124, 4, 729–40, 2006.
13. Norovirus - StatPearls - NCBI Bookshelf [Internet]. [cited 2024 Nov 24]. Available from: https://www.ncbi.nlm.nih.gov/books/NBK513265/.
14. Santiana, M., Ghosh, S., Ho, B.A., Rajasekaran, V., Du, W.L., Mutsafi, Y., *et al.*, Vesicle-Cloaked Virus Clusters Are Optimal Units for Inter-organismal Viral Transmission. *Cell Host Microbe*, 24, 2, 208–220.e8, 2018.
15. Green, K.Y., Kaufman, S.S., Nagata, B.M., Chaimongkol, N., Kim, D.Y., Levenson, E.A., *et al.*, Human norovirus targets enteroendocrine epithelial cells in the small intestine. *Nat. Commun.*, 11, 1, 2759, 2020.
16. Kirby, A.E., Streby, A., Moe, C.L., Vomiting as a symptom and transmission risk in norovirus illness: Evidence from human challenge studies. *PLoS One*, 11, 4, e0143759, 2016.
17. Division of Viral Diseases, National Center for Immunization and Respiratory Diseases C for DC and P. Updated norovirus outbreak management and disease prevention guidelines. [Internet]. *MMWR. Recommendations Reports Morbidity Mortality Weekly Rep. Recommendations Rep.*, 60, 1–18, 2011 [cited 2024 Nov 25. Available from: http://www.ncbi.nlm.nih.gov/pubmed/21368741.

18. Makison Booth, C. and Frost, G., Potential distribution of viable norovirus after simulated vomiting. *J. Hosp. Infect.*, 102, 3, 304–10, 2019 Jul.
19. Tan, M., Tian, Y., Zhang, D., Wang, Q., Gao, Z., Aerosol Transmission of Norovirus. Viruses. 16, 1, 2024 Jan.
20. Overbey, K.N., Hamra, G.B., Nachman, K.E., Rock, C., Schwab, K.J., Quantitative microbial risk assessment of human norovirus infection in environmental service workers due to healthcare-associated fomites. *J. Hosp. Infect.*, 117, 52–64, 2021.
21. NIH scientists discover norovirus and other "stomach viruses" can spread through saliva, 2022.
22. Ghosh, S., Kumar, M., Santiana, M., Mishra, A., Zhang, M., Labayo, H., *et al.*, Enteric viruses replicate in salivary glands and infect through saliva. *Nat.*, 607, 7918, 345–50, 2022.
23. De Graaf, M., van Beek, J., Koopmans, M.P., Human norovirus transmission and evolution in a changing world. *Nat. Rev. Microbiol.*, 14, 7, 421–433, 2016.
24. Pisanic, N., Ballard, S.B., Colquechagua, F.D., François, R., Exum, N., Yori, P.P., *et al.*, Minimally invasive saliva testing to monitor norovirus infection in community settings. *J. Infect. Dis.*, 219, 8, 1234–1242, 2019.
25. Teunis, P.F., Sukhrie, F.H., Vennema, H., Bogerman, J., Beersma, M.F., Koopmans, M.P., Shedding of norovirus in symptomatic and asymptomatic infections. *Epidemiol. Infect.*, 143, 8, 1710–7, 2015 Jun.
26. Qiu, Y., Freedman, S.B., Williamson-Urquhart, S., Farion, K.J., Gouin, S., Poonai, N., *et al.*, Significantly Longer Shedding of Norovirus Compared to Rotavirus and Adenovirus in Children with Acute Gastroenteritis. *Viruses*, 15, 7, 1–10, 2023.
27. Milbrath, M.O., Spicknall, I.H., Zelner, J.L., Moe, C.L., Eisenberg, J.N.S., Heterogeneity in norovirus shedding duration affects community risk. *Epidemiol. Infect.*, 141, 8, 1572–84, 2013 Aug.
28. Davis, A., Cortez, V., Grodzki, M., Dallas, R., Ferrolino, J., Freiden, P., *et al.*, Infectious norovirus is chronically shed by immunocompromised pediatric hosts. *Viruses*, 12, 6, 619, 2020 Jun.
29. Duan, L., Yang, X., Xie, J., Zhan, W., Zhang, C., Liu, H., *et al.*, Prevalence of GII.4 Sydney Norovirus Strains and Associated Factors of Acute Gastroenteritis in Children: 2019/2020 Season in Guangzhou, China. *Food Environ. Virol. [Internet]*, 13, 3, 357–67, 2021. Available from: https://doi.org/10.1007/s12560-021-09482-0.
30. Cornejo-Sánchez, T., Soldevila, N., Coronas, L., Alsedà, M., Godoy, P., Razquín, E., *et al.*, Epidemiology of GII.4 and GII.2 norovirus outbreaks in closed and semi-closed institutions in 2017 and 2018. *Sci. Rep*, 13, 1, 1–8, 2023.
31. Huhti, L., Significance of the Norovirus GII.4 Genotype as a Cause of Acute Gastroenteritis in Finnish Children, and the Production of a Candidate Virus-like Particle Vaccine. 2015

32. Woodward, J., Gkrania-Klotsas, E., Kumararatne, D., Chronic norovirus infection and common variable immunodeficiency. *Clin. Exp. Immunol.*, 188, 3, 363–70, 2017 Jun.
33. Hemarajata, P. and Versalovic, J., Effects of probiotics on gut microbiota: Mechanisms of intestinal immunomodulation and neuromodulation. *Therap. Adv. Gastroenterol.*, 6, 1, 39–51, 2013.
34. Marshall, J.A. and Bruggink, L.D., The dynamics of norovirus outbreak epidemics: recent insights. *Int. J. Environ. Res. Public Health*, 8, 4, 1141–9, 2011 Apr.
35. Teunis, P.F.M., Sukhrie, F.H.A., Vennema, H., Bogerman, J., Beersma, M.F.C., Koopmans, M.P.G., Shedding of norovirus in symptomatic and asymptomatic infections. *Epidemiol. Infect.*, 143, 8, 1710–7, 2015.
36. Phattanawiboon, B., Nonthabenjawan, N., Boonyos, P., Jetsukontorn, C., Towayunanta, W., Chuntrakool, K., Norovirus transmission mediated by asymptomatic family members in households. *PLoS One*, 15, 7, e0236502, 2020.
37. Watson, S. and Gong, YY. Risk behaviours and practices of food handlers in norovirus transmission. *Br. Food J.*, 120, 11, 2510–2523, 2018.
38. Cates, S.C., Kosa, K.M., Brophy, J.E., Hall, A.J., Fraser, A., Consumer education needed on norovirus prevention and control: findings from a nationally representative survey of US adults. *J. Food Prot.*, 78, 3, 484–490, 2015.
39. Liu, L. and Moore, M.D., A survey of analytical techniques for noroviruses. *Foods*, 9, 3, 318, 2020.
40. Nakanishi, K., Tsugawa, T., Honma, S., Nakata, S., Tatsumi, M., Yoto, Y., *et al.*, Detection of enteric viruses in rectal swabs from children with acute gastroenteritis attending the pediatric outpatient clinics in Sapporo, Japan. *J. Clin. Virol.*, 46, 1, 94–7, 2009.
41. Kirby, A., Ashton, L., Hart, I.J., Detection of norovirus infection in the hospital setting using vomit samples. *J. Clin. Virol.*, 51, 1, 86–7, 2011.
42. Kirby, A., Dove, W., Ashton, L., Hopkins, M., Cunliffe, N.A., Detection of norovirus in mouthwash samples from patients with acute gastroenteritis. *J. Clin. Virol.*, 48, 4, 285–7, 2010.
43. Centers for Disease Control and Prevention. *Norovirus Specimen Collect. [Internet]*, 2021 [cited 2024 Nov 26. Available from: https://www.cdc.gov/norovirus/lab/specimen-collection.html.
44. Umeda, Y., Norovirus infection. *Pract. Otorhinolaryngol (Basel)*, 100, 6, 497–504, 2007.
45. Atmar, R.L., Neill, F.H., Le Guyader, F.S., Detection of Human Caliciviruses in Fecal Samples by RT-PCR. *Methods Mol. Biol.*, 665, 39–50, 2011.
46. Frange, P., Touzot, F., Debré, M., Héritier, S., Leruez-Ville, M., Cros, G., *et al.*, Prevalence and clinical impact of norovirus fecal shedding in children with inherited immune deficiencies. *J. Infect. Dis. [Internet]*, 206, 8, 1269–74, 2012 Oct 15. Available from: https://doi.org/10.1093/infdis/jis498.

47. Mattioli, M.C.M., Davis, J., Mrisho, M., Boehm, A.B., Quantification of human norovirus GII on hands of mothers with children under the age of five years in Bagamoyo, Tanzania. *Am. J. Trop. Med. Hyg.*, 93, 3, 478–84, 2015 Sep.
48. Kapikian, A.Z., Wyatt, R.G., Dolin, R., Thornhill, T.S., Kalica, A.R., Chanock, R.M., Visualization by Immune Electron Microscopy of a 27-nm Particle Associated with Acute Infectious Nonbacterial Gastroenteritis. *J. Virol.*, 10, 5, 1075–81, 1972 Nov.
49. Glass, R.I., Noel, J., Ando, T., Fankhauser, R., Belliot, G., Mounts, A., *et al.*, The epidemiology of enteric caliciviruses from humans: A reassessment using new diagnostics. *J. Infect. Dis.*, 181, Suppl. 2, S254–61, 2000 May.
50. Griffin, S.M., Converse, R.R., Leon, J.S., Wade, T.J., Jiang, X., Moe, C.L., *et al.*, Application of salivary antibody immunoassays for the detection of incident infections with Norwalk virus in a group of volunteers. *J. Immunol. Methods [Internet]*, 424, 53–63, 2015. Available from: https://www.sciencedirect.com/science/article/pii/S0022175915001581.
51. Wade, T.J., Griffin, S.M., Egorov, A.I., Sams, E., Hudgens, E., Augustine, S., *et al.*, Application of a multiplex salivary immunoassay to detect sporadic incident norovirus infections. *Sci. Rep*, 9, 1, 3–9, 2019.
52. Rabenau, H.F., Stürmer, M., Buxbaum, S., Walczok, A., Preiser, W., Doerr, H.W., Laboratory diagnosis of norovirus: Which method is the best? *Intervirology*, 46, 4, 232–8, 2003.
53. Anfruns-Estrada, E., Sabrià, A., Fuentes, C., Sabaté, S., Razquin, E., Cornejo, T., *et al.*, Detection of norovirus in saliva samples from acute gastroenteritis cases and asymptomatic subjects: Association with age and higher shedding in stool. *Viruses*, 12, 12, 1369, 2020.
54. Bonifait, L., Charlebois, R., Vimont, A., Turgeon, N., Veillette, M., Longtin, Y., *et al.*, Detection and Quantification of Airborne Norovirus during Outbreaks in Healthcare Facilities. *Clin. Infect. Dis*, 61, 3, 299–304, 2015.
55. Kou, B., Crawford, S.E., Ajami, N.J., Czakó, R., Neill, F.H., Tanak, T.N., *et al.*, Characterization of cross-reactive norovirus-specific monoclonal antibodies. *Clin. Vaccine Immunol.*, 22, 2, 160–7, 2015 Feb.
56. Kim, J.H., Park, J.E., Lin, M., Kim, S., Kim, G.H., Park, S.J., *et al.*, Sensitive, Quantitative Naked-Eye Biodetection with Polyhedral Cu Nanoshells. *Adv. Mater.*, 29, 37, 1702945, 2017.
57. Hagström, A.E.V., Garvey, G., Paterson, A.S., Dhamane, S., Adhikari, M., Estes, M.K., *et al.*, Sensitive detection of norovirus using phage nanoparticle reporters in lateral-flow assay. *PLoS One*, 10, 5, e0126571, 2015.
58. Saylan, Y., Erdem, O., Ünal, S., Denizli, A., An alternative medical diagnosis method: Biosensors for virus detection. *Biosensors*, 9, 2, 65, 2019.
59. Barclay, L., Park, G.W., Vega, E., Hall, A., Parashar, U., Vinjé, J., *et al.*, Infection control for norovirus. *Clin. Microbiol. Infect.*, 20, 8, 731–40, 2014.
60. Jones, R.M. and Brosseau, L.M., Aerosol transmission of infectious disease. *J. Occup. Environ. Med.*, 57, 5, 501–8, 2015 May.

61. Després, V.R., Alex Huffman, J., Burrows, S.M., Hoose, C., Safatov, A.S., Buryak, G., *et al.*, Primary biological aerosol particles in the atmosphere: A review. . *Tellus Ser. B Chem. Phys. Meteorol*, 64, 1, 15598, 2012.
62. Alsved, M., Fraenkel, C.J., Bohgard, M., Widell, A., Söderlund-Strand, A., Lanbeck, P., *et al.*, Sources of airborne norovirus in hospital outbreaks. *Clin. Infect. Dis.*, 70, 10, 2023–8, 2020 May.
63. Verreault, D., Moineau, S., Duchaine, C., Methods for Sampling of Airborne Viruses. *Microbiol. Mol. Biol. Rev.*, 72, 3, 413–44, 2008.
64. Boles, C., Brown, G., Nonnenmann, M., Determination of murine norovirus aerosol concentration during toilet flushing. *Sci. Rep.*, 11, 1, 23558, 2021.
65. Tappes, S.P., Chaves Folly, D.C., Da Silva Santos, G., De Aquino Feijó, C., Pustiglione, M., Food handlers and foodborne diseases: Grounds for safety and public and occupational health actions. *Rev. Bras. Med. do Trab.*, 17, 3, 431–40, 2019.
66. De Graaf, M., Van Beek, J., Koopmans, M.P.G., Human norovirus transmission and evolution in a changing world. *Nat. Rev. Microbiol. [Internet]*, 14, 7, 421–33, 2016. Available from: https://doi.org/10.1038/nrmicro.2016.48.
67. Centre for Health Protection - Smoking [Internet]. 2016 [cited 2024 Nov 28]. Available from: https://www.chp.gov.hk/en/healthtopics/content/25/8806.html.
68. Moe, C.L., Preventing norovirus transmission: how should we handle food handlers? *Clin. Infect. Dis.*, 48, 1, 38–40, 2009.
69. Todd, E.C.D., Greig, J.D., Bartleson, C.A., Michaels, B.S., Outbreaks where food workers have been implicated in the spread of foodborne disease. Part 6. Transmission and survival of pathogens in the food processing and preparation environment. *J. Food Prot. [Internet]*, 72, 1, 202–19, 2009. Available from: https://doi.org/10.4315/0362-028X-72.1.202.
70. What doctors wish patients knew about the contagious norovirus | American Medical Association [Internet]. [cited 2024 Nov 29]. Available from: https://www.ama-assn.org/delivering-care/public-health/what-doctors- wish-patients-knew-about-contagious-norovirus.
71. Centres for Disease Control and Prevention. How Norovirus Spreads | CDC [Internet]. Center for Disease Control and Prevention. 2021 [cited 2024 Nov 29]. Available from: https://www.cdc.gov/norovirus/causes/index.html.
72. Hall, A.J., Noroviruses: The perfect human pathogens? *J. Infect. Dis.*, 205, 11, 1622–4, 2012.
73. Norovirus | Hartmann Science Center [Internet]. [cited 2024 Nov 29]. Available from: https://www.hartmann-science-center.com/en/hygiene-knowledge/pathogens-a-z/pathogens-14/norovirus-pathogen.
74. Norovirus Outbreaks | CDC [Internet]. [cited 2024 Dec 1]. Available from: https://www.cdc.gov/norovirus/outbreaks/index.html.
75. Dozens fall ill with norovirus on month-long cruise between Singapore and California - CBS News [Internet]. [cited 2024 Dec 1]. Available from: https://www.cbsnews.com/news/sick-pasengers-princess-cruises-coral- norovirus/.

76. Parrón, I., Álvarez, J., Jané, M., Cornejo Sánchez, T., Razquin, E., Guix, S., *et al.*, A foodborne norovirus outbreak in a nursing home and spread to staff and their household contacts. *Epidemiol. Infect.*, 147, e225, 2019 Jan.
77. Jin, T., Chen, X., Nishio, M., Zhuang, L., Shiomi, H., Tonosaki, Y., *et al.*, Interventions to prevent surface transmission of an infectious virus based on real human touch behavior: a case study of the norovirus. *Int. J. Infect. Dis. [Internet]*, 122, 83–92, 2022. Available from: https://www.sciencedirect.com/science/article/pii/S1201971222003113.
78. Reimann, H.A., Price, A.H., Hodges, J.H., The Cause of Epidemic Diarrhea, Nausea and Vomiting. (Viral Dysentery?). *Proc. Soc. Exp. Biol. Med.*, 59, 1, 8–9, 1945.
79. Tan, A., Liu, J., Liu, P., Zeng, J.M., Liu, G., Investigation on epidemic of Norovirus infection in a kindergarten in Neijiang in 2020. *J. Med. Pest Control*, 38, 2, 183–6, 2022.
80. Liu, X., Huang, Y., Chu, Y., Sun, J., Yang, X., Kong, Q., *et al.*, Epidemiological characteristics and genotype of norovirus outbreaks in schools in Xicheng District of Beijing from 2017 to 2022. *Chin. J. Sch. Heal*, 45, 5, 704–7, 2024.
81. Government of Canada. Public Health Notice — Outbreak of norovirus and gastrointestinal illnesses linked to raw oysters - Canada.ca [Internet]. 2018 [cited 2024 Dec 1]. Available from: https://www.canada.ca/en/public-health/services/public-health-notices/2018/outbreak-norovirus-infections-linked-raw-oysters.html.
82. McIntyre, L., Galanis, E., Prystajecky, N., Kosatsky, T., BC oysters and norovirus: Hundreds of cases in months with an "r" [Internet]. *Br. C. Med. J.*, 59, 326–7, 2017 [cited 2024 Dec 1. Available from: https://bcmj.org/bccdc/bc-oysters-and-norovirus-hundreds-cases-months-"r".
83. Rico, E., Pérez, C., Belver, A., Sabaté, S., Razquin, E., de Benito, J., *et al.*, Norovirus detection in environmental samples in norovirus outbreaks in closed and semi-closed settings. *J. Hosp. Infect*, 105, 1, 3–9, 2020.
84. Preventing the Spread of Norovirus | CloroxPro [Internet]. [cited 2024 Nov 29]. Available from: https://www.cloroxpro.com/resource-center/norovirus/.
85. Ciofi-Silva, C.L., Bruna, C.Q.M., Carmona, R.C.C., Almeida, A.G.C.S., Santos, F.C.P., Inada, N.M., *et al.*, Norovirus recovery from floors and air after various decontamination protocols. *J. Hosp. Infect*, 103, 3, 328–34, 2019.
86. Note, D.A., *Cleaning And Disinfection Advisory Note Guidelines for Dentists*, pp. 1–25, 2012.
87. CDC, Ncird. CS273773-A Norovirus: Facts for Food Workers. 2017 (January). Available from: https://www.cdc.gov/norovirus/downloads/foodhandlers.pdf.

10

The Role of Rainfall in the Control of Norovirus Outbreaks

Sandhya Chaudhary[1], Rahul Kumar[1], Kalpana Rahate[1], Shriyansh Srivastava[2] and Sathvik Belagodu Sridhar[3]*

[1]Department of Pharmacy, School of Medical and Allied Sciences, Galgotias University, Greater Noida, Uttar Pradesh, India
[2]Department of Pharmacology, Delhi Pharmaceutical Sciences and Research University, New Delhi, India
[3]RAK College of Pharmacy, RAK Medical & Health Sciences University, Ras Al Khaimah, UAE

Abstract

Noroviruses are non-enveloped, positive-sense, single-stranded RNA viruses classified within the *Caliciviridae* family of viruses. "Norwalk virus," initially identified and dubbed, caused a gastroenteritis outbreak in 1968 in Norwalk, Ohio. Research has demonstrated that noroviruses, which are significant human infections, cause devastating gastroenteritis epidemic outbreaks in assisted living facilities, hospitals, and communities. The environment and food are two other potential pathways for the norovirus to spread, even though direct human-to-human spread is thought to be a main method of spread for most of the occurrences. These epidemics annually cause over 267,000,000 illnesses worldwide. Climate change influences the frequency of norovirus outbreaks, a fact well acknowledged but not fully understood. Despite the widespread acceptance of the virus' seasonality and winter occurrence, the relationship between environmental factors and norovirus outbreaks remains unclear. Studies have demonstrated the crucial and intricate role of rainfall in determining the norovirus' ability to spread. With a focus on how rainfall affects both direct and indirect transmission channels, this chapter examines the dynamic role that precipitation plays in controlling norovirus outbreaks. Rainfall drastically impacts virus flow and water quality, which contributes to norovirus spread. The findings of this book chapter improve our understanding of the relationship between environmental factors and norovirus

**Corresponding author*: sathvik@rakmhsu.ac.ae

Sachin Kumar, Shriyansh Srivastava, Rishabha Malviya and Ranjit Sah (eds.) Emerging Infectious Disease: Unravelling Norovirus and Pandemic Dynamics, (275–294) © 2026 Scrivener Publishing LLC

outbreaks and provide guidance for future investigations into these outbursts. To fully comprehend the things of the other problems on the occurrence of norovirus, further study is necessary.

Keywords: Caliciviridae, gastroenteritis outbreak, humidity, norovirus, Norwalk virus, rainfall, rotavirus, seasonality, temperature

10.1 Introduction

In both industrialized and developing countries, noroviruses (NoV) are the main culprits behind viral gastroenteritis. Among nations where rotavirus childhood immunization is not available, norovirus is the second most significant nonbacterial etiological cause among children under 5 years old who have acute gastroenteritis, with rotavirus being the main viral agent [1]. Norovirus is another possible source for acute gastroenteritis in various categories of age. NoV gained notoriety after outbreaks in schools, nursing homes, cruise ships, hospital wards, and other places where people live close to one another [2]. Because of its low infectious dose, short incubation period, stability outside the host, and ease of transmission *via* a variety of pathways, norovirus may produce widespread epidemics in a couple of days. Medical facilities may experience a significant impact on both patients and employees, potentially affecting the hospital's ability to continue operations. The primary indications of a norovirus contamination are nausea and diarrhea, which usually go away on their own in a few days. On the other hand, aged and immunocompromised individuals may have chronic symptoms, and a norovirus infection may potentially be fatal for them [3, 4]. In addition to the health risks to these individuals, norovirus epidemics have an impact on the economy. Healthcare facilities use expensive outbreak control strategies, including closing units to new admissions and implementing additional sanitary precautions [5]. Norovirus outbreaks happen all year long by following different season patterns, with frequent occurrences in the midsummer and higher epidemic stages throughout the wintertime [6, 7].

10.2 Characteristics of Norovirus

10.2.1 Classification

The International Committee on Taxonomy of Viruses (ICTV) has identified four genera of the *Caliciviridae* family (Figure 10.1), which includes the genus Norovirus: Sapovirus, Vesivirus, Lagovirus, and Nebovirus are all significant viral entities worth examining. The genera Vesivirus and Lagovirus encompass significant veterinary infections including rabbit hemorrhagic sickness virus (lagovirus) and feline calicivirus (vesivirus). Humans only contract noroviruses and sapoviruses, except for an anecdotal zoonotic vesivirus infection. While noroviruses may infect persons of any age, sapoviruses mostly cause mild gastroenteritis in children under the age of 5 years. Recovirus and Valovirus, two new genera suggested for the *Caliciviridae* family, have not yet received recognition from the ICTV [9, 10] of the gene and protein associated with the capsid (GI–GV) separating the Norovirus genus into genogroups. The Norovirus genre is separated into five phylogenetic groupings, or genogroups, based on phylogenetic characteristics. Future discoveries of novel genotypes are likely to lead to the identification of viruses from new genogroups. Since 2002, genogroup II genotype 4 (abbreviated as GII.4) has consistently been reported as the source of outbreaks; it is single for this genotype that an additional subdivision of "variations" is employed. Public health places significant importance on foodborne viruses and prions. The tool for genotyping norovirus makes it simple to assign a genotype and, where applicable, a variation to a "partial" norovirus sequence [11]. The virus genotyping tool facilitates the assignment of genotypes and, where applicable, variants to partial norovirus sequences [12].

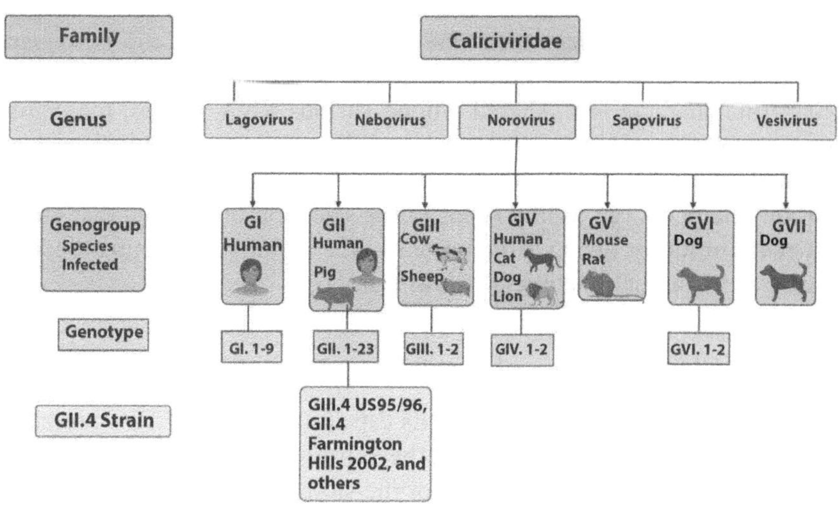

Figure 10.1 Classification of norovirus [8].

10.2.2 Structure of Norovirus

Noroviruses possess an affirmative linear genome composed of approximately 7.5-kb-sized single strand structured into three open reading frames (ORFs). ORF1 encodes a polyprotein that encompasses all seven non-structural proteins, synthesized as a single big polyprotein and subsequently split into distinct proteins. The non-structural proteins play a crucial role in the generation of multiple viruses within the diseased cells, but the components of virus particles are absent [13]. ORF2 is responsible for encoding the primary structural protein VP1, also known as the capsid protein, while ORF3 encodes the VP2 protein, which serves as the minor capsid protein. The viral protein 2 (VP2) is considered an insignificant structural component, as each norovirus virion is thought to contain only one or two copies. There is a greater understanding of the role of VP2 in other relatives within the norovirus family. The role of VP2 in feline calicivirus is crucial for generating infectious virions [14], whereas mutant rabbit hemorrhagic disease viruses lacking the VP2 protein exhibited only a minor decrease in infectivity [15]. The RNA-dependent RNA polymerase does not possess a reciprocal strand during the replication of the RNA genome, which would facilitate the proofreading activity. As a result, noroviruses display a significant mutation rate of roughly one per 103 to 104 nucleotides, in contrast to DNA-dependent DNA polymerases (present in DNA viruses or cellular organisms), which show a mutation rate of around one per 106 nucleotides. Consequently, RNA viruses exhibit significant diversity and possess an evolution rate that is nearly a million times faster than that of their host. The elevated mutation rate results in norovirus demographics within one individual host that comprises a varied assortment of almost indistinguishable strains, known as quasi species. This characteristic, shared among various RNA viruses, highlights the genetic adaptability of these viruses. When pressure is applied, such as through immunity in individuals who have been exposed, the interplay of various elements can swiftly result in the selection of variants that evade current immunity, owing to alterations in the regions concerning the VP1 that are critical for attachment to host cells [16].

10.2.3 Characteristics of Virus

Norovirus lacks a lipoidal coating; its capsid (made up of VP1 proteins) contains the genetic information. The capsid of norovirus allows them to withstand the acidic conditions of the stomach and maintain viability in the environment for prolonged periods, though the particles seem to

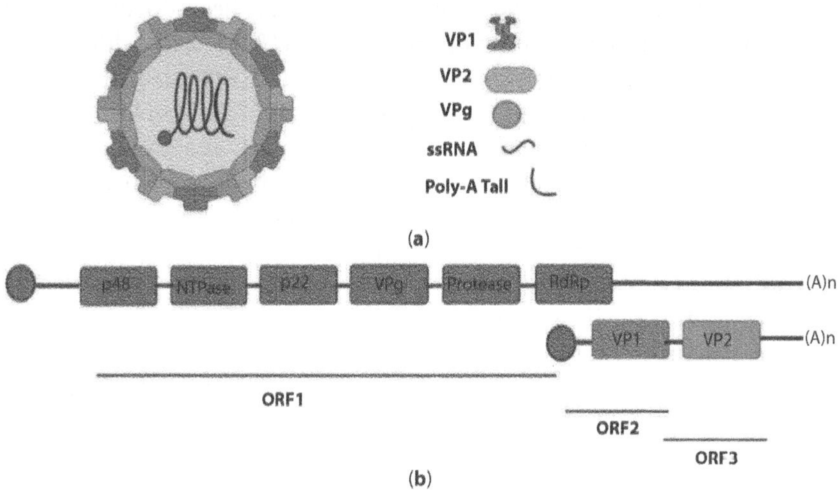

Figure 10.2 Molecular structures of norovirus [17].

exhibit reduced stability at higher pH levels [18]. Viral protein 2 (VP2) acts as the main structural element of norovirus, with each virion consisting of 180 copies (90 dimers) of the VP1 protein organized equally (Figure 10.2) [17]. The germ binds to the host cell receptor *via* the VP1 protein, highlighting the crucial role of VP1 in the relation between the virus and the host. Cryo-electron microscopic studies have demonstrated that the VP1 protein is categorized into two distinct components: the shell domain and the protruding domain (S and P domain). The S domains of the 180 VP1 copies create a shell encasing the virus RNA (icosahedron), while the P domains develop arch-like arrangements that encompass the case. The P domain is linked to the S domain through an elastic center and can be more categorized into the P1 and P2 domains, with P2 predominantly extending outward.

10.2.4 Viral Shedding and Transmission Potential

Noroviruses are ingested orally, allowing virions to traverse the sharp conditions of the stomach. Reproduction occurs in the higher duodenal tract; however, the specific chamber category remains unidentified [19]. High quantities of virions (107–1,010 RNA copies per gram) are shed in waste matter and also identified in vomitus [20]. Individuals who are infected release the virus in the greatest amounts through the severe stage, and this detaching persists through an asymptomatic phase that may last from 9 to 56 days. Continued asymptomatic discharge has been documented in children for durations reaching up to 100 days [12, 21, 22]. Additionally, immunocompromised individuals may experience extended illness and shedding, potentially lasting several years [20, 21, 23]. Norovirus exhibits a high level of infectiousness during an outbreak at a scouting camp. The estimated basic reproduction number (R0) was found to exceed 14 [22]. This indicates that, on average, each case has the potential to infect 14 additional individuals during the infection period. The implementation of sanitization actions at the scouting camp significantly impacted the R0, reducing it [24–26]. The elevated R0 can be partially attributed to the minimal dose necessary for infection. Consequently, the elimination of multiple logs of virus during infection incidents may still prove insufficient to halt the transmission of sickness [27, 28].

10.2.5 Duration of Development and Manifestations

In individuals with good health, norovirus leads to mild gastroenteritis, and the symptoms typically resolve on their own. Outbreak studies indicate that the symptoms typically persist for 1 to 5 days and encompass nausea, non-bloody diarrhea, belly pains, and fever [29]. In youngsters, nausea occurs more commonly, while adults tend to experience diarrhea more often [30]. Norovirus outbreaks generally exhibit a more extended incubation period in comparison to those caused by microbial gastroenteritis. In 85% of norovirus occurrences, the incubation period exceeded 24 h, whereas this was the case for only 39% of outbreaks associated with a microbial etiological agent [31]. In 1982, Kaplan and colleagues conducted a review of 642 cases of acute gastroenteritis outbreaks to derive criteria for the identification of outbreaks associated with Norwalk-like viruses. The criteria established by them include the following: (a) more than 50% of cases experiencing vomiting, (b) an average or median incubation time of 24 to 48 h following infection, (c) an illness duration ranging from 12 to 60 h, and (d) lack of etiological microbes in stool testers. In 2006, a re-evaluation of these criteria demonstrated that they were highly specific (99%) and temperately complex (68%), proving to be a valuable investigative tool for differentiating norovirus occurrences from bacteriological outbreaks in circumstances where alternative investigative methods are not yet accessible [32]. Norovirus mostly spreads through fecal–oral routes. Norovirus is highly contagious, but no one knows exactly how it spreads. According to literature that took into account the aggregation of particles, the infective potency of norovirus was estimated to be as low as 18 virions. Norovirus is so infectious that it can spread from one person to another directly or indirectly through contaminated food or water [33]. Vomiting can cause significant contamination of the environment, which can spread through fomites and airborne droplets. Vomiting was the main cause of norovirus transmission in long-term care facilities (LTCFs), according to literature [34].

10.3 Possible Environmental Markers for Outbreaks of the Human Norovirus

Temperature, humidity, and precipitation are among the environmental variables that have been linked to norovirus epidemics and are often mentioned in the books.

10.3.1 Temperature

Temperature is the most often mentioned ecological component which might influence norovirus outbreaks. Several experimental investigations examined the norovirus ability to survive at various temperatures. Researchers used murine norovirus as a stand for human norovirus to investigate the impact of temperature on norovirus outbreaks [35]. The findings indicated that the murine norovirus was more enduring at 4°C as opposed to 18°C and 30°C. According to a study on how temperature affects norovirus persistence on surfaces that are touched by food, such as polyvinyl chloride (PVC) and stainless steel [36], norovirus lasted slow at lower temperatures (7°C) than at room temperature (20°C). Researchers conducted a related investigation [37] on the resilience of murine norovirus to various environmental stressors in a temperature series (15°C–40°C) and comparative humidity (30%, 50%, and 70%). They discovered that compared to other temperature ranges, the murine norovirus lasted longer at the 15°C–20°C range. Another study looked at how long human norovirus (MS2) and murine norovirus (MNV) dummies could live in different climate and humidity levels, such as 4°C, 15°C, 25°C, and 40°C and 50% and 70%, respectively [38]. The results of this investigation indicated that viruses were most inactivated by the greatest temperature (40°C) and best viable at the lowest heat (4°C). The research concluded that temperature had the most impact on viral survival by following the interaction of temperature and relative moisture. Related investigations validated the harmful relationship among high temperature and norovirus persistence under a variety of experimental circumstances [39].

10.3.2 Humidity

Researchers discovered that humidity was a significant environmental element influencing norovirus survival and transmission. In fact, higher humidity could make it easier for the virus to spread *via* aerosols. However, according to the research, there was an inverse linear relationship between relative humidity and norovirus outbreaks [40]. Japanese research suggests that the rise in the norovirus incidence rate may be due to a sharp drop in humidity. Two Korean investigations statistically validated the strong negative connections between humidity and norovirus outbreaks. A few investigations experimentally examined the norovirus' ability to survive in various humidity levels. Unlike earlier research, they discovered that the norovirus remained active at lower relative humidity levels (30%) for a longer period of time than at higher relative humidity levels (70%). Additionally, they found that maintaining air humidity (AH) below 0.007 kg/kg facilitated the survival and infection of norovirus and that absolute humidity (AH) was a more significant pointer of norovirus occurrences than relative moisture [41]. Their examination of day-to-day complete moisture and heat data from Paris, France, over the previous 14 years linked low AH (less than 0.007 kg/kg air) to norovirus persistence and transmission during cold months.

10.3.3 Rainfall

Precipitation is an ecological factor that could influence the onset of norovirus outbreaks. Changes in precipitation patterns could directly affect the outbreaks of infectious disease occurrences [42]. Contaminated water is found to be the most prominent cause of norovirus infection. Prolonged and intense rainfall exceeds the wastewater storage capability and results in untreated effluent discharge. This leads to an elevated concentration of enteric virus in the aquatic species, causing norovirus gastroenteritis [43]. Precipitation has the potential to dilute or ultimately wash away bacteria retained in dusts and deposits due to earlier impurity incidents. Limited research has explored the arithmetic relationship between rainfall and norovirus outbreaks and found an observed correlation with statistical significance ($p < 0.05$) linking norovirus-associated gastroenteritis occurrences to medium monthly rainfall in Victoria, Australia, between 2002 and 2007. It was observed that a 3-month delay between the maximum typical precipitation and the outbreak of gastroenteritis was linked to norovirus. The researchers analyzed the arrangements of norovirus waves over a decade in Victoria, concluding that occurrences typically featured highest in the final part of the year. It was also assumed that sometimes the rainfall may have influenced changes in norovirus occurrence patterns, mainly in southern hemisphere nations. A comprehensive evaluation and synthesis of data examining the global seasonality of norovirus [38] observed a positive relationship between the seasonal intensity of norovirus outbreaks and the average rainfall during the wettest month. Nonetheless, there have also been reports of inverse associations between norovirus outbreaks and precipitation [44, 45].

10.4 Relationship Between Rainfall and Norovirus Outbreaks

Norovirus is also called winter vomiting disease and has been known to have a winter seasonality for a long time [44]. The relation between weather and infective illnesses may aid in determining the effects of environmental factors that raise the danger of contamination and projecting the effects of worldwide temperature variation on the spreading and transmission of contagious illnesses. Weather change may affect the transmissibility, host susceptibility, and environmental fight of norovirus, which might have an effect on seasonality. Studies have shown correlations between the seasonality of the norovirus and meteorological/climatic events. Rainfall significantly influences the seasonality of the norovirus, likely due to the virus' water-based transmission. Similarly, chilly temperatures, poor people resistance, and the introduction of novel strains are connected with elevated norovirus action. In an extensive examination conducted in England and Wales, temperature variations had the highest attributable risk for norovirus occurrence among these predictors. Researchers have

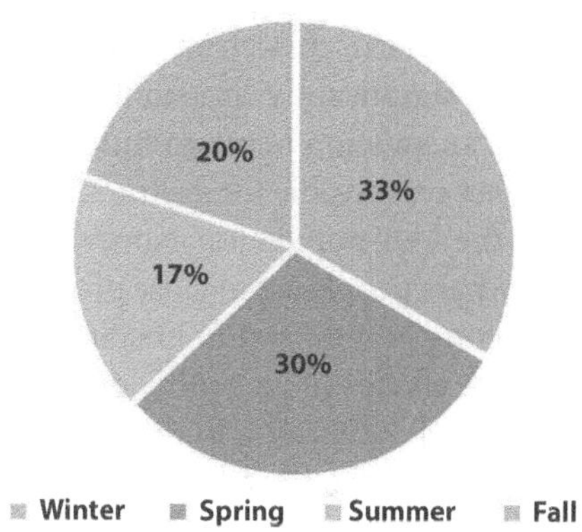

Figure 10.3 Seasonal occurrences of foodborne norovirus in different countries [43].

linked temperature, relative humidity, and rainfall to the seasonal variations of rotavirus [46], another gastrointestinal virus that is most prevalent in the winter. Host immunity, demography, and human behavior factors are responsible for the higher frequency of norovirus in the colder months. Specifically, crowded conditions and increased indoor time may exacerbate wintertime respiratory virus transmission between humans [32]. Fertility rates and regional exposure levels (the force of contamination) have been proposed as the main variables influencing the seasonal patterns of sickness caused by the rotavirus [47]. The recurring appearance of novel norovirus strains interacting with host safety in the community may potentially intermediate seasonal forms of sickness [48]. The RNA viruses known as norovirus belong to the family *Caliciviridae*. We can categorize them in five genogroups (GI–GV), with GI and GII being the most commonly occurring ones that infect humans. The majority of viruses are genotype II or genotype 4 (GGII.4), and they are evolving quickly in response to immunity in the human population. New variations of GII.4 appear every 2–4 years. The development of new GII.4 virus in 2002–2003 coincided with rare springtime/summertime and total inter-season rise in sickness, probably resulting from the initial lack of sufficient population immunity to the new variation. More than 10 years ago, we released the most current analysis of norovirus seasonality worldwide (Figure 10.3), before the general availability of PCR diagnoses [49]. In this day and age of advanced diagnostics, we did an ecological study to find parameters related to the intensity of norovirus seasonality and a thorough review of the literature to bring our knowledge of the worldwide seasonality of norovirus up to date.

10.5 Public Health and Infrastructure Challenges

Despite limited public awareness, this virus, which spreads through contaminated water and food, significantly contributes to non-bacterial gastroenteritis in Brazil [50]. Transmission between individuals occurs with ease. In contrast to other viruses responsible for gastroenteritis, such as rotavirus, norovirus often impacts the adult population. Noroviruses are strongly linked to outbreaks in settings where people are in close proximity, including within families, on ships, in nursing homes, and in hospitals. For instance, noroviruses present a worldwide challenge currently, owing to their swift spread and challenging management. According to the Laboratory of Comparative and Environmental Virology at the Oswaldo Cruz Institute (IOC), noroviruses are responsible for 99.9% of gastroenteritis outbreaks on ships [51].

10.5.1 Water and Foodborne Transmission

Norovirus can contaminate food and water supplies, which presents challenges for food safety and sanitation infrastructure. Contaminated water, particularly in underdeveloped regions or during natural disasters, can lead to widespread outbreaks. In some cases, food handlers, who are symptomatic or infected, can also contribute to the virus' spread [52].

10.5.2 Transmission Between Individuals

In Europe, it is generally observed that 74% to 85% of reported NoV infections result through human-to-human transmission, while research conducted in The Netherlands estimated that 55% (ranging from 42% to 88%) of all NoV cases arise through this pathway. On the other hand, person-to-person transmission accounts for 85% of outbreaks in England and Wales [53]. Individuals often transmit in confined environments, making it challenging to isolate infected individuals. Food can serve as a significant pathway for the initial introduction of diseases.

10.6 Prevention and Mitigation Strategies

10.6.1 Controlling the Occurrence

Healthcare institutions deliver services to at-risk populations and are frequently the primary environments for norovirus outbreaks. Consequently, our discussions will center on infection control issues within these specific environments while acknowledging that many of these principles are applicable in various other contexts as well. Managing an outbreak involves several stages: preparation, identification, response, and evaluation [54]. Public health agencies in various nations, such as Australia, Ireland, the UK, and the USA, have already issued the strategies for the control of NoV outbreaks. Certain guidelines, including those from the UK and the USA, employed organized literature reviews and subsequently assessed the strength of its endorsement. Recommendations from other countries relied on a more expert-opinion-driven method to evaluating indication [55]. Recent guidelines show a notable consistency in the recommended measures, irrespective of the methods employed. The primary plan for the management and control of norovirus outbreaks through various guidelines is mentioned in Table 10.1, encompassing the establishment of policies related to hand sanitation, the isolation of symptomatic patients, cohorting based on symptoms, exclusion of staff from work, restrictions on visitors, intensified environmental cleaning and disinfection, and the potential closure of wards [56].

Table 10.1 An overview of infection control protocols for managing and preventing norovirus outbreaks in medical facilities [57].

Sr. no.	Classification of infection control	Methods for controlling infections
1.	Identification of outbreaks	-Bacteria-negative stool. -The average length of sickness was 12–60 h. -In more than 50% of instances, vomiting. -24–48-h incubation period.
2.	Cleanliness	-Before and after interacting with patients, after using the restroom, clean hands with cleanser and warm running water. -Adopt proper PPE (personal protective equipment). -wear neat and clean gown, mask, and gloves. -Regularly replace PPE.
3.	Cleaning and sanitization	-Sanitize and disinfect places that are not impacted. - Remove any organic material from the affected areas. - Use warm water and detergent to clean the carpets. -Steam and clean any soft furniture that bleach can harm. -Throw away biohazard bags.
4.	Transferring and isolating patients	-Divide patients according to infected and not infected. -Restrict the movement and transfer of infected patients.
5.	Elimination of employees and cohorting	-Exclude the infected staff for 48 h after infection. -Allocate personnel to a specific patient group.
6.	Guests	-Limited contacts are allowed at a time. -Restrict visitors with symptoms until 48 h after they have subsided.
7.	Reports of outbreaks	-In accordance with state and municipal public health standards, notify the proper local or state health offices.
8.	Food security	-Throw away any food that is exposed. -Make sure that food is prepared, stored, and served correctly. -Avoid touching or eating ready-to-eat foods with dirty hands.

10.6.2 Sanitation

The transmission of norovirus occurs through various means, and there is no single solution to stop its spread. A wide range of suggestions is required for the inhibition and control of norovirus occurrences. In general, healthcare workers, patients, and guests in patient-care settings impacted by norovirus gastroenteritis outbreaks should be actively encouraged to practice good hand hygiene. Hands should be cleaned with cleanser and warm water for a minimum of 20 s after caring for a sick person who has a supposed or confirmed illness over occurrences [58].

10.6.3 Cleaning and Sanitization

Improved cleaning and disinfection practices can manage and stop the transmission of norovirus. This causes cleaning more often and focusing more intently on high-traffic regions and often-touched articles, including telephones and doorknobs [59]. It is recommended to prepare a new bleach solution daily, ensuring a minimum concentration of 1,000 ppm sodium hypochlorite for effective disinfection.

10.6.4 Colonization and Isolation

Other classes of suggested infection control measures are isolation, cohorting (divide patients according to symptoms), and excluding infected employees, patients, and guests [60].

10.7 Future Directions and Research Needs

The primary strategy to contain the human norovirus epidemic until the development of efficient antiviral medications or vaccinations involves the use of accurate and quick detection technologies. We should develop effective pre-treatment techniques first for complicated samples, as they can significantly impact the sensitivity and accuracy of the detection technique used to recognize human norovirus. Reverse transcription polymerase chain reaction (RT-qPCR) is now a popular and dependable technique. Identifying infectious and non-infectious viruses, processing samples from various sources, and accurately assessing disease outbreaks are the main hurdles. An inexpensive, easy-to-use, and quick diagnostic approach is the immunological colloidal gold test strip. However, promoting this technique for future applications will require increasing the detection sensitivity and prolonging the validity duration. One potential technique for detection is biosensors. Future studies must concentrate on improving fixation technology and identifying highly active and selective biometric components. Additionally, we need to investigate effective signal amplification techniques to lower the detection limit. Future research must focus on improving techniques for optimizing the concentration of viral particles from environmental or food matrices since viral contamination in these matrices is often minimal. In the interim, detection technology development will concentrate on enhancing current technology and investigating novel multidisciplinary approaches.

10.8 Conclusion

The development and implementation of strategies to reduce the risk of foodborne infections and encourage safe consumer behaviors are important tasks for food safety authorities. However, final consumers bear some responsibility for their own and their families' health and may also be a significant factor in enhancing control and prevention measures. Crucially, from sale to consumption, certain consumer behaviors raise the risk of infection while simultaneously providing chances for action to achieve considerable risk reduction and safer food, making them both a fundamental issue and a solution. A thorough and scientific understanding of the most important issues and knowledge gaps may enable the creation of novel consumer-facing educational, social, or even technical solutions aimed at lowering the risk of norovirus transmission.

Abbreviations

AH	air humidity
GII.4	genogroup II genotype 4
ICTV	International Committee on Taxonomy of Viruses
IOC	Oswaldo Cruz Institute
LTCs	long-term care facilities
NoV	norovirus
ORFs	open reading frames
ppm	parts per million
PVC	polyvinyl chloride

RO	reproduction number
RT-qPCR	reverse transcription polymerase chain reaction
VP	viral proteins

AI disclosure statement: The author acknowledges the use of AI-based tools for minor language editing or grammatical correction. However, all intellectual contributions, analyses, and conclusions presented in this chapter are entirely the author's own.

References

1. Patel, M.M., Widdowson, M.A., Glass, R.I., Akazawa, K., Vinje, J., Parashar, U.D., Systematic literature review of role of noroviruses in sporadic gastroenteritis. *Emerg. Infect. Dis.*, 14, 8, 1224, 2008.
2. Fields, B.N., *Fields' virology*, vol. 1, Lippincott Williams & Wilkins, 2007.
3. Schwartz, S., Vergoulidou, M., Schreier, E., Loddenkemper, C., Reinwald, M., Schmidt-Hieber, M., *et al.*, Norovirus gastroenteritis causes severe and lethal complications after chemotherapy and hematopoietic stem cell transplantation. *Blood J. Am. Soc. Hematol.*, 117, 22, 5850–6, 2011.
4. van Asten, L., Siebenga, J., van den Wijngaard, C., Verheij, R., van Vliet, H., Kretzschmar, M., *et al.*, Unspecified gastroenteritis illness and deaths in the elderly associated with norovirus epidemics. *Epidemiol.*, 22, 3, 336–43, 2011.
5. Johnston, C.P., Qiu, H., Ticehurst, J.R., Dickson, C., Rosenbaum, P., Lawson, P., *et al.*, Outbreak management and implications of a nosocomial norovirus outbreak. *Clin. Infect. Dis.*, 45, 5, 534–40, 2007.
6. Rohayem, J., Norovirus seasonality and the potential impact of climate change. *Clin. Microbiol. Infect.*, 15, 6, 524–7, 2009.
7. Verhoef, L., Depoortere, E., Boxman, I., Duizer, E., Van Duynhoven, Y., Harris, J., *et al.*, Emergence of new norovirus variants on spring cruise ships and prediction of winter epidemics. *Emerg. Infect. Dis.*, 14, 2, 238, 2008.
8. Ettayebi, K. and Hardy, M.E., Norwalk virus nonstructural protein p48 forms a complex with the SNARE regulator VAP-A and prevents cell surface expression of vesicular stomatitis virus G protein. *J. Virol.*, 77, 21, 11790–7, 2003.
9. Farkas, T., Sestak, K., Wei, C., Jiang, X., Characterization of a rhesus monkey calicivirus representing a new genus of Caliciviridae. *J. Virol.*, 82, 11, 5408–16, 2008.
10. L'Homme, Y., Sansregret, R., Plante-Fortier, E., Lamontagne, A.M., Ouardani, M., Lacroix, G., *et al.*, Genomic characterization of swine caliciviruses representing a new genus of Caliciviridae. *Virus Genes*, 39, 66–75, 2009.

11. Zheng, D.P., Ando, T., Fankhauser, R.L., Beard, R.S., Glass, R.I., Monroe, S.S., Norovirus classification and proposed strain nomenclature. *Virology*, 346, 2, 312–23, 2006.
12. Lee, N., Chan, M.C.W., Wong, B., Choi, K.W., Sin, W., Lui, G., *et al.*, Fecal viral concentration and diarrhea in norovirus gastroenteritis. *Emerg. Infect. Dis.*, 13, 9, 1399, 2007.
13. Lamhoujeb, S., Fliss, I., Ngazoa, S.E., Jean, J., Molecular study of the persistence of infectious human norovirus on food-contact surfaces. *Food Environ. Virol.*, 1, 51–6, 2009.
14. Sosnovtsev, S.V., Belliot, G., Chang, K.O., Onwudiwe, O., Green, K.Y., Feline calicivirus VP2 is essential for the production of infectious virions. *J. Virol.*, 79, 7, 4012–24, 2005.
15. Liu, G., Ni, Z., Yun, T., Yu, B., Chen, L., Zhao, W., *et al.*, A DNA-launched reverse genetics system for rabbit hemorrhagic disease virus reveals that the VP2 protein is not essential for virus infectivity. *J. Gen. Virol.*, 89, 12, 3080–5, 2008.
16. Hardy, M.E., Norovirus protein structure and function. *FEMS Microbiol. Lett.*, 253, 1, 1–8, 2005.
17. Campillay-Véliz, C.P., Carvajal, J.J., Avellaneda, A.M., Escobar, D., Covián, C., Kalergis, A.M., *et al.*, Human norovirus proteins: implications in the replicative cycle, pathogenesis, and the host immune response. *Front. Immunol.*, 11, 961, 2020.
18. Trivedi, T.K., Desai, R., Hall, A.J., Patel, M., Parashar, U.D., Lopman, B.A., Clinical characteristics of norovirus-associated deaths: a systematic literature review. *Am. J. Infect. Control.*, 41, 7, 654–7, 2013.
19. Tu, E.T.V., Bull, R.A., Kim, M.J., McIver, C.J., Heron, L., Rawlinson, W.D., *et al.*, Norovirus excretion in an aged-care setting. *J. Clin. Microbiol.*, 46, 6, 2119–21, 2008.
20. Ip, D.K.M., Lau, L.L.H., Leung, N.H.L., Fang, V.J., Chan, K.H., Chu, D.K.W., *et al.*, Viral shedding and transmission potential of asymptomatic and paucisymptomatic influenza virus infections in the community. *Clin. Infect. Dis.*, 64, 6, 736–42, 2017.
21. Lacey, B., Lewington, S., Clarke, R., Kong, X.L., Chen, Y., Guo, Y., *et al.*, Age-specific association between blood pressure and vascular and non-vascular chronic diseases in 0·5 million adults in China: a prospective cohort study. *Lancet Glob. Health*, 6, 6, e641–e649, 2018.
22. Ludwig, A., Adams, O., Laws, H.J., Schroten, H., Tenenbaum, T., Quantitative detection of norovirus excretion in pediatric patients with cancer and prolonged gastroenteritis and shedding of norovirus. *J. Med. Virol.*, 80, 8, 1461–7, 2008.
23. Teunis, P.F.M., Moe, C.L., Liu, P., E. Miller, S., Lindesmith, L., Baric, R.S., *et al.*, Norwalk virus: how infectious is it? *J. Med. Virol.*, 80, 8, 1468–76, 2008.

24. Murata, T., Katsushima, N., Mizuta, K., Muraki, Y., Hongo, S., Matsuzaki, Y., Prolonged norovirus shedding in infants≤ 6 months of age with gastroenteritis. *Pediatr. Infect. Dis. J.*, 26, 1, 46–9, 2007.
25. Roos-Weil, D., Ambert-Balay, K., Lanternier, F., Mamzer-Bruneel, M.F., Nochy, D., Pothier, P., *et al.*, Impact of norovirus/sapovirus-related diarrhea in renal transplant recipients hospitalized for diarrhea. *Transplant.*, 92, 1, 61–9, 2011.
26. Schorn, R., Höhne, M., Meerbach, A., Bossart, W., Wüthrich, R.P., Schreier, E., *et al.*, Chronic norovirus infection after kidney transplantation: molecular evidence for immune-driven viral evolution. *Clin. Infect. Dis.*, 51, 3, 307–14, 2010.
27. Heijne, J.C.M., Teunis, P., Morroy, G., Wijkmans, C., Oostveen, S., Duizer, E., *et al.*, Enhanced hygiene measures and norovirus transmission during an outbreak. *Emerg. Infect. Dis.*, 15, 1, 24, 2009.
28. Tuladhar, E., Hazeleger, W.C., Koopmans, M., Zwietering, M.H., Beumer, R.R., Duizer, E., Residual viral and bacterial contamination of surfaces after cleaning and disinfection. *Appl. Environ. Microbiol.*, 78, 21, 7769–75, 2012.
29. Götz, H., Ekdahl, K., Lindbäck, J., De Jong, B., Hedlund, K.O., Giesecke, J., Clinical spectrum and transmission characteristics of infection with Norwalk-like virus: findings from a large community outbreak in Sweden. *Clin. Infect. Dis.*, 33, 5, 622–8, 2001.
30. Widdowson, M.A., Sulka, A., Bulens, S.N., Beard, R.S., Chaves, S.S., Hammond, R., *et al.*, Norovirus and foodborne disease, United States, 1991–2000. *Emerg. Infect. Dis.*, 11, 1, 95, 2005.
31. Kaplan, J.E., Feldman, R., Campbell, D.S., Lookabaugh, C., Gary, G.W., The frequency of a Norwalk-like pattern of illness in outbreaks of acute gastroenteritis. *Am. J. Public Health*, 72, 12, 1329–32, 1982.
32. Turcios, R.M., Widdowson, M.A., Sulka, A.C., Mead, P.S., Glass, R.I., Reevaluation of epidemiological criteria for identifying outbreaks of acute gastroenteritis due to norovirus: United States, 1998–2000. *Clin. Infect. Dis.*, 42, 7, 964–9, 2006.
33. Barrett, M., Fitzhenry, K., O'Flaherty, V., Dore, W., Keaveney, S., Cormican, M., *et al.*, Detection, fate and inactivation of pathogenic norovirus employing settlement and UV treatment in wastewater treatment facilities. *Sci. Total Environ.*, 568, 1026–36, 2016.
34. Eregno, F.E., Tryland, I., Tjomsland, T., Myrmel, M., Robertson, L., Heistad, A., Quantitative microbial risk assessment combined with hydrodynamic modelling to estimate the public health risk associated with bathing after rainfall events. *Sci. Total Environ.*, 548, 270–9, 2016.
35. Santiago-Rodriguez, T.M., Tremblay, R.L., Toledo-Hernandez, C., Gonzalez-Nieves, J.E., Ryu, H., Santo Domingo, J.W., *et al.*, Microbial quality of tropical inland waters and effects of rainfall events. *Appl. Environ. Microbiol.*, 78, 15, 5160–9, 2012.

36. Bruggink, L.D. and Marshall, J.A., The incidence of norovirus-associated gastroenteritis outbreaks in Victoria, Australia (2002–2007) and their relationship with rainfall. *Int. J. Environ. Res. Public Health*, 7, 7, 2822–7, 2010.
37. Bruggink, L.D. and Marshall, J.A., Molecular changes associated with altered patterns of norovirus outbreak epidemics in Victoria, Australia, in 2006 to 2007. *J. Clin. Microbiol.*, 48, 3, 857–61, 2010.
38. Ahmed, S.M., Lopman, B.A., Levy, K., A systematic review and meta-analysis of the global seasonality of norovirus. *PLoS One*, 8, 10, e75922, 2013.
39. Seo, K., Lee, J.E., Lim, M.Y., Ko, G., Effect of temperature, pH, and NaCl on the inactivation kinetics of murine norovirus. *J. Food Prot.*, 75, 3, 533–40, 2012.
40. de la Noue, A., Estienney, M., Aho, S., Perrier-Cornet, J.M., de Rougemont, A., Pothier, P., *et al.*, Absolute humidity influences the seasonal persistence and infectivity of human norovirus. *Appl. Environ. Microbiol.*, 80, 23, 7196–205, 2014.
41. Wang, P. and Goggins, W.B., Chan EYY. A time-series study of the association of rainfall, relative humidity and ambient temperature with hospitalizations for rotavirus and norovirus infection among children in Hong Kong. *Sci. Total Environ.*, 643, 414–22, 2018.
42. Lopman, B.A., Reacher, M., Gallimore, C., Adak, G.K., Gray, J.J., Brown, D.W.G., A summertime peak of "winter vomiting disease": surveillance of noroviruses in England and Wales, 1995 to 2002. *BMC Public Health*, 3, 1–4, 2003.
43. Shamkhali Chenar, S. and Deng, Z., Environmental indicators for human norovirus outbreaks. *Int. J. Environ. Health Res.*, 27, 1, 40–51, 2017.
44. Greer, A.L., Drews, S.J., Fisman, D.N., Why "winter" vomiting disease? Seasonality, hydrology, and norovirus epidemiology in Toronto, Canada. *Ecohealth*, 6, 192–9, 2009.
45. Groen, A., A geospatial analysis of norovirus outbreaks in California, and an investigation of the impact of environmental variables, 2015.
46. Sumi, A., Rajendran, K., Ramamurthy, T., Krishnan, T., Nair, G.B., Harigane, K., *et al.*, Effect of temperature, relative humidity and rainfall on rotavirus infections in Kolkata, India. *Epidemiol. Infect*, 141, 8, 1652–61, 2013.
47. Leung, A.K.C., Kellner, J.D., Dele Davies, H., Rotavirus gastroenteritis. *Adv. Ther.*, 22, 476–87, 2005.
48. Kambhampati, A.K., Calderwood, L., Wikswo, M.E., Barclay, L., Mattison, C.P., Balachandran, N., *et al.*, Spatiotemporal trends in norovirus outbreaks in the United States, 2009–2019. *Clin. Infect. Dis.*, 76, 4, 667–73, 2023.
49. Khot, P.D. and Fredricks, D.N., PCR-based diagnosis of human fungal infections. *Expert Rev. Anti. Infect. Ther.*, 7, 10, 1201–21, 2009.
50. Raboni, S.M., Damasio, G.A.C., Ferreira, C.E.O., Pereira, L.A., Nogueira, M.B., Vidal, L.R., *et al.*, Acute gastroenteritis and enteric viruses in hospitalised children in southern Brazil: aetiology, seasonality and clinical outcomes. *Mem. Inst. Oswaldo Cruz*, 109, 4, 428–35, 2014.

51. Leite, J.P.G., Carvalho-Costa, F.A., Linhares, A.C., Group A rotavirus genotypes and the ongoing Brazilian experience: a review. *Mem. Inst. Oswaldo Cruz*, 103, 745–53, 2008.
52. Ushijima, H., Fujimoto, T., Müller, W.E.G., Hayakawa, S., Norovirus and foodborne disease: a review. *Food Saf.*, 2, 3, 37–54, 2014.
53. Gaythorpe, K.A.M., Trotter, C.L., Lopman, B., Steele, M., Conlan, A.J.K., Norovirus transmission dynamics: a modelling review. *Epidemiol. Infect.*, 146, 2, 147–58, 2018.
54. Eftim, S.E., Hong, T., Soller, J., Boehm, A., Warren, I., Ichida, A., *et al.*, Occurrence of norovirus in raw sewage--a systematic literature review and meta-analysis. *Water Res.*, 111, 366–74, 2017.
55. Sobolik, J.S., Newman, K.L., Jaykus, L.A., Bihn, E.A., Leon, J.S., Norovirus transmission mitigation strategies during simulated produce harvest and packing. *Int. J. Food Microbiol.*, 357, 109365, 2021.
56. Greig, J.D. and Lee, M.B., A review of nosocomial norovirus outbreaks: infection control interventions found effective. *Epidemiol. Infect.*, 140, 7, 1151–60, 2012.
57. Hall, A.J., Vinjé, J., Lopman, B., Park, G.W., Yen, C., Gregoricus, N., *et al.*, Updated norovirus outbreak management and disease prevention guidelines. *MMWR Recomm. Rep.*, 60, RR-3, 1–18, 2011.
58. Grönberg, A. and Hjorth, R.A., Cleaning-in-place and sanitization, in: *Biopharmaceutical processing*, pp. 675–99, Elsevier, 2018.
59. Schmidt, R.H., Basic elements of equipment cleaning and sanitizing in food processing and handling operations. University of Florida Cooperative Extension Service, Institute of Food and~…, 1997.
60. Orsini, L., Vanoverbeke, J., Swillen, I., Mergeay, J., De Meester, L., Drivers of population genetic differentiation in the wild: isolation by dispersal limitation, isolation by adaptation and isolation by colonization. *Mol. Ecol.*, 22, 24, 5983–99, 2013.

11

The Impact of COVID-19 on Norovirus Circulation and Research

Laxmi[1], Shriyansh Srivastava[2*], Sanjita Das[1], Malakapogu Ravindra Babu[1], Rishabha Malviya[1] and S.K. Abdul Rahaman[1]

[1]*Department of Pharmacy, School of Medical and Allied Sciences, Galgotias University, Greater Noida, India*
[2]*Department of Pharmacology, Delhi Pharmaceutical Sciences and Research University, New Delhi, India*

Abstract

The COVID-19 pandemic profoundly altered the dynamics of infectious disease transmission, particularly affecting the global circulation patterns of norovirus, a leading cause of severe gastroenteritis. This chapter examines the effects of COVID-19 prevention strategies on norovirus transmission and seasonal outbreaks, including mask-wearing, social distancing, lockdowns, and better cleanliness standards. These measures contributed to a significant drop in norovirus cases globally in the early phases of the epidemic. However, as soon as limits were relaxed, norovirus cases quickly increased again, demonstrating the virus' resilience and the challenges of maintaining control. The pandemic brought about significant changes in healthcare systems and public health procedures, which significantly affected norovirus dynamics. The complex relationship between COVID-19 interventions and norovirus circulation calls for an evaluation of its implications for future preventative measures and emphasizes the significance of strong norovirus surveillance as the world recovers from the disruptions caused by the epidemic. It also highlights the possibility of future norovirus outbreaks, especially in healthcare institutions, cruise ships, and the food sector, and talks about the long-term consequences of altered viral relationships. The pandemic highlighted the need for focused preventative strategies and vaccinations specific to noroviruses. In addition to discussing the implications of public health initiatives and the impact of COVID-19 on the virus' transmission, this research explores the virology, pathophysiology, transmission channels, epidemiological patterns, and

**Corresponding author*: shriyanshsrivastav@gmail.com

Sachin Kumar, Shriyansh Srivastava, Rishabha Malviya and Ranjit Sah (eds.) Emerging Infectious Disease: Unravelling Norovirus and Pandemic Dynamics, (295–326) © 2026 Scrivener Publishing LLC

seasonal outbreaks of norovirus. In addition, comparisons between norovirus and other viral infections offer important information about how public health interventions affect viral transmission more generally. The primary aim of this chapter is to investigate how the COVID-19 pandemic impacted research, public health initiatives, and norovirus circulation to avoid and control norovirus and related infectious disorders in the future.

***Keywords*:** Norovirus, acute gastroenteritis, coronavirus-19, genotype, transmission, vaccination, prevention strategies, surveillance

11.1 Introduction

Norovirus (NoVs) is a small, spherical virus characterized by a positive-sense, non-enveloped RNA genome, classified within the *Caliciviridae* family and the Norovirus genus [1, 2]. Noroviruses, identified in the 1970s shortly before the discovery of rotaviruses, are among the primary causes of non-bacterial gastroenteritis [3]. Although norovirus affects people of all ages and demographics globally, high-risk groups like newborns, the elderly, immunocompromised individuals, and transplant recipients have higher infection rates [4]. Currently, there are no approved treatments or vaccinations for this extremely infectious human virus [5]. It is often self-limiting, and in healthy adults, symptoms including diarrhea, vomiting, nausea, low-grade fever, and cramping in the abdomen normally go away in 2 to 4 days [6]. Noroviruses exhibit significant genetic diversity and are categorized into various genotypes, genogroups, polymerase groups (P-groups), and polymerase types (P-types). The VP1 protein's complete amino acid sequence and the RNA-dependent RNA polymerase's partial nucleotide sequence allow for the subtyping of NoVs [7]. The genus is classified into a minimum of five and potentially six genogroups (I–VI). The majority of viruses identified in genogroups I, II, and IV are linked to gastroenteritis in humans. Bovine noroviruses, murine noroviruses, and canine noroviruses represent genotypes III, V, and VI, respectively. Norovirus genogroups are subdivided into genotypes or clusters according to their genetic similarities. The majority of viruses are strains of genogroup II genotype 4 (GGII.4), which quickly change in reaction to human immunity. The novel GII.4 variations emerge biennially to quadrennially. Even so, in some regions of Asia, the recently discovered GII.P17–GII.17 genotype has lately taken over as the most common strain [8, 9]. The human norovirus is 27–35 nm in diameter and with a length of about 7.6 kb. The genome transcripts are polyadenylated at 3′ ends and then covalently linked to viral protein genomes (VPg) at 5′ ends [1]. The RNA secondary structures present in the UTRs are evolutionarily conserved and extend into the coding regions of the genome. Viral pathogenicity, translation, and replication depend on these structures [10]. Their conservation indicates that they have a high functional value across coronavirus strains, which could have consequences for the development of vaccines and therapeutic approaches. The global pandemic that resulted from the SARS-CoV-2 virus' emergence in December 2019, known as COVID-19, has resulted in millions of fatalities and enduring health complications for survivors. The World Health Organization (WHO) designated the COVID-19 pandemic as a public health emergency of international concern (PHEIC) on January 30, 2020, which was subsequently elevated to a pandemic status on March 11, 2020 [11, 12]. Although this pandemic has had a wide-ranging impact, no single worldwide strategy has been implemented to fully address the complex issues raised by such an unprecedented health disaster [11]. The pandemic caused significant disruptions to the world's economies, social institutions, and health systems, leading to a recession, a large loss of jobs, and psychological stress [13]. To curb the spread of COVID-19, governments imposed public lockdowns and vaccination campaigns [14]. This chapter examines the correlation between the COVID-19 pandemic and norovirus circulation patterns. It analyzes the influence of public health interventions implemented to curb the transmission of COVID-19 on the dissemination of norovirus and evaluates the possible long-term effects on managing outbreaks and controlling the norovirus. By comprehending these dynamics, the chapter aims to provide insightful information for upcoming public health initiatives that may minimize virus outbreaks in post-pandemic situations.

11.2 Norovirus: Characteristics and Epidemiology

Norovirus is a predominant cause of acute gastroenteritis (AGE) worldwide, accounting for approximately 60% of cases and notably impacting both young and elderly populations [15]. Epidemiological studies demonstrate that norovirus infections increase in the cooler months, especially in autumn and winter, with the highest prevalence noted in children aged 1 to 3 years. Further details regarding the features and epidemiology of norovirus are provided below [16].

11.2.1 Virology and Pathogenesis

The majority of the NoVs genome is arranged into three overlapping open reading frames (ORF) except the fourth ORF found in murine NV (MNV-1) [10, 17]. ORF1 has a length of 5,100 bases and encodes a structural polyprotein. In post-translational processing, the viral protease NS6 cleaves this polyprotein, generating six non-structural proteins. One of these proteins, NS7, functions as a 3D-like RdRp, serving a pivotal function in the replication of viral RNA [18]. ORF2 spans 1,600 bases encoding viral protein 1 (VP1), which serves as an essential capsid protein. VP1 comprises the S and P domains, and the P domain is subdivided into protruding (P1) and protruding 2 (P2). In the capsid structure, the S domain encases viral RNA and is linked to the P domain which consists of P1 and P2 through a bendable hinge [19]. One hypervariable section of the viral particle and its P2 subdomain is the most externally exposed region and is classified as a hypervariable area. ORF3, comprising 720 bases, encodes the minor capsid protein VP2. This protein resides within the viral particle and is believed to contribute to capsid stabilization, with indications that it may also facilitate capsid assemblage and enclose the genome. Moreover, during viral infection, ORF4 is translated, which is different from murine norovirus 1 and overlays with ORF2's VP1 coding area. Research employing reverse genetics has demonstrated that ORF4 is not needed for viral replication. However, the protein it encodes, known as virulence factor 1 (VF1), has been shown to enhance infection by inhibiting innate immune responses [7]. Before the viral DNA is delivered into the cytoplasm and replicates, the pathogenesis of HuNoV begins with the virus attaching itself to the target cell and progressing through receptor interaction, endocytosis, cell membrane penetration, and uncoating. The initial and frequent rate-limiting stage of virus binding to host cells is called viral entry, and it is facilitated by both host attachment factors and receptors [20]. Human noroviruses attach themselves to mucosal epithelial cells' surface expression of histo-blood group antigens (HBGAs). HBGAs are neutral carbohydrates that are connected to the P2 region of the VP1 proteins' P domain or lipids on cell surfaces to allow norovirus entrance [21]. Receptor engagement is the second step of viral pathogenesis, during which the human receptor is still unknown. The virus enters the cell through endocytosis after receptor engagement, where receptor interaction causes the minor capsid protein VP2 to generate a membrane portal that facilitates the liberation of viral DNA into the cytoplasm [22]. The VPg protein, which is covalently bound to the 5′ end of the viral genome, forms a translation complex with cellular translation initiation factors like eIF3 when the viral genome is released. Following this, the recruitment of the main and minor ribosomal subunits causes the non-structural polyprotein containing the virus' non-structural proteins to be translated [9]. The protease is automatically cleaved following the production of the non-structural polyprotein by HuNoV. The protease then cleaves the residual polyprotein both co- and post-translationally to produce three protein precursors: p48/NTPase, p22/VPg, and Pro/Pol [23]. The enzymatic function of the p48/NTPase and p22/VPg complexes is currently unknown. The Pol, VPg, and NTPase proteins work together to carry out the replication process. The latter possesses three functions: chaperone, NTP hydrolase, and helicase. After the three proteins are produced, the virions are assembled and released *via* a mechanism that has not yet been determined [24].

11.2.2 Transmission Routes

The primary means of norovirus transmission are person-to-person transmission (82%–85%), oral–fecal transmission (by the ingestion of contaminated water or food), as well as by oral contact with contaminated objects (fomites) in the environment, such as shared restrooms or elevator buttons [25, 26]. Outbreaks are more frequent in semi-enclosed and confined environments, including facilities for child care, restaurants, cruise ships, elderly homes, hospitals, and social gatherings, as well as in the military and navy, due to the low infectious dosage of norovirus [4, 27]. NoV can contaminate food products by contact with vomit or fecal material, which can happen at any point during the food preparation process. Food contamination by NoV can happen either before or after harvest. For example, fresh produce and shellfish can be irrigated with NoV-contaminated water or contaminated manure or foods like strawberries and raspberries can be picked, processed, and prepared by hand [28]. Additionally, people can also contract NoV *via* drinking water or by coming into touch with recreational surface water [28, 29]. Particles of aerosolized vomit are contagious and can spread viruses. Furthermore, persistent norovirus shedding after infection has been shown as a nosocomial transmission mechanism and may further aid in the spread of the virus [4, 30]. The prevalence of infecting genotypes varies among individuals and transmission techniques; nonetheless, GII.4 genotype strains constituted 70%–80% of all recorded outbreaks in the past 13 years. In comparison to GII.6 and GII.12, which are more commonly associated with foodborne transmission, GII.4 is more frequently connected to person-to-person transmission. Additionally, genotype GI strains are more related to aquatic transmission than GII infections (Figure 11.1) [26].

11.2.3 Global Epidemiology of Norovirus

Noroviruses are a predominant etiological agent of acute gastroenteritis (AGE) across all age demographics, accounting for almost one-fifth (20%) of all AGE incidences globally [31, 32]. About 685 million AGE instances and more than 200,000 fatalities each year have been reported, with the severity of symptoms increasing dramatically in young children, the elderly, organ transplant patients, and persons who are immunocompromised [33]. According to estimates, AGE causes 1.5 million deaths worldwide each year, with 136,000 to 278,000 of those fatalities being linked to norovirus infection. Children residing in low-income countries face a higher risk of death from norovirus gastroenteritis (AGE) due to dehydration and severe diarrhea. In contrast, fatalities due to this disease are less frequent in high-income countries and more often seen in the elderly as a result of complications like sepsis, heart problems, malnourishment, and colon perforation. Between 2012 and 2013, the prevalence of norovirus-related outpatient medically attended gastroenteritis was 5.6 cases per 1,000 person-years (PYs), whereas the overall community incidence of norovirus AGE was 68.9 cases per 1,000 PYs [34]. In 2015, three out of four cases (75.0%) had GII.17 as the predominant genotype. In 2016, five out of 19 patients (26.3%) and seven out of 19 cases (36.7%) who had GII.2, GII.4, and GII.17 were discovered, respectively [35]. In 2019, NoV killed over 130,000 people worldwide, with over 43,000 of those fatalities being in children under five and another 54,000 among those 70 and older [36]. Currently, the most commonly found genomes globally are noroviruses GI and GII variants. Specifically, the most common variation is GII.4, a genotype of GII [37].

11.2.4 Seasonal Patterns and Outbreaks

Seasonal patterns and climatic sensitivities are linked to the activities of many of the most prevalent infectious diseases, and identifying these changes in the climate or weather may have an impact on the spread of norovirus infection [38, 39]. Noroviruses are known to cause infections all year round, but they also exhibit annual fluctuations in incidence and winter seasonality in the Northern Hemisphere. The norovirus season lasted an average of 6 months nationwide, with onset taking place between October to December and offset taking place between April to May [40]. The seasonality of norovirus is linked to variations in climatic factors such as humidity, temperature cycles, wind patterns, and precipitation [41]. These factors may affect norovirus resistance, host susceptibility, and transmissibility, which in turn may affect seasonality. Rainfall has been emphasized as a significant influence in norovirus seasonality, most likely because the virus is spread through water. Similarly, low population immunity, the introduction of novel variations, and cool temperatures are linked to higher norovirus activity. Both an overall increase in sickness during the winter months and unusual spring/summer seasonality were linked to the introduction of a novel GII.4 genotype in 2002/2003 [8]. Infections with GII.4 were observed more frequently in the fall, with a double peak between 2018 and 2019. Since norovirus has long been recognized to be seasonal in the winter, it has been referred to as "winter vomiting disease" [42], with outbreaks predominantly transmitted from person to person. NoV infection has been linked to eating contaminated food from sources like salad, berries, and seafood or as a result of a food handler contaminating the food in some way while it was being processed or served. The most defecating foodborne outbreaks were linked to shellfish out of all food kinds. In a wide range of circumstances, including in military installations and weddings, food handlers contributed to defiant food-handler outbreaks using a variety of commodities. Large outbreaks of NoV have

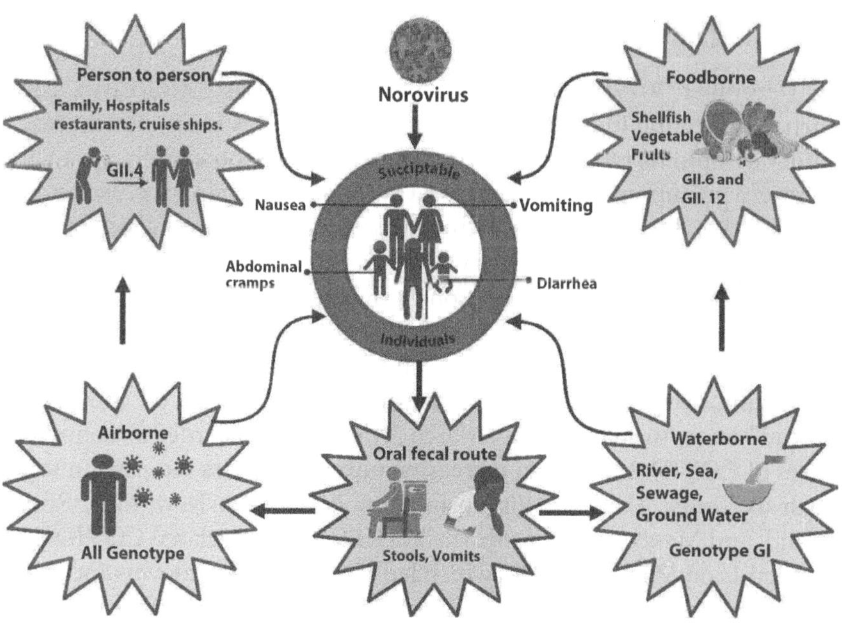

Figure 11.1 The multifaceted transmission of norovirus, including person-to-person transmission, contaminated water and food, fomites, and aerosolized particles from vomitus or feces.

been linked to sewage-contaminated water systems [43]. In total, 1,008 norovirus outbreaks (23%) were documented, with contaminated transmission serving as the primary mechanism. Among the 4,318 norovirus outbreaks, there were also reports of person-to-person transmission (2,976 [69%], environmental transmission (15 [0.35%]), waterborne transmission (11 [0.26%]), and unknown transmission (308 [7%]). Wintertime saw the highest frequency of norovirus outbreaks, with 2,394 (55%) [44].

11.3 The COVID-19 Pandemic: An Overview

The COVID-19 pandemic is a worldwide health emergency initiated by the emergence of SARS-CoV2. On March 11, 2020, the WHO designated it as a pandemic, resulting in significant health, economic, and social upheavals globally, causing millions of infections and considerable mortality. As it spreads, COVID-19 caused a variety of symptoms, ranging from minor to severe, largely affecting respiratory health and leading to numerous public health initiatives, such as mask requirements, social distancing, and immunizations [45].

11.3.1 Timeline of COVID-19 Spread

In December 2019, a new coronavirus identified as SARS-CoV2 commenced infecting individuals in Wuhan, China. This virus induces respiratory symptoms ranging from moderate to severe, potentially advancing to severe pneumonia, referred to as COVID-19. The aggregate case-fatality rate for this disease is 2.3%, with 49.0% of the cases progressing to critical conditions [46]. On January 30, 2020, the WHO declared it a public health emergency of international concern (PHEIC) [47]. During the week of February 23, 2020, the Centers for Disease Control and Prevention (CDC) in the United States reported fresh confirmed COVID-19 cases in California, Oregon, and Washington, indicating the onset of "community spread" across the nation. As of March 2, the US had recorded 33 confirmed cases of COVID-19, with an increasing number of cases reported in states such as Texas, Arizona, Wisconsin, Illinois, Florida, New York, Rhode Island, and Massachusetts [46]. The WHO formally declared COVID-19 a pandemic on March 11, 2020, which led to extensive lockdowns and travel restrictions around the world. On April 8, 2020, Wuhan formally reopened after a 76-day lockdown. The purpose of this lockdown was to stop the spread of COVID-19 [47]. A major turning point in the pandemic was reached on May 27, 2020, when the US announced that COVID-19 mortality had surpassed 100,000 [48]. Globally, there were 11 million COVID-19 instances as of July 3, 2020, demonstrating the virus' quick spread throughout several nations [49]. By mid-2021, limitations had gradually been loosened as a result of vaccination campaigns that started in late 2020 [50]. By early 2023, the Biden administration had announced intentions to end the public health emergency. The WHO declared the end of the global health emergency on May 5, 2023, which was a major turning point in the pandemic's trajectory [51].

11.3.2 Public Health Interventions and Their Impact

The ultimate purpose of public health interventions, which are control measures for public health emergencies, is to stop the epidemic from spreading and to lessen the outbreak's size. When safe and effective vaccines or antivirals are not available, public health measures like quarantine and isolation, city lockdowns, mass gathering bans, school closures, contact tracing, travel and entry restrictions, and protective masks are the main tools used to manage and lessen the pandemic pressure [52, 53]. Contact and close contact isolation and quarantine are essential for managing the COVID-19 outbreak and minimizing human-to-human transmission [54]. People who have been recognized as having contact with confirmed or suspected cases, as well as those who have recently traveled to places impacted by COVID-19, are placed in residential treatment centers or placed in quarantine at home for 2 weeks to stop the virus from spreading across the community. The spread of SARSCoV2 can also be stopped by city lockdowns, physical distancing measures, and stay-at-home directives [55]. Closing schools was intended to lower population contact rates and, consequently, the spread of the pandemic-causing coronavirus known as SARSCoV-2 [53]. Any confirmed COVID-19 case is immediately subject to epidemiological investigation and contact quarantine in order to facilitate contact tracing. When contact tracing is challenging, public announcements are used in conjunction with medical records, credit card information, closed-circuit television (CCTV), and mobile GPS [56]. Border controls, travel and entrance restrictions, and quarantine of visitors from impacted nations are all effective ways to stop the spread of the SARS-CoV-2 infection. In addition to other public health measures, these measures included imposing a 14-day quarantine on passengers upon arrival. The general public and healthcare professionals can effectively prevent SARS-CoV-2 infections by wearing masks [55].

11.3.3 Changes in Healthcare Systems

Global healthcare systems have seen substantial alterations stemming from the COVID-19 pandemic, which has shown weaknesses and forced adjustments. The quick uptake of telehealth services, which increased from almost non-existent to over 40% of outpatient visits by April 2020, is one of the most noteworthy changes. This has improved access for patients who face obstacles to receiving care in person [57]. However, the pandemic also caused significant stress in the healthcare field, with many medical personnel considering early retirement and experiencing burnout as a result of elevated stress levels [57, 58]. Additionally, there were sharp drops in the use of basic treatments like routine vaccines and cancer screenings, with a median decrease in healthcare use of 37% during the pandemic. Factors such as lockdown procedures, fear of contagion, and the reallocation of medical resources to handle COVID-19 cases were blamed for this disruption [59]. As a result, there is a rising realization of the need for systemic reforms to strengthen resilience and ensure that healthcare systems can effectively respond to future public health emergencies while sustaining vital services [58, 60].

11.4 Effects of COVID-19 on Norovirus Circulation

Norovirus circulation has been significantly impacted by the COVID-19 pandemic, which has resulted in a notable decline in outbreaks during the peak of public health strategies to control the spread of SARS-CoV2. A reported 63% decrease in outbreak notifications occurred in Germany during the pandemic, demonstrating the remarkable suppression of norovirus activity brought about by improved hygiene measures, social distancing, and gathering limitations [61].

11.4.1 Changes in Reporting and Surveillance

Changes in reporting and surveillance have been significantly influenced by the COVID-19 pandemic and have altered healthcare-seeking habits, disrupted service delivery, and altered local capacities to detect, test, and report norovirus cases or outbreaks. Nevertheless, the decline in norovirus reporting is believed to stem from multiple factors. Furthermore, non-pharmaceutical interventions (NPIs) enacted in England to mitigate the transmission of SARS-CoV-2 have likely led to an actual reduction in norovirus transmission [62]. There was a significant decrease in norovirus cases during the pandemic due to stringent public health efforts; laboratory-confirmed reports decreased by more than 90% from pre-pandemic levels. To promote a more thorough public health response, the new system seeks to improve the integration of data on several respiratory viruses, including norovirus, influenza, and respiratory syncytial virus (RSV) [63]. Sustained investment in reliable surveillance systems is necessary to track these trends and efficiently inform public health actions [64].

11.4.2 Impact of Non-Pharmaceutical Interventions (NPIs)

Non-pharmaceutical interventions (NPIs) have been widely used in the United States and other countries since March 2020 to prevent the spread of SARS-CoV-2. These include travel restrictions both domestically and abroad, lockdowns and stay-at-home guidelines, universal handwashing instructions, physical distancing, and face covering [65]. These changes may also have an impact on the spread of other infections, such as norovirus, and probably decrease the prevalence of influenza in other nations. NPIs for SARS-CoV-2 prevention started close to the conclusion of the main wintertime disease transmission season, which includes norovirus [66]. A significant decrease in COVID-19 hospitalizations and fatalities has been noted since the introduction of the COVID-19 vaccine in England, which has replaced the ineffective NPIs [67]. Because NPIs are not pathogen-specific, their use can also impact the spread of other infectious agents. This effect is best demonstrated by the 2020–2021 influenza season being shortened or even nonexistent, as well as a change in the seasonal peak of respiratory syncytial virus (RSV) cases, which was postponed by the strict NPIs but recovered in 2021 once the strictness of these measures was reduced [68]. These trends were noted for respiratory syncytial virus (RSV), rhinovirus infections, hand, foot, and mouth disease, norovirus gastroenteritis, and influenza, all of which saw reductions attributed to strategies including mask usage, social distancing and hand hygiene [69]. NPIs are a useful tool for reducing the spread of infectious diseases, and risk patterns differ according to age, socioeconomic position, and NPI levels. These results have significant ramifications for developing focused infectious disease prevention strategies [70].

11.4.3 Predictions for Norovirus Resurgence Post-Pandemic

Predictions regarding the resurgence of norovirus post-pandemic indicate a significant increase in infection cases due to population susceptibility. According to research, non-pharmaceutical interventions (NPIs) significantly decreased norovirus transmission during the COVID-19 pandemic, and several vulnerable individuals accumulated. According to a mathematical modeling study, the incidence of norovirus might quadruple in comparison to normal years if contact patterns return to pre-pandemic levels [67, 71]. This risk is increased by the temporary nature of norovirus immunity, which is mostly obtained by infection, as well as the possibility of new variations developing [72]. Since the reappearance may result in more hospital admissions and strain on healthcare systems, especially during the winter months when norovirus usually people, it is thought that ongoing nationwide surveillance is crucial for public health preparation [73]. To mitigate the effects of upcoming outbreaks, the implications of these findings highlight the necessity of preventative actions and strong monitoring systems. Knowing the dynamics of norovirus transmission will be essential for allocating resources and implementing effective public health responses as communities adjust to shifting contact patterns following COVID-19 [67, 71].

11.5 Research Trends and Findings

Recent studies on norovirus underscore escalating concerns about its prevalence and genetic development, especially in the outcome of the COVID-19 pandemic. The primary circulating strains predominantly belong to genogroup 2, particularly genotype GII.17, which has been linked to numerous outbreaks. Continuous molecular surveillance is essential for monitoring these alterations and preventing future outbreaks, as well as comprehending their interaction with COVID-related changes in infection patterns [74].

11.5.1 Changes in Norovirus Research Focus During COVID-19

The focus of norovirus research during the COVID-19 pandemic changed dramatically from developing vaccines and conducting conventional epidemiological studies to comprehending the pandemic's effects on the dynamics of norovirus transmission and getting ready for any resurgences. Prior to the pandemic, studies focused mostly on the epidemiology, genetic diversity, and prevention of norovirus, specifically examining different genotypes such as GII.4 and the difficulties in developing a vaccine because of their high range [75]. At first, strict public health measures significantly decreased the number of norovirus cases worldwide. Reduced mobility and improved hygiene standards, for example, were responsible for a reported 90.4% decrease in norovirus cases in the United States between April 2020 and April 2021 when compared to the prior year [71, 76]. The pandemic also made it clear how important it is to keep up strong norovirus epidemiological surveillance. Although the number of cases initially decreased, the research cautioned that populations were more susceptible because of decreased exposure during lockdowns, which could result in higher incidence rates of norovirus-related acute gastroenteritis (AGE) once regular contact patterns resumed [77, 78]. Additionally, because the pandemic disrupted conventional surveillance systems, researchers have adopted sophisticated statistical modeling techniques to forecast norovirus activity. These algorithms improved prediction accuracy by using external data sources like internet searches and mobility patterns. As society returns to normal, this change emphasizes the necessity for public health organizations to modify their surveillance plans and get ready for any spikes in norovirus cases [78].

11.5.2 Advances in Norovirus Vaccine Development

Numerous researchers are working on a number of NoV subunit vaccine candidates derived from virus-like particles (VLPs) and particles that are undergoing clinical and preclinical studies.

11.5.2.1 Vaccine in Human Clinical Trials

Two vaccines just received approval for clinical trial testing, and two more are presently undergoing clinical trials.

a. Monovalent GI.1 oral vaccine (Vaxart oral vaccine): Vaxart, a biotechnology company, is working on a recombinant vaccine designed to prevent norovirus infections. This vaccine will be administered to adults in the form of an oral tablet and to children as a liquid. Since these formulations are stable even when kept at ambient temperature, a "cold chain" is unnecessary to keep them in quality. Using a non-replicating adenovirus 5 vector, it expresses the VP1 gene of the gastrointestinal virus; it was created as a monovalent vaccination. Additionally, it includes adjuvant double-stranded RNA and a single strain of norovirus [79, 80]. This monovalent norovirus vaccine was found to be safe and well-tolerated in the phase 1 clinical study. It effectively generated systemic, mucosal, and memory IgA/IgG antibodies (Table 11.1) [81].

Table 11.1 Advances in norovirus vaccine development with their target genotypes.

Vaccine type	Manufacturer	Target genotype	Study phase	Notable finding	References
Monovalent GI.1 oral vaccine	Vaxart Biotechnology Company	GI.1	Phase 1	Well-tolerated, safe, producing systemic mucosal and IgA/ IgG antibodies	[79, 81]
Bivalent GI.1/GII.4 vaccine	Takeda Pharmaceutical Company Limited	GI.1, GII.4	Clinical	Reduced norovirus-associated AGE in trials; reformulated from monovalent to bivalent formulation	[81]
Bivalent GI.1/GII.4 vaccine	National Vaccine and Serum Institute (NVSI)	GI.1, GII.4	Clinical	Generated IgG antibody and HBGA-blocking antibody with noted safety outcomes	[84]
Tetravalent vaccine	Institute Pasteur of Shanghai (IPS) (Chinese Academy of Science)	GII.4, GII.2, GII.3, and GI.1	Clinical	Currently under study; formulation details pending	[81]
HuNoV + rotavirus (trivalent)	-	HuNoV GI.3, GII.4-1999, RV VP6	Preclinical	Induced immune response for HuNoV and RV; designed for broader AGE protection	[85, 86]
Combined norovirus + enterovirus vaccine	-	GII.4 (NoV), EV71 (enterovirus)	Preclinical	Induced comparable immune responses in combined vs. monovalent vaccines without immune interference	[88]
HuNoV P particle-based vaccines	-	Varies	Preclinical	Induced higher T-cell responses (CD4+, IFN-γ+ CD8+, Tregs, TGF-β+) compared to VLP-based vaccines	[87]

b. Bivalent GI.1/GII.4 vaccine (Takeda vaccine): Takeda Pharmaceutical Company Limited is developing a bivalent vaccine targeting GI.1 and GII.4 norovirus strains. Initially, this vaccine was created as an intranasal virus-like particle (VLP) vaccine with a GI.1 genotype and adjuvants of chitosan and monophosphoryl lipid A [82]. A monovalent intranasal vaccine has no adverse effects and decreased AGE related to norovirus following exposure to a similar virus, as indicated by challenging research findings in a human clinical trial (Table 11.1) [83]. Following this, Takeda used an amino acid consensus from the GII.4 Houston, Yerseke, and Den Haag variant strains for the GII.4 component to assemble the bivalent GI.1/GII.4 VLP vaccine [81].

c. Bivalent GI.1/GII.4 vaccine (National Vaccine and Serum Institute): The NVSI of China created a bivalent VLP base vaccine for Hansenulapolymorpha, which is the third norovirus vaccine undergoing clinical trials. It is composed of two recombinant VLPs, one for the GI.1 genotype and one for the GII.4 genotype. The vaccination was injected into

Table 11.2 Comparative analysis of norovirus with other viruses based on their hosts, transmission, incubation period, symptoms, and prevention strategies.

Virus	Host(s)	Transmission	Incubation period	Primary symptoms	Prevention methods	References
Norovirus	Humans	Fecal–oral, contaminated food/water, human-to-human	12–48 h	Nausea, vomiting, diarrhea, stomach pain	Hand hygiene, disinfection, safe food handling	[99]
Sapovirus	Humans, pigs, and other animals	Fecal–oral, contaminated food/water	1–4 days	Mild vomiting, diarrhea, stomach cramps	Hand hygiene, sanitation	[100]
Lagovirus	Rabbits and hares	Contaminated food/water, direct contact	1–3 days	Hemorrhagic disease, liver damage	Vaccination in certain areas	[101]
Vesivirus	Cats, dogs, marine mammals	Fecal–oral respiratory droplets, contaminated object	2–7 days	Respiratory symptoms, mouth ulcer, fever	Sanitation, quarantine measures	[102]
Nebovirus	Young calf	Fecal–oral, contaminated water/food	12 to 48 h	Diarrhea, mild gastrointestinal symptoms	Improved sanitation, water treatment	[103]

the muscle in two or three doses. The outcomes were measured by looking at things like GI.1 and GII.4 norovirus-specific IgG antibody titers and positive rates as well as HBGA-blocking antibody titers and matching positive rates (Table 11.1) [84].

d. Tetravalent vaccine (Chinese Academy of Sciences): The Chinese National Medical Product Administration granted the Chinese Academy of Sciences' Institute Pasteur of Shanghai (IPS) a clinical research permit in May 2019 to investigate a tetravalent vaccination that had undergone 4 years of preclinical development before a human clinical trial (Table 11.1). This tetravalent vaccine's formulation has not yet been released [81].

11.5.2.2 Vaccines in Pre-Clinical Trials

a. Vaccines based on HuNoV VLPs: Several norovirus vaccines in preclinical trials have been developed as multivalent immunizations to protect against human norovirus as well as other viruses including rotavirus and enterovirus. AGE induced by HuNoV and RV can be mitigated through the integration of norovirus and rotavirus vaccinations; formulations incorporating both HuNoV and RV immunogens have been investigated. Initially, only human norovirus GII.4 virus-like proteins were chosen for the combination vaccine, however, the central capsid VP6 of rotavirus was nominated because of its notable conservation among group A rotaviruses, as well as its impressive immunogenic properties and adjuvant capabilities [85, 86]. A trivalent vaccine was developed that includes human norovirus GII.4-1999 and GI.3 VLPs along with oligomeric rotavirus VP6, as the initial combination vaccination did not successfully stimulate the production of neutralizing antibodies against heterologous human norovirus genotypes [87]. A novel vaccination combines enterovirus and NoV, demonstrating efficacy against both pathogens, potentially benefiting regions such as the Asia-Pacific where these diseases are more prevalent. A formulation containing GII.4 and EV71 VLPs was developed to assess immunogenicity in mouse models, and it was compared to monovalent GII.4 and EV71 VLPs (Table 11.1) [88].

b. Vaccine based on HuNoV P particles: HuNoV combination vaccines employing P particles have been designed to avert several possible viral infections, integrating immunogens from other viruses. P particles induced more robust immune responses compared to VLPs, as the significantly increased counts of stimulated CD4+ T cells in every organ, IFN-γ+ CD8+ T cells in the duodenum, regulatory T cells in the circulation, and TGF-β+, CD4+, CD25−FoxP3+ cells are evident, T regulatory cells within the spleen (Table 11.1) [87].

11.5.3 Comparative Studies of Norovirus and Other Viruses

According to the International Committee on Taxonomy of Viruses (ICTV), the family *Caliciviridae* includes four other recognized genera in addition to noroviruses: Sapovirus, Lagovirus, Vesivirus, and Nebovirus [89]. Although the genomic architectures of all *Caliciviridae* are closely related, they differ greatly in terms of genetics and antigens and infect a variety of mammalian host species, including humans [90].

11.5.3.1 Sapoviruses

Both sapoviruses and noroviruses are members of the *Caliciviridae* family, but despite their similarities, they have different epidemiological and clinical traits. Noroviruses are known to be the main causes of non-bacterial pandemic gastroenteritis in humans, and they frequently cause severe outbreaks, especially in enclosed spaces like hospitals [91, 92]. Conversely, sapoviruses are less commonly linked to clinical symptoms; many infections happen without any symptoms, and they mainly affect elderly people and children under five (Table 11.2) [101].

11.5.3.2 Lagovirus

Although both norovirus and lagovirus belong to the *Caliciviridae* family, their host specificity, clinical effects, and epidemiology differ greatly. Mostly affecting humans, norovirus is a major cause of AGE, accounting for around 23 million incidences in the US alone each year. Symptoms include diarrhea, vomiting, and cramping in the abdomen which usually lasts 1 to 3 days [1, 93]. In contrast, lagovirus, which contains the rabbit hemorrhagic sickness virus (RHDV), mostly affects lagomorphs like hares and rabbits and causes high mortality rates and severe hemorrhagic sickness (Table 11.2) [94].

11.5.3.3 *Vesivirus*

Vesivirus and norovirus have quite different effects on human health. A well-known human pathogen that causes acute gastroenteritis outbreaks and is mainly spread by the oral–fecal route is norovirus, which is a member of the Norovirus genus [95]. In contrast, Vesivirus, which comprises viruses like vesicular exanthema of swine virus, mostly infects animals and has a varied host range, including some cases of mild infections in humans (Table 11.2) [96].

11.5.3.4 *Nebovirus*

Both nebovirus and norovirus mainly infect cattle and cause gastroenteritis, especially in young calves. Bovine norovirus, or BoNoV, is a type of norovirus that is known to cause acute gastroenteritis, which is typified by vomiting and severe diarrhea. It is widespread throughout different regions and has a genome of about 7.5 kb. Research suggests that it is a major pathogen that causes calf diarrhea worldwide. Conversely, nebovirus, which has been found less frequently than norovirus and has a slightly smaller genome (roughly 7.4 kb), nevertheless plays a significant role in calf diarrhea (Table 11.2) [97, 98].

11.6 Public Health Implications

Norovirus presents considerable public health concerns due to its extreme contagiousness and the considerable morbidity it inflicts, especially among at-risk groups like small children and the elderly [104].

11.6.1 Preparedness for Future Outbreaks

Preparedness for future outbreaks, particularly for norovirus, is essential to minimize their effect on public health. The norovirus a major reason of gastroenteritis, is notoriously hard to control because of its short-lived immunity and high contagiousness. Effective management techniques include the creation of outbreak control teams capable of carrying out quick response actions, like cleaning and decontaminating impacted areas right away, preventing employees with symptoms from returning to work for at least 48 h after they appear and strictly adhering to hand hygiene guidelines [105, 106]. During outbreaks, ongoing risk assessment aids in identifying the appropriate control measures, such as patient testing with sensitive PCR techniques to verify infections [107]. Furthermore, healthcare facilities must to implement uniform infection control protocols and guarantee that all employees receive prompt training on identifying and reporting outbreaks [106]. Strong monitoring systems must be integrated for early detection and reaction in order to enable prompt actions that can drastically lower transmission rates [108]. Overall, controlling norovirus outbreaks and strengthening resistance to future public health risks require a multidisciplinary strategy that incorporates preparedness planning, community involvement, and education [109, 110]. Since there is presently no approved norovirus vaccine, vaccine development is also essential. However, advancements in this field could greatly lessen the severity of outbreaks. AI, predictive analytics, and international cooperation can help coordinate responses even more, maximizing resource allocation and improving outbreak preparedness [111, 112].

11.6.2 Strategies for Mitigating Norovirus Spread

Norovirus mitigation necessitates a multifaceted and deliberate strategy, especially in healthcare settings. Important tactics include putting in place stringent infection control protocols, such as hand washing, enhanced ward closures, visitor restrictions, staff absence from work, patient isolation and cohorts, and environmental cleansing and disinfecting [112, 113]. It has been shown that practicing good hand hygiene can effectively stop the spread of the norovirus during foodborne outbreaks [114]. Hand washing thoroughly with running water and plain or antiseptic soap for 20 s is the best way to reduce any norovirus that may be on hands. There is ongoing debate on the effectiveness of alcohol-based and alternative hand sanitizers against norovirus [115]. The spread of norovirus may be managed and stopped by improved cleaning and disinfection practices. This means cleaning more often and paying close attention to areas that see a lot of traffic and surfaces that are touched frequently, such door handles and telephones. To disinfect, use a bleach solution that is made fresh each day and contains at least 1,000 parts per million sodium hypochlorite. There are EPA-registered compounds, notably List G, that can effectively combat norovirus surrogates in locations where bleach is not accessible or where objects corrode [113]. It is recommended that patients experiencing norovirus gastroenteritis remain on contact precautions for at least 48 h following the onset of symptoms. When symptomatic patients cannot be placed in single occupancy rooms, it is crucial to attempt to categorize them into three different groups: symptomatic, exposed but asymptomatic, and unexposed [116]. To stop the virus from spreading further, other tactics include putting a 48-h return-to-work policy in place after symptoms have subsided and immediately excluding employees who show symptoms [105]. Strict visiting limits and ward closures are necessary in healthcare settings in order to prevent the transmission of infection. Only essential personnel should be permitted to reach impacted areas during an outbreak, and non-essential visitors should be subject to a temporary suspension of visiting (TSV). This action lowers the chance of additional transmission among susceptible groups, including the elderly and individuals with compromised immune systems [117, 118]. Another crucial tactic is ward closures, which keep the outbreak under control by limiting patient transfers to other wards or facilities and preventing new admissions [105, 119].

11.6.3 Lesson Learned from COVID-19

Lessons learned from the COVID-19 pandemic include the necessity of producing and distributing point-of-care diagnostic tests quickly and on a large scale, the necessity of addressing the trade of wildlife and ecosystem loss as major contributors to the emergence of infectious diseases, and more. Future public health challenges, particularly the development of novel diseases, will require the application of the lessons learned from the COVID-19 pandemic [120]. Firstly, the requirement of early and decisive action was underlined, as countries that enacted rapid public health measures, such as lockdowns and travel restrictions, managed to prevent the virus' spread more successfully than those that delayed action [121]. Secondly, the pandemic highlighted significant healthcare inequities, underlining the need for equitable access to healthcare services, particularly for marginalized communities that faced higher chances of severe outcomes [122, 123]. Thirdly, the pandemic made clear how crucial it is to maintain investments in research and development, especially in the field of vaccine technology, which allowed for the quick creation of potent vaccines [124, 125]. Furthermore, the pandemic hastened the use of telehealth services, indicating a possible change in the way medical treatment is provided going forward [126]. Last but not least, the crisis demonstrated the importance of community resilience and adaptability, as people and organizations came up with creative ways to help one another through trying times. Taken together, these lessons emphasize the necessity of readiness, equity, and cooperation in tackling future public health challenges [127].

Conclusion

In conclusion, this chapter has explored the complex relationship between COVID-19 and norovirus transmission, highlighting the unique traits and epidemiology of norovirus and the important impacts the COVID-19 pandemic has had on its monitoring, research, and transmission. The decline in norovirus outbreaks during the initial phases of the pandemic—driven by extensive public health measures such as heightened hygiene practices, social distancing, and mask-wearing—has offered a temporary respite. However, as these measures are gradually relaxed, there are increased worries about a potential resurgence of norovirus, which poses important challenges for public health preparedness. While research efforts predominantly shifted towards COVID-19 during the pandemic, this period also catalyzed advancements in virology, diagnostics, and vaccine development that could enhance norovirus research moving forward. As we look to the future, it is essential to reassess public health strategies aimed at reducing the spread of norovirus in light of the lessons derived from the COVID-19 pandemic. Improving surveillance systems, increasing preparedness for possible outbreaks, and refocusing on norovirus vaccine development are all critical steps to lessen the effect of future epidemics. The combination of pandemic experiences with the persistent threats of gastrointestinal viruses such as norovirus underscores the importance of constant attention and flexible approaches to public health protection as the world community develops.

Abbreviations

COVID-19	coronavirus disease-19
NoVs	norovirus
RNA	ribonucleic acid
P-groups	polymerase groups
P-types	polymerase types
RdRp	RNA-dependent RNA polymerase
VP1	viral protein 1
VPg	viral protein genome
UTRs	untranslated regions
WHO	World Health Organization
PHEIC	public health emergency of international concern
SARS-CoV-2	severe acute respiratory syndrome coronavirus

AGE	acute gastroenteritis
ORFs	open reading frames
MNV-1	murine NV
NS7	non-structural protein 7
P1	protruding 1
P2	protruding 2
VF1	virulence factor 1
DNA	deoxyribonucleic acid
HuNoV	human norovirus
HBGAs	histo-blood group antigens
NTPase	nucleoside triphosphatase
Pro/Pol	protease–polymerase
PYs	person-years
CDC	Centers for Disease Control and Prevention
CCTV	closed-circuit television
GPS	global positioning system
NPIs	non-pharmaceutical interventions
RSV	respiratory syncytial virus
VLPs	virus-like particles
IgA	immunoglobulin A
IgG	immunoglobulin G
NVSI	National Vaccine and Serum Institute
IPS	Institute Pasteur of Shanghai
RV	rotavirus
EV71	enterovirus 71
IFN-γ+	interferon-gamma
TGF-β+	transforming growth factor-beta
CD25	cluster of differentiation 25
FoxP3+	forkhead box P3
Tregs	regulatory T cell
ICTV	International Committee on Taxonomy of Viruses
RHDV	rabbit hemorrhagic sickness virus
BoNoV	bovine norovirus
PCR	polymerase chain reaction
EPA	Environmental Protection Agency
TSV	temporary suspension of visiting

AI disclosure statement: The author acknowledges the use of AI-based tools for minor language editing or grammatical correction. However, all intellectual contributions, analyses, and conclusions presented in this chapter are entirely the author's own.

References

1. Robilotti, E., Deresinski, S., Pinsky, B.A., Norovirus. *Clin. Microbiol. Rev.*, 28, 1, 134–64, 2015.
2. Karst, S.M. and Tibbetts, S.A., Recent advances in understanding norovirus pathogenesis. *J. Med. Virol.*, 88, 11, 1837–43, 2016.
3. Weinberg, G.A., *Outbreak epidemiology: one of many new frontiers of norovirus biology*, pp. 1349–52, Oxford University Press, US, 2019. https://doi.org/10.1093/infdis/jiy570
4. Netzler, N.E., Enosi Tuipulotu, D., White, P.A., Norovirus antivirals: Where are we now? *Med. Res. Rev.*, 39, 3, 860–86, 2019.
5. Lee, S., Liu, H., Wilen, C.B., Sychev, Z.E., Desai, C., Hykes, B.L., *et al.*, A secreted viral nonstructural protein determines intestinal norovirus pathogenesis. *Cell Host Microbe*, 25, 6, 845–57.e5, 2019.
6. Matthews, J., Dickey, B., Miller, R., Felzer, J., Dawson, B., Lee, A., *et al.*, The epidemiology of published norovirus outbreaks: a review of risk factors associated with attack rate and genogroup. *Epidemiol. Infect.*, 140, 7, 1161–72, 2012.
7. Winder, N., Gohar, S., Muthana, M., Norovirus: an overview of virology and preventative measures. *Viruses*, 14, 12, 2811, 2022.
8. Ahmed, S.M., Lopman, B.A., Levy, K., A systematic review and meta analysis of the global seasonality of norovirus. *PLoS One*, 8, 10, e75922, 2013.
9. Karst, S.M., Wobus, C.E., Goodfellow, I.G., Green, K.Y., Virgin, H.W., Advances in norovirus biology. *Cell Host Microbe*, 15, 6, 668–80, 2014.
10. Thorne, L.G. and Goodfellow, I.G., Norovirus gene expression and replication. *J. Gen. Virol.*, 95, 2, 278–91, 2014.
11. De Bruin, Y.B., Lequarre, A.-S., McCourt, J., Clevestig, P., Pigazzani, F., Jeddi, M.Z., *et al.*, Initial impacts of global risk mitigation measures taken during the combatting of the COVID-19 pandemic. *Saf. Sci.*, 128, 104773, 2020.
12. Srivastava, S., Kumar, S., Ashique, S., Satapathy, P., Mehta, R., Mohanty, A., Sah, R., Fast-Spreading JN. 1: What You Need to Know About the Latest COVID-19 Subvariant. *J. Infect. Public Health*, 17, 7, 102451, 2024 May 14.
13. Clemente-Suárez, V.J., Navarro-Jiménez, E., Jimenez, M., Hormeño-Holgado, A., Martinez-Gonzalez, M.B., Benitez-Agudelo, J.C., *et al.*, Impact of COVID-19 pandemic in public mental health: An extensive narrative review. *Sustainability*, 13, 6, 3221, 2021.
14. Arbel, R. and Pliskin, J., Vaccinations versus lockdowns to prevent COVID-19 mortality. *Vaccines*, 10, 8, 1347, 2022.
15. Ahmed, S.M., Hall, A.J., Robinson, A.E., Verhoef, L., Premkumar, P., Parashar, U.D., Koopmans, M., Lopman, B.A., Global prevalence of norovirus in cases of gastroenteritis: a systematic review and meta-analysis. *Lancet Infect. Dis.*, 14, 8, 725–30, 2014 Aug 1.
16. Yang, L., Shi, S., Na, C., Li, B., Zhao, Z., Yang, T., Yao, Y., Rotavirus and norovirus infections in children under 5 years old with acute gastroenteritis in

Southwestern China, 2018–2020. *J. Epidemiol. Global Health*, 12, 3, 292–303, 2022 Sep.

17. Sarmento, S.K., JdSR, .A., MP, M., Fumian, T.M., Virological and Epidemiological Features of Norovirus Infections in Brazil, 2017–2018. *Viruses*, 13, 9, 1724, 2021.
18. Roth, A.N. and Karst, S.M., Norovirus mechanisms of immune antagonism. *Curr. Opin. Virol.*, 16, 24–30, 2016.
19. Lee, C.-C., Chiu, C.-H., Lee, H.-Y., Tsai, C.-N., Chen, C.-L., Chen, S.-Y., Clinical and virological characteristics of viral shedding in children with norovirus gastroenteritis. *J. Microbiol. Immunol. Infect.*, 55, 6, 1188–94, 2022.
20. Graziano, V.R., Wei, J., Wilen, C.B., Norovirus attachment and entry. *Viruses*, 11, 6, 495, 2019.
21. Karst, S.M., Pathogenesis of noroviruses, emerging RNA viruses. *Viruses*, 2, 3, 748–81, 2010.
22. Vera-Garcia, M.E., Martinez-Acuña, N., Lozano-Sepulveda, S.A., Arellanos-Soto, D., Galan-Huerta, K.A., Merino-Mascorro, J.A., *et al.*, Unraveling the mechanisms involved in human norovirus pathogenesis: An update in norovirus pathogenesis. *Med. Universitaria*, 23, 4, 142–7, 2021.
23. Belliot, G., Sosnovtsev, S.V., Chang, K.-O., Babu, V., Uche, U., Arnold, J.J., *et al.*, Norovirus proteinase-polymerase and polymerase are both active forms of RNA-dependent RNA polymerase. *J. Virol.*, 79, 4, 2393–403, 2005.
24. Campillay-Véliz, C.P., Carvajal, J.J., Avellaneda, A.M., Escobar, D., Covián, C., Kalergis, A.M., *et al.*, Human norovirus proteins: implications in the replicative cycle, pathogenesis, and the host immune response. *Front. Immunol.*, 11, 961, 2020.
25. de Graaf, M., Villabruna, N., Koopmans, M.P., Capturing norovirus transmission. *Curr. Opin. Virol.*, 22, 64–70, 2017.
26. Gaythorpe, K., Trotter, C.L., Lopman, B., Steele, M., Conlan, A., Norovirus transmission dynamics: a modelling review. *Epidemiol. Infect.*, 146, 2, 147–58, 2018.
27. Towers, S., Chen, J., Cruz, C., Melendez, J., Rodriguez, J., Salinas, A., *et al.*, Quantifying the relative effects of environmental and direct transmission of norovirus. *R. Soc. Open Sci.*, 5, 3, 170602, 2018.
28. Mathijs, E., Stals, A., Baert, L., Botteldoorn, N., Denayer, S., Mauroy, A., *et al.*, review of known and hypothetical transmission routes for noroviruses. *AFood Environ. Virol.*, 4, 131–52, 2012.
29. De Graaf, M., van Beek, J., Koopmans, M.P., Human norovirus transmission and evolution in a changing world. *Nat. Rev. Microbiol.*, 14, 7, 421–33, 2016.
30. Karst, S.M., Zhu, S., Goodfellow, I.G., The molecular pathology of noroviruses. *J. Pathol.*, 235, 2, 206–16, 2015.
31. Cannon, J.L., Bonifacio, J., Bucardo, F., Buesa, J., Bruggink, L., Chan, M.C.-W., *et al.*, Global trends in norovirus genotype distribution among children with acute gastroenteritis. *Emerging Infect. Dis.*, 27, 5, 1438, 2021.

32. Wei, N., Ge, J., Tan, C., Song, Y., Wang, S., Bao, M., *et al.*, Epidemiology and evolution of Norovirus in China. *Hum. Vaccin. Immunother.*, 17, 11, 4553–66, 2021.
33. Jahun, A.S. and Goodfellow, I.G., Interferon responses to norovirus infections: current and future perspectives. *J. Gen. Virol.*, 102, 10, 001660, 2021.
34. Carlson, K.B., Dilley, A., O'Grady, T., Johnson, J.A., Lopman, B., Viscidi, E., A narrative review of norovirus epidemiology, biology, and challenges to vaccine development. *npj Vaccines*, 9, 1, 94, 2024.
35. Honjo, S., Kuronuma, K., Fujiya, Y., Nakae, M., Ukae, S., Nihira, H., *et al.*, Genotypes and transmission routes of noroviruses causing sporadic acute gastroenteritis among adults and children, Japan, 2015–2019. *Infection Genet. Evol.*, 104, 105348, 2022.
36. Zhang, P., Hao, C., Di, X., Chuizhao, X., Jinsong, L., Guisen, Z., *et al.*, Global prevalence of norovirus gastroenteritis after emergence of the GII. 4 Sydney 2012 variant: a systematic review and meta-analysis. *Front. Public Health*, 12, 1373322, 2024.
37. Wang, J., Gao, Z., Z-r, Y., Liu, K., Zhang, H., Global prevalence of asymptomatic norovirus infection in outbreaks: a systematic review and meta-analysis. *BMC Infect. Dis.*, 23, 1, 595, 2023.
38. Yadav, N. and Upadhyay, R.K., Global effect of climate change on seasonal cycles, vector population and rising challenges of communicable diseases: a review. *J. Atmos. Sci. Res.*, 6, 1, 21–59, 2023.
39. Al Meslamani, A.Z., How climate change influences pathogen transmission. *Pathog. Global Health*, 118, 5, 450–2, 2024.
40. Bezirtzoglou, C., Dekas, K., Charvalos, E., Climate changes, environment and infection: facts, scenarios and growing awareness from the public health community within Europe. *Anaerobe*, 17, 6, 337–40, 2011.
41. Kambhampati, A.K., Calderwood, L., Wikswo, M.E., Barclay, L., Mattison, C.P., Balachandran, N., *et al.*, Spatiotemporal trends in norovirus outbreaks in the United States, 2009–2019. *Clin. Infect. Dis.*, 76, 4, 667–73, 2023.
42. Chiu, S.-C., Hu, S.-C., Liao, L.-M., Chen, Y.-H., Lin, J.-H., Norovirus Genogroup II Epidemics and the Potential Effect of Climate Change on Norovirus Transmission in Taiwan. *Viruses*, 14, 3, 641, 2022.
43. Hardstaff, J.L., Clough, H.E., Lutje, V., McIntyre, K.M., Harris, J.P., Garner, P., *et al.*, Foodborne and food-handler norovirus outbreaks: a systematic review. *Foodborne Pathog. Dis.*, 15, 10, 589–97, 2018.
44. Hall, A.J., Wikswo, M.E., Pringle, K., Gould, L.H., Parashar, U.D., Vital signs: foodborne norovirus outbreaks—United States, 2009–2012. *MMWR Morb. Mortal Wkly Rep.*, 63, 22, 491–5, 2014.
45. Ferrer, R., COVID-19 Pandemic: the greatest challenge in the history of critical care. *Med. Intensiva*, 44, 6, 323, 2020 Aug.
46. Lin L., Zhao Y., Chen B., He D. Multiple COVID-19 waves and vaccination effectiveness in the United States. *Int. J. Environ. Res. Public Health*, 19, 4, 2022.

47. Carvalho, T., Krammer, F., Iwasaki, A., The first 12 months of COVID-19: a timeline of immunological insights. *Nat. Rev. Immunol.*, 21, 4, 245–56, 2021.
48. News a. COVID-19 timeline: How the deadly virus and the world's response have evolved over 4 years 2024 [Available from: https://abcnews.go.com/Health/covid-19-timeline-deadly-virus-worlds-response-evolved/story?id=107880313.
49. Zhou, Y., Li, J., Chen, Z., Luo, Q., Wu, X., Ye, L., *et al.*, The global COVID-19 pandemic at a crossroads: relevant countermeasures and ways ahead. *J. Thoracic Dis.*, 12, 10, 5739, 2020.
50. Basu, S., Ashok, G., Debroy, R., Ramaiah, S., Livingstone, P., Anbarasu, A., Impact of the COVID-19 pandemic on routine vaccine landscape: A global perspective. *Hum. Vaccin. Immunother.*, 19, 1, 2199656, 2023.
51. News A. Biden Administration Announces End of COVID-19 Public Health Emergency 2023 [Available from: https://news.thoracic.org/washington-letter/2023/biden-administration-announces-end-of-covid-19-public-health-emergency.php.
52. Xiang, Y., Jia, Y., Chen, L., Guo, L., Shu, B., Long, E., COVID-19 epidemic prediction and the impact of public health interventions: A review of COVID-19 epidemic models. *Infect. Dis. Modell.*, 6, 324–42, 2021.
53. Jüni, P., Rothenbühler, M., Bobos, P., Thorpe, K.E., da Costa, B.R., Fisman, D.N., *et al.*, Impact of climate and public health interventions on the COVID-19 pandemic: a prospective cohort study. *Cmaj*, 192, 21, E566–E73, 2020.
54. Kucharski, A.J., Klepac, P., Conlan, A.J., Kissler, S.M., Tang, M.L., Fry, H., *et al.*, Effectiveness of isolation, testing, contact tracing, and physical distancing on reducing transmission of SARS-CoV-2 in different settings: a mathematical modelling study. *Lancet Infect. Dis.*, 20, 10, 1151–60, 2020.
55. Ayouni, I., Maatoug, J., Dhouib, W., Zammit, N., Fredj, S.B., Ghammam, R., *et al.*, Effective public health measures to mitigate the spread of COVID-19: a systematic review. *BMC Public Health*, 21, 1, 1015, 2021.
56. Lee, H., Lee, H., Song, K.-H., Kim, E.S., Park, J.S., Jung, J., *et al.*, Impact of public health interventions on seasonal influenza activity during the COVID-19 outbreak in Korea. *Clin. Infect. Dis.*, 73, 1, e132–e40, 2021.
57. Cutler, D.M. (Ed.), Health System Change in the Wake of COVID-19, in: *JAMA Health Forum*, American Medical Association, 2023. doi:10.1001/jamahealthforum.2023.4355
58. Filip, R., Gheorghita Puscaselu, R., Anchidin-Norocel, L., Dimian, M., Savage, W.K., Global challenges to public health care systems during the COVID-19 pandemic: a review of pandemic measures and problems. *J. Pers. Med.*, 12, 8, 1295, 2022.
59. Moynihan, R., Sanders, S., Michaleff, Z.A., Scott, A.M., Clark, J., To, E.J., *et al.*, Impact of COVID-19 pandemic on utilisation of healthcare services: a systematic review. *BMJ Open*, 11, 3, e045343, 2021.

60. Tracie. COVID-19 Healthcare Delivery Impacts 2021 [Available from: https://files.asprtracie.hhs.gov/documents/covid-19-healthcare-delivery-impacts-quick-sheet.pdf.
61. Jacobsen, S., Faber, M., Altmann, B., Marques, A.M., Bock, C.T., Niendorf, S., Impact of the COVID-19 pandemic on norovirus circulation in Germany. *Int. J. Med. Microbiol.*, 314, 151600, 2024 Mar 1.
62. Douglas, A., Sandmann, F.G., Allen, D.J., Celma, C.C., Beard, S., Larkin, L., Impact of COVID-19 on national surveillance of norovirus in England and potential risk of increased disease activity in 2021. *J. Hosp. Infect.*, 112, 124–6, 2021.
63. Organization WH. Coronavirus disease (COVID-19) pandemic 2020 [Available from: https://www.who.int/europe/emergencies/situations/covid-19.
64. Tsai, P., Scott, K.A., González, M.C., Pappaioanou, M., Keusch, G.T. (Eds.), Sustaining global surveillance and response to emerging zoonotic diseases.
65. Pijnacker, R., Mughini-Gras, L., Verhoef, L., van den Beld, M., Franz, E., Friesema, I., Impact of non-pharmaceutical interventions during the COVID-19 pandemic on pathogens transmitted via food in the Netherlands. *Epidemiol. Infect.*, 152, e130, 2024.
66. Skarp, J.E., Downey, L.E., Ohrnberger, J.W., Cilloni, L., Hogan, A.B., Sykes, A.L., Wang, S.S., Shah, H.A., Xiao, M., A systematic review of the costs relating to non-pharmaceutical interventions against infectious disease outbreaks. *Appl. Health Econ. Health Policy*, 19, 673–97, 2021. 2021 Sep.
67. O'Reilly, K.M., Sandman, F., Allen, D., Jarvis, C.I., Gimma, A., Douglas, A., *et al.*, Predicted norovirus resurgence in 2021–2022 due to the relaxation of nonpharmaceutical interventions associated with COVID-19 restrictions in England: a mathematical modeling study. *BMC Med.*, 19, 1–10, 2021.
68. Oh, K.-B., Doherty, T.M., Vetter, V., Bonanni, P., Lifting non-pharmaceutical interventions following the COVID-19 pandemic–the quiet before the storm? *Expert Rev. Vaccines*, 21, 11, 1541–53, 2022.
69. Hausmann, J., Dörre, A., Katz, K., van de Berg, S., Lifting COVID-19-associated non-pharmaceutical interventions: potential impact on notifications of infectious diseases transmitted from person to person in 2022 in Bavaria, Germany. *Front. Public Health*, 12, 1437485, 2024.
70. Zhang, W., Wu, Y., Wen, B., Zhang, Y., Wang, Y., Yin, W., *et al.*, Non-pharmaceutical interventions for COVID-19 reduced the incidence of infectious diseases: a controlled interrupted time-series study. *Infect. Dis. Poverty*, 12, 02, 60–71, 2023.
71. Lappe, B.L., Wikswo, M.E., Kambhampati, A.K., Mirza, S.A., Tate, J.E., Kraay, A.N., *et al.*, Predicting norovirus and rotavirus resurgence in the United States following the COVID-19 pandemic: a mathematical modelling study. *BMC Infect. Dis.*, 23, 1, 254, 2023.

72. Lucero, Y., Matson, D.O., Ashkenazi, S., George, S., O'Ryan, M., Norovirus: facts and reflections from past, present, and future. *Viruses*, 13, 12, 2399, 2021 Nov 30.
73. Clark, T., Academy of Medical Science's Report: Preparing for a challenging winter 2020/21: Preparing for a challenging winter 2020/2.
74. Kumthip, K., Khamrin, P., Thongprachum, A., Malasao, R., Yodmeeklin, A., Ushijima, H., Maneekarn, N., Diverse genotypes of norovirus genogroup I and II contaminated in environmental water in Thailand during the COVID-19 outbreak from 2020 to 2022. *Virologica Sin.*, 39, 556–564, 2024 May 30.
75. Yasmin, F., Ali, S.H., Ullah, I., Norovirus outbreak amid COVID-19 in the United Kingdom; priorities for achieving control. *J. Med. Virol.*, 94, 3, 1232–5, 2022.
76. Lu, Y., Zhang, Z., Xie, H., Su, W., Wang, H., Wang, D., *et al.*, The rise in Norovirus-related Acute Gastroenteritis during the fight against the COVID-19 pandemic in Southern China. *Front. Public Health*, 9, 785373, 2022.
77. Sarmento, S.K., JdSR, .A., FC, M., AM, F., Mello, MS., FM, B., *et al.*, Norovirus Epidemiology and Genotype Circulation during the COVID-19 Pandemic in Brazil, 2019–2022. *Pathog.*, 13, 1, 3, 2023.
78. Ondrikova, N., Clough, H., Douglas, A., Vivancos, R., Itturiza-Gomara, M., Cunliffe, N., *et al.*, Comparison of statistical approaches to predicting norovirus laboratory reports before and during COVID-19: insights to inform public health surveillance. *Sci. Rep.*, 13, 1, 21457, 2023.
79. Kim, L., Liebowitz, D., Lin, K., Kasparek, K., Pasetti, M.F., Garg, S.J., *et al.*, Safety and immunogenicity of an oral tablet norovirus vaccine, a phase I randomized, placebo-controlled trial. *JCI Insight*, 3, 13, e121077, 2018.
80. Ghosh, S., Malik, Y.S., Kobayashi, N., Therapeutics and immunoprophylaxis against noroviruses and rotaviruses: the past, present, and future. *Curr. Drug Metab.*, 19, 3, 170–91, 2018.
81. Cates, J.E., Vinjé, J., Parashar, U., Hall, A.J., Recent advances in human norovirus research and implications for candidate vaccines. *Expert Rev. Vaccines*, 19, 6, 539–48, 2020.
82. Esposito, S. and Principi, N., Norovirus vaccine: priorities for future research and development. *Front. Immunol.*, 11, 1383, 2020.
83. Atmar, R.L., Bernstein, D.I., Harro, C.D., Al-Ibrahim, M.S., Chen, W.H., Ferreira, J., *et al.*, Norovirus vaccine against experimental human Norwalk Virus illness. *New Engl. J. Med.*, 365, 23, 2178–87, 2011.
84. Tan, M., Norovirus vaccines: current clinical development and challenges. *Pathog.*, 10, 12, 1641, 2021.
85. Caddy, S.L., Vaysburd, M., Wing, M., Foss, S., Andersen, J.T., Mayes, K., *et al.*, Intracellular neutralisation of rotavirus by VP6-specific IgG. *PLoS Pathog.*, 16, 8, e1008732, 2020.
86. Lappalainen, S., Pastor, A.R., Malm, M., López-Guerrero, V., Esquivel-Guadarrama, F., Palomares, L.A., *et al.*, Protection against live rotavirus

challenge in mice induced by parenteral and mucosal delivery of VP6 subunit rotavirus vaccine. *Arch. Virol.*, 160, 2075–8, 2015.

87. Zhang, M., Fu, M., Hu, Q., Advances in human norovirus vaccine research. *Vaccines*, 9, 7, 732, 2021.
88. Esposito, S. and Principi, N., Norovirus vaccine: priorities for future research and development. Frontiers in immunology. 2020 Jul 7;11:1383.
89. Turan, T., Işıdan, H., Atasoy, M.O., Irehan, B., Detection and molecular analysis of bovine enteric norovirus and nebovirus in Turkey. *J. Vet. Res.*, 62, 2, 129–35, 2018.
90. Desselberger, U., Caliciviridae other than noroviruses. *Viruses*, 11, 3, 286, 2019.
91. Logan, C. and O 'Sullivan, N., Detection of viral agents of gastroenteritis: Norovirus, Sapovirus and Astrovirus. *Future Virol.*, 3, 1, 61–70, 2008.
92. Tatusov, R.L., Chhabra, P., Diez-Valcarce, M., Barclay, L., Cannon, J.L., Vinjé, J., Human Calicivirus Typing tool: A web-based tool for genotyping human norovirus and sapovirus sequences. *J. Clin. Virol.*, 134, 104718, 2021.
93. CDC. Infection Control 2011 [Available from: https://www.cdc.gov/infection-control/hcp/norovirus-guidelines/background.html.
94. Abrantes, J., Droillard, C., Lopes, A.M., Lemaitre, E., Lucas, P., Blanchard, Y., *et al.*, Recombination at the emergence of the pathogenic rabbit haemorrhagic disease virus Lagovirus europaeus/GI. 2. *Sci. Rep.*, 10, 1, 14502, 2020.
95. Plavsic, M., Qiu, Y., Jones, N., Keegan, J., Woodcock, D., Morris, J., *et al.*, Caliciviridae and vesivirus 2117. *Bioprocess. J.*, 9, 2, 6 12, 2011.
96. Kang, S. and Kim, Y.-C., Identification of Viral Taxon-Specific Genes (VTSG): Application to Caliciviridae. *Genomics Inf.*, 16, 4, e23, 2018.
97. Guo, Z., He, Q., Yue, H., Zhang, B., Tang, C., First detection of Nebovirus and Norovirus from cattle in China. *Arch. Virol*, 163, 475–8, 2018.
98. Karayel-Hacioglu, I. and Alkan, F., Molecular characterization of bovine noroviruses and neboviruses in Turkey: detection of recombinant strains. *Arch. Virol*, 164, 1411–7, 2019.
99. DiCaprio, E., Ma, Y., Hughes, J., Li, J., Epidemiology, prevention, and control of the number one foodborne illness: human norovirus. *Infect. Dis. Clin.*, 27, 3, 651–74, 2013 Sep 1.
100. Sajewski, E.T., Vinjé, J., Glass, R.I., Lopman, B.A., Noroviruses, sapoviruses, and astroviruses, in: *Viral Infections of Humans: Epidemiology and Control*, pp. 1–46, Springer US, New York, NY, 2022 Mar 5. https://doi.org/10.1007/978-1-4939-9544-8_20-1
101. Hardy, S.A., Spread of rabbit hemorrhagic disease virus 2 (rhdv2) in wild and domestic lagomorphs in the western United States.
102. EFSA Panel on Biological Hazards (BIOHAZ), Scientific opinion on an update on the present knowledge on the occurrence and control of foodborne viruses. *EFSA J.*, 9, 7, 2190, 2011 Jul.
103. Mischenko, V.A., Mischenko A.V., Nikeshina T.B., Brovko Y.V., Kushlubaeva, A.I., Bovine nebovirus infection. *Vet. Sci. Today*, 12, 278, 2023.

104. Lanrewaju, A.A., Enitan-Folami, A.M., Sabiu, S., Edokpayi, J.N., Swalaha, F.M., Global public health implications of human exposure to viral contaminated water. *Front. Microbiol.*, 13, 981896, 2022 Aug 30.
105. Chadwick, P.R., Trainor, E., Marsden, G.L., Mills, S., Chadwick, C., O'Brien, S.J., Evans, C.M., Mullender, C., Strazds, P., Turner, S., Weston, V. Guidelines for the management of norovirus outbreaks in acute and community health and social care settings. *J. Hosp. Infect.*, 1, 136, 127–91, 2023.
106. Scotland HP., General information to prepare for and manage norovirus in care settings 2016 [Available from: https://www.nss.nhs.scot/media/4082/general-information-to-prepare-for-and-manage-norovirus-in-care-settings-v1.pdf.
107. Yao, L., Zhu, W., Shi, J., Xu, T., Qu, G., Zhou, W., Yu, X.F., Zhang, X., Jiang, G., Detection of coronavirus in environmental surveillance and risk monitoring for pandemic control. *Chem. Soc. Rev.*, 50, 6, 3656–76, 2021.
108. Zhang, L., Guo, W., Lv, C., Modern technologies and solutions to enhance surveillance and response systems for emerging zoonotic diseases. *Sci. One Health*, 3, 100061, 2024.
109. Bedi, J.S., Vijay, D., Dhaka, P., Gill, J.P.S., Barbuddhe, S.B., Emergency preparedness for public health threats, surveillance, modelling & forecasting. *Indian J. Med. Res.*, 153, 3, 287–98, 2021.
110. Singh, S., Sharma, P., Pal, N., Sarma, D.K., Tiwari, R., Kumar, M., Holistic one health surveillance framework: synergizing environmental, animal, and human determinants for enhanced infectious disease management. *ACS Infect. Dis.*, 10, 3, 808–26, 2024.
111. Wang, X., Wu, T., Oliveira, L.F., Zhang, D., Sheet, Surveillance, Strategy, Salvage and Shield in global biodefense system to protect the public health and tackle the incoming pandemics. *Sci. Total Environ.*, 822, 153469, 2022.
112. Steffen, R., Baños, A., debernardis, C., Vaccination priorities. *Int. J. Antimicrob. Agents*, 21, 2, 175–80, 2003.
113. Barclay, L., Park, G., Vega, E., Hall, A., Parashar, U., Vinjé, J., *et al.*, Infection control for norovirus. *Clin. Microbiol. Infect.*, 20, 8, 731–40, 2014.
114. Sobolik, J.S., Newman, K.L., Jaykus, L.-A., Bihn, E.A., Leon, J.S., Norovirus transmission mitigation strategies during simulated produce harvest and packing. *Int. J. Food Microbiol.*, 357, 109365, 2021.
115. Hall, A.J., Vinjé, J., Lopman, B., Park, G.W., Yen, C., Gregoricus, N., *et al.*, Updated norovirus outbreak management and disease prevention guidelines. *MMWR Recomm. Rep.*, 60, RR-3, 1–18, 2011.
116. Kambhampati, A., Koopmans, M., Lopman, B.A., Burden of norovirus in healthcare facilities and strategies for outbreak control. *J. Hosp. Infect.*, 89, 4, 296–301, 2015.
117. Currie, K., Curran, E., Strachan, E., Bunyan, D., Price, L., Temporary suspension of visiting during norovirus outbreaks in NHS Boards and the independent care home sector in Scotland: a cross-sectional survey of practice. *J. Hosp. Infect.*, 92, 3, 253–8, 2016.

118. Smallwood, N., Harrex, W., Rees, M., Willis, K., Bennett, CM., COVID-19 infection and the broader impacts of the pandemic on healthcare workers. *Respirology*, 27, 6, 411–26, 2022.
119. Trust, N.F., Ward closures [Available from: https://roh.nhs.uk/supporting-services/ipc/ward-closures.
120. Khanna, R.C., Cicinelli, M.V., Gilbert, S.S., Honavar, S.G., Murthy, G.V., COVID-19 pandemic: Lessons learned and future directions. *Indian J. Ophthalmol.*, 68, 5, 703–10, 2020
121. Stenseth, N.C., Dharmarajan, G., Li, R., Shi, Z.-L., Yang, R., Gao, G.F., Lessons learnt from the COVID-19 pandemic. *Front. Public Health*, 9, 694705, 2021.
122. Organization, W.H., Lessons learned from COVID-19 pandemic: virtual regional consultation with informal expert group. Lessons learned from COVID-19 pandemic: virtual regional consultation with informal expert group, 2022.
123. Assefa, Y., Gilks, C.F., Reid, S., van de Pas, R., Gete, D.G., Van Damme, W., Analysis of the COVID-19 pandemic: lessons towards a more effective response to public health emergencies. *Globalization Health*, 18, 1, 10, 2022.
124. Fauci, A.S. and Folkers, G.K., Pandemic preparedness and response: lessons from COVID-19. *J. Infect. Dis.*, 228, 4, 422–5, 2023.
125. Buchy, P., Buisson, Y., Cintra, O., Dwyer, D.E., Nissen, M., de Lejarazu, R.O., *et al.*, COVID-19 pandemic: lessons learned from more than a century of pandemics and current vaccine development for pandemic control. *Int. J. Infect. Dis.*, 112, 300–17, 2021.
126. Medicine, Y., 8 Lessons We Can Learn From the COVID-19 Pandemic 2021 [Available from: https://www.yalemedicine.org/news/8-lessons-covid-19-pandemic.
127. Fleming, M.D., Safaeinili, N., Knox, M., Brewster, A.L., Organizational and community resilience for COVID-19 and beyond: leveraging a system for health and social services integration. *Health Serv. Res.*, 59, e14250, 2024.

Part 3

CROSS-SPECIES STUDIES AND THERAPEUTIC DEVELOPMENT

12

Norovirus in Animals: Bovine, Murine, and Other Hosts

Aritra Banerjee[1], K.V. Leela[1], Shriyansh Srivastava[2], Pranav Gupta[3], Malakapogu Ravindra Babu[3] and Ranjit Sah[4*]

[1]*Department of Microbiology, S.R.M. Medical College Hospital & Research Centre, Potheri, India*
[2]*Department of Pharmacology, Delhi Pharmaceutical Sciences and Research University, New Delhi, India*
[3]*Department of Pharmacy, School of Medical and Allied Sciences, Galgotias University, Greater Noida, India*
[4]*BMC South, Boston Medical Centre Health System, MA, USA*

Abstract

One of the four acknowledged genera in the family Caliciviridae, enteric caliciviruses belong to genus Norovirus. These single-stranded, positive-sense RNA viruses exhibit a significant degree of genetic and antigenic diversity. Numerous animal enteric caliciviruses are genetically related to human noroviruses and have the same physical characteristics. Newbury Agent 2 refers to one of the initial bovine enteric noroviruses identified in Great Britain. Porcine noroviruses belonging to genogroup II combine at least three genomic clusters. The most common cause of acute gastroenteritis in humans of all ages is human noroviruses. They are linked to between 30% and 50% of all food-borne outbreaks in the United States. Noroviruses have not yet been linked to epidemics of gastroenteritis in animals with immune competence. A sophisticated three-dimensional culture system can facilitate the development of human norovirus. Nevertheless, neither bovine nor porcine noroviruses are capable of replicating in cell culture. The recently found murine noroviruses serve as model viruses for the study of human noroviruses because of their capacity to reproduce in cell culture. Unlike human noroviruses, the main topics of this research are animal norovirus taxonomy, virion structure, pathophysiology, epidemiology, immune response, and diagnostics. The genus Norovirus includes animal enteric caliciviruses, which increases the risk of

**Corresponding author*: ranjitsah57@gmail.com

Sachin Kumar, Shriyansh Srivastava, Rishabha Malviya and Ranjit Sah (eds.) Emerging Infectious Disease: Unravelling Norovirus and Pandemic Dynamics, (329–358) © 2026 Scrivener Publishing LLC

human transmission from an animal reservoir. In order to assess the likelihood of cross-species infections, which can significantly affect the epidemiology and development of these viruses and make treating human norovirus infections more difficult, the answer to this issue is essential.

Keywords: Norovirus, Calicivirus, zoonotic, host specificity, contagious

12.1 Introduction

The norovirus family is classified as Caliciviridae. Caliciviruses are diminutive, non-enveloped viruses characterized by single-stranded, positive-sense RNA, measuring between 27 and 35 nm in size. They impact numerous hosts, resulting in various diseases and lesions, such as upper respiratory tract and systemic diseases in felines, vesicular lesions, vesicular lesions and reproductive failures in swine and in other marine mammals including sea lions, as well as hemorrhagic disease in rabbits [1]. Furthermore, the investigation identified caliciviruses in calves with clinical respiratory symptoms [2]. The initial norovirus (NoV) was identified and linked to a human gastroenteritis outbreak in Norwalk, Ohio, and the prototype strain was designated as Norwalk virus (NV) in 1968 [3]. Subsequently, in 1972, immunological electron microscopy (IEM) was employed on stool samples from volunteers who received fecal filtrates from children impacted by the outbreak, thereby confirming the presence of the virus [4]. The Caliciviridae family was established nearly simultaneously [5]. Following observation using electron microscopy, numerous enteric viruses were classified as "small round-structured viruses" (SRSV). This occurred following the identification of rotavirus [6–8] and Norwalk virus [9] as intestinal pathogens by immunoelectron microscopy (IEM) [10–12]. The availability of sequencing an cloning data for the whole NV genome (strain FIIa) in the early 1990s initiated a new era of research on these viruses [13, 14]. NoV virions are excreted in the feces of gastroenteritis patients, with a buoyant density of 1.33–1.41 g/cm^3 in CsCl, and possess a ss-RNA genome. SRSV are categorized into astroviruses, sapoviruses (sometimes referred to as Sapporo-like viruses or SLVs), and noroviruses (also known as Norwalk-like viruses or NLVs). Electron microscopy revealed several viruses exhibiting the characteristic calicivirus morphology in fecal samples from domestic animals, especially in calves [15] and pigs [16]. Porcine NoV, strain SW918, was identified in fecal material of fully grown pig in Japan in 1997. Consequently, numerous pig genotypes were identified on different continents [17, 18]. The Jena agent, which was molecularly defined in 1999 and was isolated from cow in Germany over the 1980s, was initially discovered in the feces of diarrheal calves in

1978, as well as the Newbury Agent 2 which represents the two prototype strains of bovine noroviruses. Two enteric bovine caliciviruses, namely, Nebraska (NB) and Newbury Agent 1 strain, were found in British diarrheal calf samples in 1984 and in cattle in the USA, respectively. The two viruses exhibit 98% similarity in the amino acid composition of their whole sequence of capsid protein [19]. Moreover, murine noroviruses (murine NoV-1, -2, -3, and -4) have been identified in both immunocompetent and immunodeficient laboratory mice [20, 21]. Furthermore, a deceased lion juvenile in Italy was found to have contracted norovirus [22]. Canine feces have been identified to have some caliciviruses (genus Vesivirus); however, no instances of norovirus infection have been documented in other animal species [23–25]. Additionally, vesivirus-specific antibodies have been associated with equine abortion [26]. The study indicated the presence of vesivirus viremia in human sera [27]. Caliciviruses discovered in mink are classified within the Sapovirus genus (2001). Furthermore, antibodies targeting human norovirus have been detected in certain non-human primates [28].

12.2 Genomic Organization and Virus

Electron microscopy demonstrated that both human and animal noroviruses are spherical, non-enveloped entities exhibiting a frothy appearance and indistinct surface morphology. Research on human norovirus indicates that the capsid has 180 copies of a singular protein, organized in 90 dimers, exhibiting T = 3 icosahedral symmetry. Furthermore, the surface features protruding arches and 32 cup-shaped depressions [29]. Structural changes have been identified, and their functional consequences have been examined [30]. The NoV genome has three open reading frames (ORFs) and is classified as a positive-sense, single-stranded RNA with a length of 7.5 kb. A proposed genome-associated viral protein (VPg) is situated at the 5′ terminus [31]. NoV lacks a ribosomal entry site and a cap structure; these features are present in other animal caliciviruses, namely, at the N-terminal genomic extremity associated with VPg [32–35]. The expected role of VPg may include initiating viral RNA translation *in vitro* by interacting with translation machinery components (ribosomal protein, eIF3, eIF4E, and eIF4GI) through a particular protein–protien interaction [31]. In the case of mouse NoV, the VPg is associated with genomic NoV RNA [35, 36]. The 3C-like viral protease cleaves the 195-kDa polyprotein. ORF1 encodes at least six non-structural proteins at the 5′ terminal of the genomic RNA: RNA-dependent RNA polymerase; protein p48, indicating intracellular protein trafficking; nucleoside triphosphate (NTPase); protein p22, presumably enabling cellular membrane mobility and propagation; protease; protein p52 and protease. ORF2 encodes the major capsid protein (VP1), which plays a role in host specificity, strain variety, immunogenicity, receptor binding, capsid formation, and self-assembly [30]. The highly conserved genomic sequence of the GI and GII strains of NoV extends 18 nucleotides from the polymerase gene's C-terminal to the capsid coding region's N-terminal. This region may facilitate recombination [37, 38] or function as a packing signal for norovirus genome or a transcription initiation point [39]. The VP1 subunit contains a protrusion (P) domain and a shell (S) domain. The P domain, comprising the P1 and P2 subunits, exhibits the most significant structural alterations, whereas the hypervariable domain, P2, is situated on the surface of intestinal cells and functions as a ligand for cell receptors [40]. VP2, a 20-kDa protein, is encoded by ORF3, situated at the 3′ terminus of the chromosome. It aids in the stability and production of the VP1 capsid protein [41]. Mouse and human noroviruses may produce and accumulate subgenomic RNAs during propagation in infected and cDNA-transfected cells [42, 43]. This method facilitates the control and adequate synthesis of structural proteins by other positive-strand RNA viruses [44].

12.3 Interactions Between Viruses and Cells

Despite the presence of infection in the small intestine, norovirus has not been detected in the enterocytes, as it is probable that NoV replication is not confined to this area. The receptors for animal noroviruses remain unidentified. The study revealed that the mouse norovirus had an unanticipated affinity for hematopoietic cell lineages, particularly macrophages and dendritic cells [43]. Investigations on other caliciviruses also indicate a similar conclusion. The rabbit hemorrhagic disease virus attaches to antigen from ABH histo-blood group family [45]. It is noteworthy that oysters, often implicated in human foodborne norovirus outbreaks, have carbohydrates in their digestive tissues that are strongly associated with specific histo-blood group antigens. These carbohydrates, when linked to human NoVs, enhance their concentration and enable their dissemination post-consumption [46]. *In vitro*, human NoV-like particles may bind to pig gastrointestinal washings coated on plaques [47]. Furthermore, *in vivo* studies reveal that human NoVs were observed to cling to pig intestinal tissues, exhibiting some symptoms of replication [48]. Nevertheless, human saliva specimens from the primary histo-blood group types did not adhere to virus-like particles (VLP) produced from SW918 strain

(GII.11) [49]. This research suggests that this NoV has a unique receptor type, and human NoVs may occasionally bind to naturally occurring carbohydrates on porcine gastrointestinal tissues associated with human HBGAs [48, 50]. The absence of supplementary components required for NoV replication *in vitro* and *in vivo*, which are present in cellular culture in "3D conformation," may also be a contributing factor [51]. The replication of the sapovirus porcine enteric calicivirus (PEC)/Cowden strain in cell culture requires bile acids, increased cAMP levels, and the inhibition of interferon-induced phosphorylation of signal transducer and activator of transcription 1 (STAT1), a crucial component of innate immunity [52]. Furthermore, recent data indicate that trypsin-mediated proteolysis may facilitate human NoV replication within the host [53]. Animal NoVs may replicate and disrupt internal membranes upon engagement, similar to membrane-associated transport seen in other members of the Caliciviridae family and positive-strand RNA viruses [54–56]. In an experiment, the complete NV N-terminal protein (amino acid 1–398 of ORF1 polyprotein) was demonstrated to co-localize with Golgi apparatus proteins in cells transfected with a vector plasmid, leading to its production [57]. The complete lack of a Golgi apparatus is distinctly observed in cells transfected and infected with NoV [43]. It should be noted that VPg may inhibit host protein production. Experimental experiments including Proteinase K treatment of animal calicivirus RNA demonstrated a reduction in its infectivity across many assays, hence underscoring the essential involvement of VPg [32, 33]. Moreover, viral proteinase can obstruct cellular translation by cleaving the polyA binding protein [58].

12.4 Norovirus in Bovines

Human norovirus was first identified in 1972 [9], and 4 years later, in 1976, bovine norovirus was also discovered. Bovine noroviruses are classified into two genotypes, GIII.1 and GIII.2, with GIII.2 being the more prevalent type [59, 69]. The bovine Pdimer features a flattened P2 subdomain and shorter extended loops that are reminiscent of the GI.1P domain. Investigations have shown that a monoclonal antibody (MAb) is capable of binding to five distinct norovirus genogroups. Animal noroviruses are categorized into several groups: GV for mice, GVI for dogs, GIII for ruminants, GIV for cats, dogs, and lions, and GII for pigs [15, 60–66]. As per Vinje *et al.* (2015), a study has found that third group of canine noroviruses has been found in fecal specimens [67] from a dog in Hong Kong and has been designated as a novel potential genogroup (GVII). Two different genotypes make up GIII: genotype 1 and genotype 2 are represented by the progenitor strains

Bo/Newbury2/76/UK and Bo/Jena/80/DE, respectively. An analysis of genomic sequences obtained from original BoNoVs [15, 68–71] grounds this categorization. Recombination at the ORF1/ORF2 junction occurs often; genogroups and genotypes are established based on amino acid variety throughout the whole VP1 sequence. Consequently, based on the evolutionary links of both partial RdRp (P) and capsid coding sequences, a dual taxonomy system has been developed [72]. This nomenclature identifies the GIII.1 prototype strain Bo/Jean/80/DE as GIII/Bo/DE/1980/GIII.P1-GIII.1/Jean and the GIII.2 prototype strain Bo/Newbury2/76/UK as GIII/Bo/UK/1976/GIII.2/Newbury2. Mauroy *et al.* (2012) [73] assert that a thorough phylogenetic study and evaluation of evolutionary rates for BoNoVs is unattainable due to the insufficient number of completely sequenced BoNoV strains and the lack of accessible ORF2 sequences.

12.5 Bovine Norovirus Distribution

GIII.2 is the predominant strain of norovirus globally. According to Ike *et al.* (2007), Mijovski *et al.* (2010), and Van der Poel *et al.* (2003), the molecular frequency of GIII.2 in Europe varies from 1.8% to 50.3%. In the Netherlands, 4.2% of individual feces samples from milking cows and 31.6% of aggregated stool specimens from veal cow herds were found positive for GIII.2 BoNoVs through real-time polymerase chain reaction. In the United Kingdom, BoNoVs were identified in 11% of diarrheal specimens [59]. In Belgium, IgG antibodies against GIII.2 BoNoVs were examined in 93.2% of serum samples from adult cattle and calves. GIII was identified in 7.5% of stool specimens from diarrheal cattle aged 1 week to 6 months [74]. Italy has investigated GIII.2 BoNoVs in 8.6% of medically healthy calf, and 7.9% of cattle with mixed rotavirus infections had diarrhea. GIII.2 BoNoVs were mostly discovered in 72% of veal cattle in Ohio, followed by 80% in Michigan calves exhibiting diarrhea, and 25% in Wisconsin calves with diarrhea. Thomas *et al.* (2014) [75] report a notable frequency of IgG antibodies against GIII.2 BoNoVs in calves from Ohio, Arkansas, and New Mexico with rates between 94% and 100%. Moreover, BoNoVs were identified in 1.6% of Canadian bovine fecal specimens [76]. Rotavirus, nebovirus, BVDV, bovine coronavirus, *Escherichia coli*, and bovine torovirus were identified as causal agents of enteric diseases, representing 5.9% of the 645 fecal samples in South Korea that tested the presence of BoNoVs with nested RT-PCR [77]. In this work, one of the 12 South Korean BoNoV genomes was genetically aligned with GIII.1, while the remaining 11 exhibited the most amino acid (93.5%–99.1%) and nucleotide (88.0%–90.5%) similarity with the strain of GIII.2. In Africa, 16.6% of Tunisian calves with diarrhea tested positive for bovine noroviruses [78]. GIII.1 BoNoVs are encountered with lesser frequency, indicating the formation of a moderate cluster. BoNoVs were found in 1% of animals in Venezuela in South America [79]. The GIII.1 BoNoVs strain has been recorded at 1/300 (0.3%) in Belgium, 1/47 (2.1%) in Hungary [80], 3/101 (2.9%), in France [81], and 25/456 (5%) [82]. In the United States, five out of 74 (6.8%) dairy cows with diarrhea in Wisconsin and Michigan tested positive for GIII.1 BoNoV sequences [83]. In Germany, 99.1% of cattle exhibit IgG antibodies against GIII.1 BoNoV [84]. In the identical experiment, GIII.1 BoNoVs were detected in 8.9% of fecal samples utilizing an Ag ELISA. Multiple molecular investigations conducted in various countries have produced unique recombinant strains, the majority of which demonstrate genetic similarity to the prototype strain Bo/Thirsk 10/00/UK [85]. Research on the ORF1/ORF2 region identified recombinant type GIII.P1/GIII.2 in Belgium [80], Norway [82], and Italy [81]. The recombinant GIII.P1/GIII.2 strain CV521-OH/02/US was identified in United States, Ohio by Han *et al.* (2004). Bull *et al.* 2007 [86] identified recombinant strain B-1SV/03/US characterized by GIII.2 polymerase and GIII.1 capsid (GIII.P1/GIII.2) during their investigation of NoV recombinants across all genogroups globally. The necessity of analyzing these strains for BoNoV detection is shown by the high detecting level of the recombinant Bo/Thirsk 10/00/UK-like (GIII.P1GIII.2) sequence. It also promotes compelling inquiries concerning the genesis of these viruses within the framework of restricted circulation of GIII.1 viruses.

12.6 The Pathophysiology and Clinical Characteristics of Bovine Norovirus

While the pathophysiology of BoNoVs remains incompletely recognized, their resemblance to HuNoVs is a basis for certain conclusions to be drawn. Human noroviruses are extremely infectious and are mostly transmitted from patient to patient by contaminated food or beverages and by the fecal–oral route. The primary clinical symptom in cattle infected with BoNoVs is diarrhea, accompanied by transient anorexia and malabsorption syndrome, as indicated by several experimental studies [15, 87–89]. Günther and Otto (1987) [87] state that diarrhea may persist for 3 to 4 days and is more severe in 3-week-old animals compared to newborn cattle. Anorexia, diarrhea, intestinal epithelium necrosis, and villous atrophy were observed in genotobiotic cattle affected with GIII.1 BoNoV strain Bo/Jean/80/DE. Following inoculation with

Bo/Jena/80/DE, immunocompetent calves that were deprived of colostrum experienced diarrhea lasting 2 to 3 days [88]. The GIII.2 BoNoV strain CV186-OH/00/US caused severe diarrhea in gnotobiotic calves aged 4–7 days, resulting in prolonged fecal shedding (median 28 days) and seroconversion; however, no significant intestinal lesions were seen [89]. Molecular methods have been used to find BoNoVs in the feces of calves that have diarrhea. These viruses have been found alone or with other intestinal viruses like nebovirus, bovine viral diarrhea virus (BVDV), rotavirus, and coronavirus [77, 90–93]. Mixed infections likely affect the severity of BoNoV infections. Calves in Italy and The Netherlands have been documented to possess asymptomatic infections [81, 94]. Quantitative reverse transcriptase polymerase chain reaction (qPCR) was employed to detect GIII BoNoVs in 272 fecal specimens from 190 herds in Norway. The incidence of these viruses was approximately the same in healthy calves (50.3%) and diarrheal calves (49.3%) [92]. BoNoVs were detected more frequently and in higher numbers in diarrheal calves compared to the healthy ones in a case study of calf diarrhea conducted in the United States of America [90]. More research needs to be done on both adult cattle and kids to find out how vulnerable cattle become to BoNoVs as they age. Figure 12.1 shows the pathophysiology and clinical characteristics of bovine norovirus.

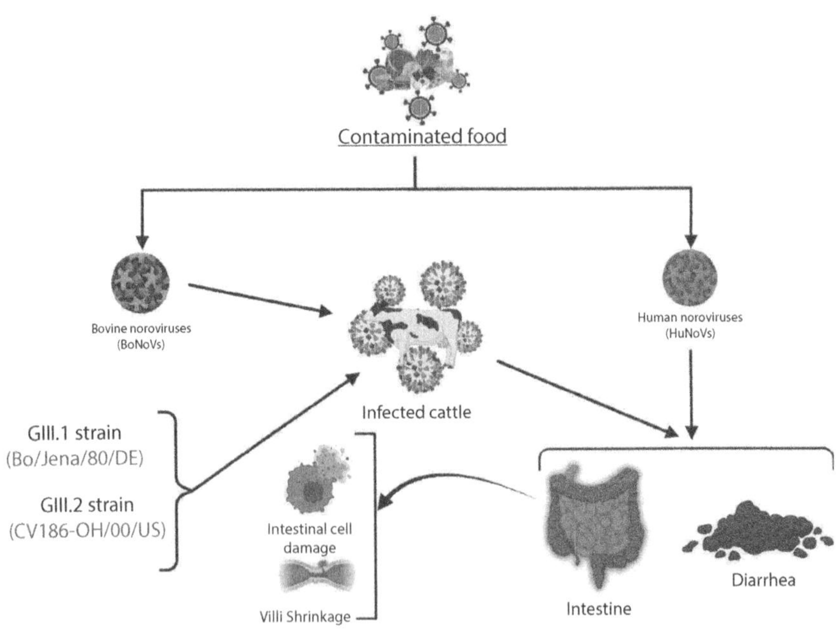

Figure 12.1 The pathophysiology and clinical characteristics of bovine norovirus.

12.7 Diagnosis of the Bovine Norovirus

12.7.1 Electron Microscopy

Electron microscopy (EM) was the initial diagnostic method for noroviruses (NoVs) [95]. Nonetheless, owing to its low sensitivity, with a detection threshold of around 106 virus particles/mL of feces, immunoelectron microscopy (IEM) is employed to enhance the detection sensitivity; however, it remains inadequate for routine diagnosis.

12.7.2 Molecular Methods

RT-PCR has enhanced the ability to detect NoVs in feces [96]. RT-PCR was created by De Leon *et al.* in 1992 as a way to find viruses that are genetically related to the Norwalk virus [97]. The main approach for diagnosing BoNoV infections has developed into several RT-PCR formats [98–101]. The highly conserved regions located at the termini of RdRp gene, initiation of ORF2 gene, and NTPase/RNA helicase (NS3) coding sequence were successfully amplified by the majority of RTPCR oligonucleotide primers [102]. A combination of multiple primer sets has been employed to generate complete genomes [71, 73, 96]. For the detection of HuNoV cases, new molecular methods have come out that are more sensitive and specific [103–106]. Differentiating infectious from non-infectious particles can be accomplished by RT-PCR combined with binding. Putting together binding RT-PCR and long-range RT-PCR can be used to check the stability of the genome. These techniques can be employed to identify BoNoV infections [107–109].

12.7.3 Antigen ELISAs

The capsid proteins of noroviruses self-assemble into virus-like particles (VLPs) that mimic infectious viral particles both visually and antigenically when generated by baculovirus. Antigens derived from VLPs can be utilized in monoclonal or polyclonal antibodies for antigen Ag ELISAs and in serological assays. While RT-PCR and qPCR demonstrate superior sensitivity and specificity compared to Ag ELISAs for HuNoV detection, Ag ELISAs may prove advantageous for cost-effectiveness and high-throughput screening of numerous fecal samples for HuNoVs. Retesting using molecular methods is crucial to confirm the infection in samples from probable cases showing negative findings in the Ag ELISA. Deng *et al.* [127] in 2003 developed a BoNoV Ag ELISA to identify capsid antigens associated with BoNoV Bo/Jean/80/DE virus (GIII.1) in fecal specimens of diarrheal cows.

12.7.4 Antibody ELISAs

The evaluation of seroprevalence for noroviruses in humans and veterinary species has been performed using many formats of antibody ELISAs [49, 84, 110, 111]. Ab ELISAs have been created to assess bovine blood specimen utilizing GIII.1 or GIII.2 BoNoV VLPs produced in baculovirus system [49, 74, 75, 84, 112]. Identifying heterologous reactions among genetically related strains can be difficult, as antibody ELISAs often exhibit wider reactivity than antigen ELISAs. The antibodies identified by these Ab ELISAs do not clearly show the specific response to a particular strain of illness.

12.8 Norovirus in Murine

Though noroviruses from cattle, pigs, and mice have been discovered, the majority of noroviruses cause human gastrointestinal disorders [61, 69, 113]. The only norovirus capable of replicating in both small animal and cell culture with some potential experimental models is the murine norovirus (MNV) [61, 114]. Researchers in the US and Canada found that, out of serum samples of numerous mice from research colonies, 22.1% of the samples had MNV-1 reactive antibodies [115]. Also, over 35 new MNV isolates have been identified in studies. Hence, it is suggestive that MNV is one of the most common potent pathogens for illnesses in modern research mice.

12.8.1 Murine Norovirus Identification

Murine norovirus-1, or MNV-1, was the initial norovirus to infect mice. It was found in 2003 using representational difference analysis, a method that can give sequences that are similar to but different from the sequences of other noroviruses. The full 7,382-bp polyadenylated MNV-1 genome was obtained by cloning and sequencing using a mix of PCR and fast amplification of cDNA ends. It was then connected to norovirus genomes. Based on a phylogenetic study of the capsid protein and genome, MNV-1 is a new type of norovirus that has not been looked into much. A recent work employing PCR amplicons has confirmed the precision of the average sequencing of MNV-1 brain homogenate from IFN-αβYR-/- mice [61].

12.8.2 Structural Features of MNV

From a molecular perspective, MNV-1 and human noroviruses exhibit numerous genetic and metabolic similarities. MNV-1 exhibits an icosahedral structure, measuring between 28 and 35 nm in diameter, with a buoyant density of 1.36 ± 0.04 g/cm^3, which is consistent with that of human noroviruses [61, 116]. Furthermore, the study of MNV-1 genome uncovers the three open reading frames (ORF) that are common among the Caliciviridae genera, specifically in vesiviruses and noroviruses. The 187.5-kDa polyprotein, characterized by a projection, encodes three polymerase, 2C helicase, and 3C protease domains, which are also present in other picornaviruses and caliciviruses, as detailed in MNV-1's ORF1. ORF2 of MNV-1 encodes a 58.9-kDa capsid protein capable of self-assembling into virus-like particles, generated inside a baculovirus production system. Concurrently, ORF3 encodes the expected 22.1-kDa simple protein [61].

The investigation into norovirus replication remains in its preliminary phases. The genomes of human and murine noroviruses exhibit numerous molecular similarities, as evidenced by the acquisition of capsid protein sequences from murine norovirus 1, various human norovirus strains, multiple strains of porcine noroviruses, VP1 of poliovirus 1 Mahoney, numerous human sapovirus strains, rabbit hemorrhagic illness virus, feline calicivirus, and European brown hare syndrome virus. Therefore, the study of MNV could assist in uncovering the mechanisms that drive viral replication. To date, the use of MNV-1 has led to the proposal that subgenomic RNA synthesis and intracellular membrane reorganization during viral replication are two crucial steps in the replication process of norovirus. It has been established that animal caliciviruses can be cultured in tissue systems, accompanied by the generation of subgenomic RNA during the viral replication process. The incorporation of the 5′ end of both genomic and subgenomic RNA is attributed to the capsid protein associated with these viruses [32–35, 117, 118]. The conserved nucleotides at the 5′ end of the genome and region directly upstream of open reading frame 2 point to subgenomic RNA maybe expressing the capsid and basic protein, particularly open reading frame 3 (ORF3) of noroviruses. Three latest studies show that during norovirus replication a subgenomic RNA is generated. RNA was isolated from MNV-1-infected cells, and a Northern blot analysis was conducted, revealing that the quantity of a subgenomic RNA increased over time [114]. Unpublished observations indicate that MNV-1 infecting wild-type and STAT1/macrophages display comparable subgenomic RNA species. Furthermore, cell transfection using the cDNA clone of either U201 or Norwalk virus (NV), which are two human noroviruses, results in the detection of a subgenomic-length RNA [42, 119].

Like certain other RNA viruses, noroviruses employ a rich subgenomic message to control the synthesis of structural proteins. The existence of subgenomic RNA species during human and murine norovirus replication serves as evidence for this.

12.8.3 MNV Infection in Wild-Type Mice

Recent investigations indicate that MNV-1 can infect both wild-type mice and those with compromised immune systems. Observational studies indicated that both inbred 129 and outbred CD1 mice, when either i.n. or p.o. with MNV-1 [131, 135], demonstrate a notable seroconversion to the capsid protein. Furthermore, at 1 day post-inoculation, 129 grown mice (8 weeks old) exhibited considerable quantities of viral RNA in the liver, spleen, and proximal intestine, except brain, blood, feces and lung following the injection of MNV-1 brain homogenate at a dosage of 3×10^4 PFU [131]. MNV RNA is present in the spleen jejunum, mesenteric lymph node, and feces of young CD1 mice at least 8 weeks post-injection with 1×10^6 PFU of three distinct murine norovirus isolates: MNV3, MNV4, and MNV2. This concept suggests that MNV and other noroviruses in wild-type mice may serve as a reservoir for norovirus outbreaks [120, 121].

12.8.4 To Control Norovirus Infection, Innate Immunity is Necessary

Prior *in vivo* studies have demonstrated that IFN-αβϒR--/-- animals exhibit increased susceptibility to lethal MNV-1 infection, similar to STAT-/- mice. A further study is necessary to clarify the unique roles of INF-ϒ and INF-αβ after MNV-1 infection *in vivo* [27]. The preliminary animal studies exclusively examined the lethal impact of the virus, and the role of IFN-αβ and IFN-ϒ in the more nuanced suppression of MNV infection remains unclear. Subsequent investigations on MNV indicated that the rapid resolution of human norovirus illness may be attributed to innate immunity, particularly immunological responses reliant on IFN or STAT1. Additionally, the presence of an innate illness in immunocompromised mice following the injection of MNV-1 suggests that individuals with a weakened innate immunity may be at a higher risk of contracting a more severe and pervasive form of norovirus disease, which may have unidentified medical consequences [122].

12.8.5 To Eradicate a Norovirus Infection, Adaptive Immunity is Required

A lot of research has already been done on the adaptable immune response and how it helps viruses get rid of themselves. The study demonstrated that RAG1-/- and RAG2-/- animals exhibit resistance to MNV-1-induced mortality, in stark contrast to wild-type mice [131]. It implies that, unlike the STAT1-dependent innate immune response, protection against mortality from murine norovirus infection does not need the adaptive immune response. In RAG-/- mice, elevated viral RNA levels have been detected in the blood and feces at 90 days post-exposure, as well as in the liver, proximal intestine, spleen, brain, and lung [20]. Unpublished results also indicate that infectious MNV-1 is present in various organs of RAG-/- mice, accompanied by substantial amounts of viral RNA [20]. Unlike the swift elimination of MNV-1 in wild-type mice, MNV-1 replication continues in RAG-/- animals, suggesting that T and B cells are essential for controlling and resolving MNV-1 infections. These investigations further indicate that MNV-1 can persistently proliferate in tissues without inducing an apparent disease or mortality. This trait also elucidates the persistence of norovirus infections in wild-type hosts. Preliminary research indicates that the development of a neutralizing antibody response may provide some *in vivo* protection. While the results of MNV-1 infection in RAG-/-mice indicate that B and T cells contribute to viral control and clearance, more investigation is required to clarify the specific roles and mechanisms of action of these cells [114].

12.9 Porcine Norovirus

In The Netherlands, norovirus RNA was identified in two out of 100 pooled pig fecal samples and in four out of 1,017 normal slaughtered pigs in Japan. The introduction of the RT-PCR technique facilitated this classification of porcine NoVs as GII, revealing genomic similarities [18, 113]. Moreover, it has been noted that antibodies directed against human GII, but not GI NoVs, demonstrate cross-reactivity with virus-like particles (VLPs) of the Sw918 strain [49, 123–125]. Due to the genetic and antigenic similarities between human and porcine noroviruses, there is an imperative to enhance public health awareness about the potential of zoonotic transmission and the emergence of novel pandemic human strains. It is important to highlight that pigs exhibiting no clinical signs were the source of all porcine NoVs [18, 113]. Porcine GII NoVs have genetic and antigenic similarities to human NoVs, prompting concerns over their potential for zoonotic persistence. Additionally, subclinically infected pigs could serve as natural reservoirs for NoVs. The transmission of human NoV strains associated with QW101-like porcine NoVs among individuals working with pigs remains uncertain. However, these informative studies could yield valuable insights into the zoonotic potential of these swine NoVs. In pigs from the United States, three genotypes of porcine NoVs were identified; the genotype that exhibited similarities to the human GII NoVs, both genetically and antigenically, was QW101-like, GII-18. Additionally, norovirus particles were detected in the digestive content of gnotobiotic pigs infected with QW101-like NoVs, and recombinant strains of NoV were also identified in the pigs. The results suggest the possibility of porcine/human GII recombinants or that pigs may serve as reservoirs for the formation of novel human noroviruses [37, 126–129].

12.10 In Lions, Norovirus

The feline parvovirus, feline herpesvirus, feline calicivirus (FCV), feline retrovirus, and canine distemper virus can impact lions (*Panthera leo*) [130–133]. In the autumn of 2006, a 4-week-old cub at the Pistoia Zoo in Italy succumbed to acute hemorrhagic enteritis. The prevalence of viral infections in felines and canines was assessed by examining the tissues and intestinal contents. The incidence of viral infections in cats and dogs was evaluated by the analysis of tissues and digestive contents. The tests for recognized canine infections (parvovirus, adenoviruses type 1 and type 2, and distemper virus) and feline pathogens (parvovirus, coronaviruses, herpesvirus, and retroviruses) yielded negative results. However, calicivirus was detected in the digestive contents [134]. A phylogenetic analysis was conducted utilizing the capsid protein of various human and animal noroviruses from distinct genogroups (I to V) after extensive experimentation with the sample [17, 135]. The analysis revealed that the genogroup IV human noroviruses were associated with the lion calicivirus strain. Furthermore, the capsid protein analysis indicated that lion NoV 387/06 had a genetic similarity to human genogroup IV NoVs, with an amino acid identity ranging from 69.3% to 70.1%. Consequently, although human genogroup IV noroviruses are classified as genotype IV.1, the virus may be regarded as a distinct genotype (IV.2) within genogroup IV. Finally, the great genetic similarity between human genogroup IV NoVs and lion NoV strain supports the theory that human NoV evolution is entwined with that of animal NoVs [136].

12.11 Conclusion

Animal noroviruses, like their human counterparts, pose significant challenges in veterinary and public health due to their high infectivity and resilience in the environment. These viruses can cause gastrointestinal illness in a wide range of animal species, impacting animal welfare, agricultural productivity, and, in some cases, zoonotic disease transmission. Although the molecular biology, pathophysiology, and epidemiology of animal noroviruses are now better understood, major gaps still exist in the development of sensible vaccines and antiviral treatments. Continued research is essential to improve diagnostic tools, implement preventive measures, and assess the potential risk of cross-species transmission. Enhancing surveillance and biosecurity measures in animal populations will also play a critical role in controlling outbreaks. By addressing these challenges, we can mitigate the impact of animal noroviruses and safeguard both animal and human health.

AI disclosure statement: The author acknowledges the use of AI-based tools for minor language editing or grammatical correction. However, all intellectual contributions, analyses, and conclusions presented in this chapter are entirely the author's own.

References

1. Torrence, M.E. and Isaacson, R.E., *Microbial Food Safety in Animal Agriculture: Current Topics*, p. 436, 2008, [cited 2025 Jan 13]. Available from: https://books.google.com/books/about/Microbial_Food_Safety_in_Animal_Agricult.html?id=kKotTIISeNMMC.
2. Smith, A.W., Mattson, D.E., Skilling, D.E., Schmitz, J.A., Isolation and partial characterization of a calicivirus from calves. *Am. J. Vet. Res. [Internet]*, 44, 5, 851–5, 1983 May 1. [cited 2025 Jan 13]. Available from: https://europepmc.org/article/med/6307089.
3. Winter Vomiting Disease on JSTOR [Internet],. [cited 2025 Jan 13]. Available from: https://www.jstor.org/stable/30102351.
4. Kapikian, A.Z., The Discovery of the 27-nm Norwalk Virus: An Historic Perspective. *J. Infect. Dis. [Internet]*, 181, Supplement_2, S295–302, 2000 May 1. [cited 2025 Jan 13] Available from: https://dx.doi.org/10.1086/315584.
5. Matthews, R.E.F., Classification and Nomenclature of Viruses Second Report of the International Committee on Taxonomy of Virology-nomenclature I Matthews. *R E F II Title QW*, 12, 5, 163, 1979.

6. Adams, W.R. and Kraft, L.M., Epizootic Diarrhea of Infant Mice: Identification of the Etiologic Agent. *Sci. (1979) [Internet]*, 141, 3578, 359–60, 1963 Jul 26. [cited 2025 Jan 13] Available from: https://www.science.org/doi/10.1126/science.141.3578.359.
7. Bridger, J.C. and Woode, G.N., Neonatal Calf Diarrhoea: Identification of a Reovirus-Like (Rotavirus) Agent in Faeces by Immunofluorescence and Immune Electron Microscopy. *Br. Vet. J.*, 131, 5, 528–35, 1975 Sep 1.
8. Bishop, R.F., Davidson, G.P., Holmes, I.H., Ruck, B.J., Virus Particles In Epithelial Cells Of Duodenal Mucosa From Children With Acute Non-Bacterial Gastroenteritis. *Lancet*, 302, 7841, 1281–3, 1973 Dec 8.
9. Kapikian, A.Z., Wyatt, R.G., Dolin, R., Thornhill, T.S., Kalica, A.R., Chanock, R.M., Visualization by Immune Electron Microscopy of a 27-nm Particle Associated with Acute Infectious Nonbacterial Gastroenteritis. *J. Virol. [Internet]*, 10, 5, 1075–81, 1972 Nov. [cited 2025 Jan 13] Available from: https://journals.asm.org/doi/10.1128/jvi.10.5.1075-1081.1972.
10. Appleton, H. and Higgins, P.G., Viruses And Gastroenteritis In Infants. *Lancet*, 305, 7919, 1297, 1975 Jun 7. [cited 2025 Jan 13] Available from: http://www.thelancet.com/article/S0140673675925817/fulltext.
11. Chiba, S., Nakata, S., Numata-Kinoshita, K., Honma, S., Sapporo Virus: History and Recent Findings. *J. Infect. Dis. [Internet]*, 181, Supplement_2, S303–8, 2000 May 1. [cited 2025 Jan 13] Available from: https://dx.doi.org/10.1086/315574.
12. Srivastava, S., Kumar, S., Sharma, P.K., Rustagi, S., Mohanty, A., Donovan, S., *et al.*, Control strategies for emerging infectious diseases: Crimean-Congo hemorrhagic fever management. *Health Sci. Rep. [Internet]*, 7, 9, e70053, 2024 Sep 1. [cited 2025 Jan 13] Available from: https://onlinelibrary.wiley.com/doi/full/10.1002/hsr2.70053.
13. Jiang, X., Wang, M., Wang, K., Estes, M.K., Sequence and Genomic Organization of Norwalk Virus. *Virology.*, 195, 1, 51–61, 1993 Jul 1.
14. Jiang, X., Graham, D.Y., Wang, K., Estes, M.K., Norwalk Virus Genome Cloning and Characterization. *Sci. (1979) [Internet]*, 250, 4987, 1580–3, 1990 Dec 14. [cited 2025 Jan 13] Available from: https://www.science.org/doi/10.1126/science.2177224.
15. Woode, G.N. and Bridger, J.C., Isolation of small viruses resembling astroviruses and caliciviruses from acute enteritis of calves. *J. Med. Microbiol. [Internet]*, 11, 4, 441–52, 1978 Nov 1. [cited 2025 Jan 13] Available from: https://www.microbiologyresearch.org/content/journal/jmm/10.1099/00222615-11-4-441.
16. Saif, L.J., Bohl, E.H., Theil, K.W., Cross, R.F., House, J.A., Rotavirus-Like, Calicivirus-Like, and 23-nm Virus-Like Particles Associated with Diarrhea in Young Pigs. *J. Clin. Microbiol. [Internet]*, 12, 1, 105–11, 1980. [cited 2025 Jan 13] Available from: https://journals.asm.org/doi/10.1128/jcm.12.1.105-111.1980.

17. Wang, Q.H., Myung, G.H., Cheetham, S., Souza, M., Funk, J.A., Saif, L.J., Porcine Noroviruses Related to Human Noroviruses. *Emerg. Infect. Dis. [Internet]*, 11, 12, 1874, 2005. [cited 2025 Jan 13] Available from: https://pmc.ncbi.nlm.nih.gov/articles/PMC3367634/.
18. Van Der Poel, W.H.M., Vinjé, J., Van Der Heide, R., Herrera, M.I., Vivo, A., Koopmans, M.P.G., Norwalk-like calicivirus genes in farm animals. *Emerg. Infect. Dis. [Internet]*, 6, 1, 36, 2000. [cited 2025 Jan 13] Available from: https://pmc.ncbi.nlm.nih.gov/articles/PMC2627973/.
19. Oliver, S.L., Asobayire, E., Dastjerdi, A.M., Bridger, J.C., Genomic characterization of the unclassified bovine enteric virus Newbury agent-1 (Newbury1) endorses a new genus in the family Caliciviridae. *Virology.*, 350, 1, 240–50, 2006 Jun 20.
20. Karst, S.M., Wobus, C.E., Lay, M., Davidson, J., Virgin IV, H.W., STAT1-dependent innate immunity to a norwalk-like virus. *Sci. (1979) [Internet]*, 299, 5612, 1575–8, 2003 Mar 7. [cited 2025 Jan 13] Available from: https://www.science.org/doi/10.1126/science.1077905.
21. Hsu, C.C., Riley, L.K., Wills, H.M., Livingston, R.S., Persistent Infection with and Serologic Crossreactivity of Three Novel Murine Noroviruses.
22. Martella, V., Campolo, M., Lorusso, E., Cavicchio, P., Camero, M., Bellacicco, A.L., *et al.*, Norovirus in Captive Lion Cub (Panthera leo). *Emerg. Infect. Dis. [Internet]*, 13, 7, 1071, 2007. [cited 2025 Jan 13] Available from: https://pmc.ncbi.nlm.nih.gov/articles/PMC2878247/.
23. Matsuura, Y., Tohya, Y., Nakamura, K., Shimojima, M., Roerink, F., Mochizuki, M., *et al.*, Complete nucleotide sequence, genome organization and phylogenic analysis of the canine calicivirus. *Virus Genes [Internet]*, 25, 1, 67–73, 2002. [cited 2025 Jan 13] Available from: https://link.springer.com/article/10.1023/A:1020174225622.
24. Mochizuki, M., Hashimoto, M., Roerink, F., Tohya, Y., Matsuura, Y., Sasaki, N., Molecular and seroepidemiological evidence of canine calicivirus infections in Japan. *J. Clin. Microbiol. [Internet]*, 40, 7, 2629–31, 2002. [cited 2025 Jan 13] Available from: https://journals.asm.org/doi/10.1128/jcm.40.7.2629-2631.2002.
25. Pratelli, A., Greco, G., Camero, M., Corrente, M., Normanno, G., Buonavoglia, C., Isolation and identification of a calicivirus from a dog with diarrhoea. *New Microbiol. [Internet]*, 23, 3, 257–60, 2000 Jul 1. [cited 2025 Jan 13] Available from: https://europepmc.org/article/med/10939040.
26. Kurth, A., Skilling, D.E., Smith, A.W., Serologic evidence of vesivirus-specific antibodies associated with abortion in horses. *Am. J. Vet. Res. [Internet]*, 67, 6, 1033–9, 2006 Jun 1. [cited 2025 Jan 13] Available from: https://avmajournals.avma.org/view/journals/ajvr/67/6/ajvr.67.6.1033.xml.
27. Smith, A.W., Iversen, P.L., Skilling, D.E., Stein, D.A., Bok, K., Matson, D.O., Vesivirus viremia and seroprevalence in humans. *J. Med. Virol. [Internet]*, 78, 5, 693–701, 2006 May 1. [cited 2025 Jan 13] Available from: https://onlinelibrary.wiley.com/doi/full/10.1002/jmv.20594.

28. Jiang, B., McClure, H.M., Fankhauser, R.L., Monroe, S.S., Glass, R.I., Prevalence of rotavirus and norovirus antibodies in non-human primates. *J. Med. Primatol. [Internet]*, 33, 1, 30–3, 2004 Feb 1. [cited 2025 Jan 13] Available from: https://onlinelibrary.wiley.com/doi/full/10.1111/j.1600-0684.2003.00051.x.
29. Prasad, B.V.V., Hardy, M.E., Dokland, T., Bella, J., Rossmann, M.G., Estes, M.K., X-ray Crystallographic Structure of the Norwalk Virus Capsid. *Sci. (1979) [Internet]*, 286, 5438, 287–90, 1999 Oct 8. [cited 2025 Jan 13] Available from: https://www.science.org/doi/10.1126/science.286.5438.287.
30. Chen, R., Neill, J.D., Noel, J.S., Hutson, A.M., Glass, R.I., Estes, M.K., *et al.*, Inter- and Intragenus Structural Variations in Caliciviruses and Their Functional Implications. *J. Virol. [Internet]*, 78, 12, 6469–79, 2004 Jun 15. [cited 2025 Jan 13] Available from: https://journals.asm.org/doi/10.1128/jvi.78.12.6469-6479.2004.
31. Daughenbaugh, K.F., Fraser, C.S., Hershey, J.W.B., Hardy, M.E., The genome-linked protein VPg of the Norwalk virus binds eIF3, suggesting its role in translation initiation complex recruitment. *EMBO J. [Internet]*, 22, 11, 2852–9, 2003 Jun 1. [cited 2025 Jan 13] Available from: https://www.embopress.org/doi/10.1093/emboj/cdg251.
32. Burroughs, J.N. and Brown, F., Presence of a covalently linked protein on calicivirus RNA. *J. Gen. Virol. [Internet]*, 41, 2, 443–6, 1978 Nov 1. [cited 2025 Jan 13] Available from: https://www.microbiologyresearch.org/content/journal/jgv/10.1099/0022-1317-41-2-443.
33. Dunham, D.M., Jiang, X., Berke, T., Smith, A.W., Matson, D.O., Genomic mapping of a calicivirus VPg. *Arch. Virol. [Internet]*, 143, 12, 2421–30, 1998 May 16. [cited 2025 Jan 13] Available from: https://link.springer.com/article/10.1007/s007050050471.
34. Herbert, T.P., Brierley, I., Brown, T.D.K., Identification of a protein linked to the genomic and subgenomic mRNAs of feline calicivirus and its role in translation. *J. Gen. Virol. [Internet]*, 78, 5, 1033–40, 1997 May 1. [cited 2025 Jan 13] Available from: https://www.microbiologyresearch.org/content/journal/jgv/10.1099/0022-1317-78-5-1033.
35. Schaffer, F.L., Ehresmann, D.W., Fretz, M.K., Soergel, M.E., A protein, VPg, covalently linked to 36S calcivirus RNA. *J. Gen. Virol. [Internet]*, 47, 1, 215–20, 1980 Mar 1. [cited 2025 Jan 13] Available from: https://www.microbiologyresearch.org/content/journal/jgv/10.1099/0022-1317-47-1-215.
36. Daughenbaugh, K.F., Wobus, C.E., Hardy, M.E., VPg of murine norovirus binds translation initiation factors in infected cells. *Virol. J. [Internet]*, 3, 1, 1–7, 2006 May 23. [cited 2025 Jan 13] Available from: https://link.springer.com/articles/10.1186/1743-422X-3-33.
37. Katayama, K., Shirato-Horikoshi, H., Kojima, S., Kageyama, T., Oka, T., Hoshino, F.B., Fukushi, S., Phylogenetic analysis of the complete genome of 18 Norwalk-like viruses. *Virol. 2002•Elsevier [Internet]*, 2002. [cited 2025 Jan 14]; Available from: https://www.sciencedirect.com/science/article/pii/S0042682202915680.

38. Bull, R.A., Hansman, G.S., Clancy, L.E., Tanaka, M.M., Rawlinson, W.D., White, P.A., Norovirus Recombination in ORF1/ORF2 Overlap. *Emerg. Infect. Dis. [Internet]*, 11, 7, 1079, 2005. [cited 2025 Jan 14] Available from: https://pmc.ncbi.nlm.nih.gov/articles/PMC3371806/.
39. Lambden, P.R., Liu, B., Clarke, I.N., A conserved sequence motif at the 5′ terminus of the Southampton virus genome is characteristic of the Caliciviridae. *Virus Genes.*, 10, 2, 149–52, 1995.
40. Tan, M., Hegde, R.S., Jiang, X., The P Domain of Norovirus Capsid Protein Forms Dimer and Binds to Histo-Blood Group Antigen Receptors. *J. Virol.*, 78, 12, 6233–42, 2004 Jun 15.
41. Bertolotti-Ciarlet, A., Crawford, S., Hutson, A.M., Estes, M.K., The 3′ end of Norwalk virus mRNA contains determinants that regulate the expression and stability of the viral capsid protein VP1: a novel function for the VP2. *J. Virol. 2003•Am. Soc. Microbiol. [Internet]*, 77, 21, 11603–15, 2003 Nov. [cited 2025 Jan 14] Available from: https://journals.asm.org/doi/abs/10.1128/jvi.77.21.11603-11615.2003.
42. Asanaka, M., Atmar, R.L., Ruvolo, V., Crawford, S.E., Neill, F.H., Estes, M.K., Replication and packaging of Norwalk virus RNA in cultured mammalian cells. *Proc. Natl. Acad. Sci. U.S.A.*, 102, 29, 10327–32, 2005 Jul 19.
43. Wobus, C.E., Karst, S.M., Thackray, L.B., Chang, K.O., Sosnovtsev, S.V., Belliot, G., *et al.*, Krug, A., Replication of Norovirus in Cell Culture Reveals a Tropism for Dendritic Cells and Macrophages. *PLoS Biol. 2004•journals.plos.org> [Internet]*, 2, 12, 2004 Dec. [cited 2025 Jan 14] Available from: https://journals.plos.org/plosbiology/article?id=10.1371/journal.pbio.0020432.
44. Miller, W. and Koev, G., Synthesis of subgenomic RNAs by positive-strand RNA viruses. *Virol. 2000•Elsevier [Internet]*, 2000. [cited 2025 Jan 14]; Available from: https://www.sciencedirect.com/science/article/pii/S0042682200904215.
45. Ruvoën-Clouet, N., Ganière, J.P., André-Fontaine, G., Blanchard, D., Le Pendu, J., Binding of Rabbit Hemorrhagic Disease Virus to Antigens of the ABH Histo-Blood Group Family. *J. Virol.*, 74, 24, 11950–4, 2000.
46. Guyader, F.L., Loisy, F., Atmar, R.L., Hutson, A.M., Estes, M.K., Ruvoën-Clouet, N., Pommepuy, M., Norwalk virus–specific binding to oyster digestive tissues. *Emerging Infect. Dis. 2006•ncbi.nlm.nih.gov [Internet]*, 2006. [cited 2025 Jan 14]; Available from: https://www.ncbi.nlm.nih.gov/pmc/articles/PMC2596755/.
47. Tian, P., Jiang, X., Zhong, W., Jensen, H., Binding of recombinant norovirus like particle to histo-blood group antigen on cells in the lumen of pig duodenum. *Elsevier [Internet]*, 2007. [cited 2025 Jan 14]; Available from: https://www.sciencedirect.com/science/article/pii/S0034528807000434.
48. Cheetham, S., Souza, M., McGregor, R., Meulia, T., Wang, Q., Saif, L.J., Binding Patterns of Human Norovirus-Like Particles to Buccal and Intestinal Tissues of Gnotobiotic Pigs in Relation to A/H Histo-Blood Group Antigen Expression. *J. Virol.*, 81, 7, 3535–44, 2007 Apr.

49. Farkas, T., Nakajima, S., Sugieda, M., Deng, X., Zhong, W., Jiang, X., Seroprevalence of noroviruses in swine. *J. Clin. Microbiol. 2005•Am. Soc. Microbiol. [Internet]*, 43, 2, 657–61, 2005 Feb. [cited 2025 Jan 14] Available from: https://journals.asm.org/doi/abs/10.1128/jcm.43.2.657-661.2005.
50. Cheetham, S., Souza, M., Meulia, T., Grimes, S., Han, M.G., Saif, L.J., Pathogenesis of a genogroup II human norovirus in gnotobiotic pigs. *J. Virol. 2006•Am. Soc. Microbiol. [Internet]*, 80, 21, 10372–81, 2006 Nov. [cited 2025 Jan 14] Available from: https://journals.asm.org/doi/abs/10.1128/jvi.00809-06.
51. Straub, T., Bentrup, K.Z., Coghlan, P.O., Dohnalkova, A., Mayer, B.K., Bartholomew, R.A., *In vitro* cell culture infectivity assay for human noroviruses. *Emerging Infect. Dis. 2007•ncbi.nlm.nih.gov [Internet]*, 2007. [cited 2025 Jan 14]; Available from: https://www.ncbi.nlm.nih.gov/pmc/articles/PMC2725917/.
52. Chang, K.O., Sosnovtsev, S.V., Belliot, G., Kim, Y., Saif, L.J., Green, K.Y., Bile acids are essential for porcine enteric calicivirus replication in association with down-regulation of signal transducer and activator of transcription 1. *Proc. Natl. Acad. Sci. U.S.A.*, 101, 23, 8733–8, 2004 Jun 8.
53. Tan, M., Meller, J., Jiang, X., C-terminal arginine cluster is essential for receptor binding of norovirus capsid protein. *J. Virol. 2006•Am. Soc. Microbiol. [Internet]*, 80, 15, 7322–31, 2006 Aug. [cited 2025 Jan 14] Available from: https://journals.asm.org/doi/abs/10.1128/jvi.00233-06.
54. Green, K.Y., Mory, A., Fogg, M.H., Weisberg, A., Belliot, G., Wagner, M., *et al.*, Mitra, T., Ehrenfeld, E., Isolation of enzymatically active replication complexes from feline calicivirus-infected cells. *J. Virology 2002•Am. Soc. Microbiol. [Internet]*, 76, 17, 8582–95, 2002 Sep. [cited 2025 Jan 14] Available from: https://journals.asm.org/doi/abs/10.1128/jvi.76.17.8582-8595.2002.
55. Schwartz, M., Chen, J., Lee, W.M., Janda, M., Ahlquist, P., Alternate, virus-induced membrane rearrangements support positive-strand RNA virus genome replication. *Proc. Natl. Acad. Sci. U.S.A.*, 101, 31, 11263–8, 2004 Aug 3.
56. Studdert, M.J. and O'Shea, J.D., Ultrastructural studies of the development of feline calicivirus in a feline embryo cell line. *Arch. Virol*, 48, 4, 317–25, 1975 Dec.
57. Fernandez-Vega, V., Sosnovtsev, S.V., Belliot, G., King, A.D., Mitra, T., Gorbalenya, A., *et al.*, Green, K.Y., Norwalk virus N-terminal nonstructural protein is associated with disassembly of the Golgi complex in transfected cells. *J. Virol. 2004•Am. Soc. Microbiol. [Internet]*, 78, 9, 4827–37, 2004 May. [cited 2025 Jan 14] Available from: https://journals.asm.org/doi/abs/10.1128/jvi.78.9.4827-4837.2004.
58. Kuyumcu-Martinez, M., Belliot, G., Sosnovtsev, S.V., Chang, K.O., Green, K.Y., Lloyd, R.E., Calicivirus 3C-like proteinase inhibits cellular translation by cleavage of poly (A)-binding protein. *J. Virol. 2004•Am. Soc. Microbiol.*

[Internet], 78, 15, 8172–82, 2004 Aug. [cited 2025 Jan 14] Available from: https://journals.asm.org/doi/abs/10.1128/jvi.78.15.8172-8182.2004.

59. Milnes, A.S., Binns, S.H., Oliver, S.L., Bridger, J.C., Retrospective study of noroviruses in samples of diarrhoea from cattle, using the Veterinary Laboratories Agency's Farmfile database. *Vet. Rec.*, 160, 10, 326–30, 2007 Mar 10.
60. Saif, L.J., Bohl, E.H., Theil, K.W., Rotavirus-like, calcivirus-like, and 23nm virus-like particles associated with diarrhea in young pigs. *J. Clin. Microbiol.*, 12, 1, 105–11, 1980.
61. Karst, S.M., Wobus, C.E., Lay, M., Davidson, J., Virgin, H.W., STAT1-Dependent Innate Immunity to a Norwalk-Like Virus. *Sci. (1979)*, 299, 5612, 1575–8, 2003 Mar 7.
62. Norovirus in captive lion cub (Panthera leo). *ncbi.nlm.nih.gov [Internet]*. [cited 2025 Jan 14]; Available from: https://www.ncbi.nlm.nih.gov/pmc/articles/PMC2878247/.
63. Martella, V., Lorusso, E., Decaro, N., Elia, G., Radogna, A., D'Abramo, M., Desario, C., Cavalli, A., Detection and molecular characterization of a canine norovirus. *Emerging Infect. Dis. 2008•ncbi.nlm.nih.gov [Internet]*, 2008. [cited 2025 Jan 14]; Available from: https://www.ncbi.nlm.nih.gov/pmc/articles/PMC2600395/.
64. Mesquita, J., Barclay, L., Nascimento, M.S.J., Vinjé, J., Novel norovirus in dogs with diarrhea. *Emerging Infect. Dis. 2010•ncbi.nlm.nih.gov [Internet]*, 2010. [cited 2025 Jan 14]; Available from: https://www.ncbi.nlm.nih.gov/pmc/articles/PMC3086253/.
65. Pinto, P., Wang, Q., Chen, N., Dubovi, E., Daniels, J.B., Millward, L.M., Buonavoglia, C., Martella, V., Discovery and genomic characterization of noroviruses from a gastroenteritis outbreak in domestic cats in the US. *PloS One 2012•journals.plos.org [Internet]*, 7, 2, 32739, 2012 Feb 28. [cited 2025 Jan 14] Available from: https://journals.plos.org/plosone/article?id=10.1371/journal.pone.0032739.
66. Srivastava, S., Kumar, S., Kumar Sharma, P., Mehta, R., Habeeb Naser, I., Mohanty, A., *et al.*, Managing and mitigating Crimean-Congo haemorrhagic fever outbreaks during mass Gatherings: A Case study of the Arbaeen pilgrimage in Iraq. *Clin. Infect. Pract.*, 22, 100356, 2024 Apr.
67. Vinjé J. Advances in laboratory methods for detection and typing of norovirus. *J. Clin. Microbiol.*, 53, 2, 373–81, 2015 Feb.
68. Günther, H. and Otto, P., Diarrhea in young calves. 6. Determination of the pathogenicity of a bovine coronavirus and an unidentified icosahedral virus, 1984. pubmed.ncbi.nlm.nih.gov [Internet]. [cited 2025 Jan 14]; Available from: https://pubmed.ncbi.nlm.nih.gov/6099109/.
69. Liu, B.L., Lambden, P.R., Günther, H., Günther, G., Otto, P., Elschner, M., *et al.*, Molecular Characterization of a Bovine Enteric Calicivirus: Relationship to the Norwalk-Like Viruses. *J. Virol. [Internet]*, 73, 1, 819–25, 1999 Jan. [cited 2025 Jan 14] Available from: https://journals.asm.org/journal/jvi.

70. Oliver, S.L., Dastjerdi, A.M., Wong, S., El-Attar, L., Gallimore, C., Brown, D.W.G., *et al.*, Molecular characterization of bovine enteric caliciviruses: a distinct third genogroup of noroviruses (Norwalk-like viruses) unlikely to be of risk to humans. *J. Virol. 2003•Am. Soc. Microbiol. [Internet]*, 77, 4, 2789–98, 2003 Feb 15. [cited 2025 Jan 14] Available from: https://journals.asm.org/doi/abs/10.1128/jvi.77.4.2789-2798.2003.
71. Oliver, S., Asobayire, E., Charpilienne, A., Cohen, J., Bridger, J.C., Complete genomic characterization and antigenic relatedness of genogroup III, genotype 2 bovine noroviruses. *Arch. Virol. 2007•Springer [Internet]*, 152, 2, 257–72, 2007 Feb. [cited 2025 Jan 14] Available from: https://link.springer.com/article/10.1007/s00705-006-0856-2.
72. Kroneman, A., Vega, E., Vennema, H., Vinjé, J., White, P.A., Hansman, G., *et al.*, Proposal for a unified norovirus nomenclature and genotyping. *Arch. Virol. [Internet]*, 158, 10, 2059–68, 2013 Oct 25. [cited 2025 Jan 14] Available from: https://link.springer.com/article/10.1007/s00705-013-1708-5.
73. Mauroy, A., Scipioni, A., Mathijs, E., Ziant, D., Daube, G., Thiry, E., Complete genome sequence of a novel bovine norovirus: evidence for slow genetic evolution in genogroup III genotype 2 noroviruses. *J. Virol. [Internet]*, 86, 22, 12449–50, 2012 Nov 15. [cited 2025 Jan 14] Available from: https://journals.asm.org/doi/abs/10.1128/jvi.02251-12.
74. Mauroy, A., Scipioni, A., Mathijs, E., Epidemiological study of bovine norovirus infection by RT-PCR and a VLP-based antibody ELISA. *Elsevier [Internet]*, 2009. [cited 2025 Jan 14]; Available from: https://www.sciencedirect.com/science/article/pii/S0378113509000522.
75. Thomas, C., Jung, K., Han, M.G., Hoet, A., Scheuer, K., Wang, Q., *et al.*, Retrospective serosurveillance of bovine norovirus (GIII. 2) and nebovirus in cattle from selected feedlots and a veal calf farm in 1999 to 2001 in the United States. *Arch. Virol. 2014•Springer [Internet]*, 159, 1, 83–90, 1999 Jan 1. [cited 2025 Jan 14] Available from: https://link.springer.com/article/10.1007/s00705-013-1795-3.
76. Mattison, K., Shukla, A., Cook, A., Pollari, F., Friendship, R., Kelton, D., Bidawid, S., Farber, J.M., Human noroviruses in swine and cattle. *Emerging Infect. Dis. 2007•ncbi.nlm.nih.gov [Internet]*, 2007. [cited 2025 Jan 14]; Available from: https://www.ncbi.nlm.nih.gov/pmc/articles/PMC2828081/.
77. Park, S., Jeong, C., Kim, H., Park, S., Molecular epidemiology of bovine noroviruses in South Korea. *Elsevier [Internet]*, 2007. [cited 2025 Jan 14]; Available from: https://www.sciencedirect.com/science/article/pii/S0378113507001472.
78. Hassine-Zaafrane, M., Kaplon, J., Sdiri-Loulizi, K., Aouni, Z., Pothier, P., Aouni, M., *et al.*, Molecular prevalence of bovine noroviruses and neboviruses detected in central-eastern Tunisia. *Arch. Virol*, 157, 8, 1599–604, 2012 Aug.

79. Alcalá, A.C., Hidalgo, M.A., Obando, C., Vizzi, E., Liprandi, F., Ludert, J.E., [Molecular identification of bovine enteric calciviruses in Venezuela]. *Acta. Cient. Venez. [Internet]*, 54, 2, 148–52, 2003 Jan 1. [cited 2025 Jan 14] Available from: https://europepmc.org/article/med/14976786.
80. Mauroy, A., Scipioni, A., Mathijs, E., Thys, C., *et al.*, Thiry, E., Molecular detection of kobuviruses and recombinant noroviruses in cattle in continental Europe. *Arch. Virol 2009•Springer [Internet]*, 154, 11, 1841–5, 2009 Oct. [cited 2025 Jan 14] Available from: https://link.springer.com/article/10.1007/s00705-009-0518-2.
81. Di Martino, B., Di Profio, F., Di Felice, E., Melegari, I., Ceci, C., Mauroy, A., *et al.*, Genetic heterogeneity of bovine noroviruses in Italy. *Arch. Virol*, 159, 10, 2717–22, 2014 Sep 1.
82. Kaplon, J., Guenau, E., Asdrubal, P., Pothier, P., Ambert-Balay, K., Possible novel nebovirus genotype in cattle, France. *Emerging Infect. Dis. 2011•ncbi.nlm.nih.gov [Internet]*, 2011. [cited 2025 Jan 14]; Available from: https://www.ncbi.nlm.nih.gov/pmc/articles/PMC3358183/.
83. Wise, A., Monroe, S., Hanson, L., Molecular characterization of noroviruses detected in diarrheic stools of Michigan and Wisconsin dairy calves: circulation of two distinct subgroups. *Elsevier [Internet]*, 2004. [cited 2025 Jan 14]; Available from: https://www.sciencedirect.com/science/article/pii/S0168170203003484.
84. Deng, Y., Batten, C.A., Liu, B.L., Lambden, P.R., Elschner, M., Günther, H., *et al.*, Otto, P., Schnurch, P., Studies of epidemiology and seroprevalence of bovine noroviruses in Germany. *J. Clin. Microbiol. 2003•Am. Soc. Microbiol. [Internet]*, 41, 6, 2300–5, 2003 Jun 1. [cited 2025 Jan 14] Available from: https://journals.asm.org/doi/abs/10.1128/jcm.41.6.2300-2305.2003.
85. Oliver, S., Brown, D., Green, J., Bridger, J.C., A chimeric bovine enteric calicivirus: evidence for genomic recombination in genogroup III of the Norovirus genus of the Caliciviridae. *Virol. 2004•Elsevier [Internet]*, 2004. [cited 2025 Jan 14]; Available from: https://www.sciencedirect.com/science/article/pii/S0042682204003976.
86. Bull, R.A., Tanaka, M.M., Correspondence, P.A.W., White, P.A., Norovirus recombination. *J. Gen. Virol. 2007•microbiologyresearch.org [Internet]*, 88, 12, 3347–59, 2007 Dec. [cited 2025 Jan 14] Available from: https://www.microbiologyresearch.org/content/journal/jgv/10.1099/vir.0.83321-0.
87. Günther, H., Diarrhea in young calves. 7." Zackenvirus"(Jena agent 117/80)–a new diarrhea pathogen in calves, 1987. [cited 2025 Jan 14]; Available from: https://pubmed.ncbi.nlm.nih.gov/2831845/.
88. Otto, P.H., Clarke, I.N., Lambden, P.R., Salim, O., Reetz, J., Liebler-Tenorio, E.M., Infection of calves with bovine norovirus GIII. 1 strain Jena virus: an experimental model to study the pathogenesis of norovirus infection. *J. Virol. 2011•Am. Soc. Microbiol. [Internet]*, 85, 22, 12013–21, 2011 Nov 15. [cited 2025 Jan 14] Available from: https://journals.asm.org/doi/abs/10.1128/jvi.05342-11.

89. Jung, K., Scheuer, K., Zhang, Z., Wang, Q., Pathogenesis of GIII. 2 bovine norovirus, CV186-OH/00/US strain in gnotobiotic calves. *Elsevier [Internet]*, 2014. [cited 2025 Jan 14]; Available from: https://www.sciencedirect.com/science/article/pii/S0378113513005075.
90. Cho, Y., Han, J., Wang, C., Cooper, V., Case–control study of microbiological etiology associated with calf diarrhea. *Elsevier [Internet]*, 2013. [cited 2025 Jan 14]; Available from: https://www.sciencedirect.com/science/article/pii/S0378113513003337.
91. Di Bartolo, I., Ponterio, E., Monini, B.M., Ruggeri, F.M., Bartolo, I.D., Monini, M., Ponterio, E., Monini, M., A pilot survey of bovine norovirus in northern Italy. *Arch. Virol. 2011•Citeseer [Internet]*. [cited 2025 Jan 14]; Available from: https://citeseerx.ist.psu.edu/document?repid=rep1&type=pdf&doi=bfe5d-7c1c5ac827bda3a912156a2ef0fc10f20b0.
92. Jor, E. and Myrmel, M., SYBR Green based real-time RT-PCR assay for detection and genotype prediction of bovine noroviruses and assessment of clinical significance in Norway. *Elsevier [Internet]*, 2010. [cited 2025 Jan 14]; Available from: https://www.sciencedirect.com/science/article/pii/S0166093410001254.
93. Smiley, J.R., Hoet, A.E., Trávén, M., Tsunemitsu, H., Saif, L.J., Reverse transcription-PCR assays for detection of bovine enteric caliciviruses (BEC) and analysis of the genetic relationships among BEC and human caliciviruses. *J. Clin. Microbiol. 2003•Am. Soc. Microbiol. [Internet]*, 41, 7, 3089–99, 2003 Jul 1. [cited 2025 Jan 14] Available from: https://journals.asm.org/doi/abs/10.1128/jcm.41.7.3089-3099.2003.
94. Poel, W.. and Heide, R.., Epidemiology of Norwalk-like virus infections in cattle in The Netherlands. *Elsevier [Internet]*, 2003. [cited 2025 Jan 14]; Available from: https://www.sciencedirect.com/science/article/pii/S0378113502004212.
95. Doane, F.W., Electron microscopy for the detection, in: *Google Scholar [Internet]*, 1994, [cited 2025 Jan 14]. Available from: https://scholar.google.com/scholar?hl=en&as_sdt=0%2C5&scioq=Alcala%2C+A.C.%2C+Hidalgo%2C+M.A.%2C+Obando%2C+C.%2C+Vizzi%2C+E.%2C+Liprandi%2C+F.%2C+Ludert%2C+J.E.%2C+2003.+Deteccion+molecular+de+calicivirus+entericos+de+bovinos+en+Venezuela.+Acta+Cientifica+Venezolana+54%2C+148%E2%80%93152.+&q=-Doane%2C+F.W.%2C+1994.+Electron+microscopy+-for+the+detection+of+gastroenteritis+viruses.+In%3A+Kapikian%2C+A.Z.+%28Ed.%29%2C+Viral+Infections+of+the+Gastro-intestinal+Tract.+Marcel+Dekker%2C+New+York%2C+USA%2C+pp.+101%E2%80%93130.&btnG=.
96. Atmar, R.L. and Estes, M.K., Diagnosis of noncultivatable gastroenteritis viruses, the human caliciviruses. *Clin. Microbiol. Rev.*, 14, 1, 15–37, 2001.

97. De Leon, R., Matsui, S.M., Baric, R.S., Herrmann, J.E., Blacklow, N.R., Greenberg, H.B., *et al.*, Detection of Norwalk virus in stool specimens by reverse transcriptase- polymerase chain reaction and nonradioactive oligoprobes. *J. Clin. Microbiol.*, 30, 12, 3151–7, 1992.
98. Scipioni, A., Mauroy, A., Ziant, D., Saegerman, C., Thiry, E., A SYBR Green RT-PCR assay in single tube to detect human and bovine noroviruses and control for inhibition. *Virol. J. 2008•Springer [Internet]*, 5, 2008. [cited 2025 Jan 14] Available from: https://link.springer.com/article/10.1186/1743-422X-5-94.
99. Scipioni, A., Mauroy, A., Vinje, J., Animal noroviruses. *Elsevier [Internet]*, 2008. [cited 2025 Jan 14]; Available from: https://www.sciencedirect.com/science/article/pii/S1090023307003966.
100. Scipioni, A., Bourgot, I., Mauroy, A., Detection and quantification of human and bovine noroviruses by a TaqMan RT-PCR assay with a control for inhibition. *Elsevier [Internet]*, 2008. [cited 2025 Jan 14]; Available from: https://www.sciencedirect.com/science/article/pii/S0890850808000261.
101. Wolf, S., Williamson, W.M., Hewitt, J., Rivera-Aban, M., Lin, S., Ball, A., *et al.*, Scholes, P., Greening, G.E., Sensitive multiplex real-time reverse transcription-PCR assay for the detection of human and animal noroviruses in clinical and environmental samples. *Appl. Environ. Microbiol. 2007•Am. Soc. Microbiol. [Internet]*, 73, 17, 5464–70, 2007 Sep. [cited 2025 Jan 14] Available from: https://journals.asm.org/doi/abs/10.1128/aem.00572-07.
102. Dastjerdi, A., Green, J., Gallimore, C., Brown, D.W.G., Bridger, J.C., The bovine Newbury agent-2 is genetically more closely related to human SRSVs than to animal caliciviruses. *Virol. 1999•Elsevier [Internet]*, 1999. [cited 2025 Jan 14]; Available from: https://www.sciencedirect.com/science/article/pii/S0042682298995149.
103. Liu, C.J., Lien, K.Y., Weng, C.Y., Shin, J.W., Chang, T.Y., Lee, G.B., Magnetic-bead-based microfluidic system for ribonucleic acid extraction and reverse transcription processes. *Biomed. Microdevices*, 11, 2, 339–50, 2009.
104. Fukuda, S., Takao, S., Kuwayama, M., Shimazu, Y., Miyazaki, K., Rapid detection of norovirus from fecal specimens by real-time reverse transcription-loop-mediated isothermal amplification assay. *J. Clin. Microbiol. 2006•Am. Soc. Microbiol. [Internet]*, 44, 4, 1376–81, 2006 Apr. [cited 2025 Jan 14] Available from: https://journals.asm.org/doi/abs/10.1128/jcm.44.4.1376-1381.2006.
105. Moore, C., Clark, E., Gallimore, C., Evaluation of a broadly reactive nucleic acid sequence based amplification assay for the detection of noroviruses in faecal material. *Elsevier [Internet]*, 2004. [cited 2025 Jan 14]; Available from: https://www.sciencedirect.com/science/article/pii/S1386653203001707.
106. Notomi, T., Okayama, H., Masubuchi, H., Yonekawa, T., Watanabe, K., Amino, N., *et al.*, Hase, T., Loop-mediated isothermal amplification of DNA. *Nucleic Acids Res. 2000•academic.oup.com [Internet]*, 28, 12, 63, 2000. [cited 2025 Jan 14] Available from: https://academic.oup.com/nar/article-abstract/28/12/e63/2359194.

107. Li, D., De Keuckelaere, A., Uyttendaele, M., Application of long-range and binding reverse transcription-quantitative PCR to indicate the viral integrities of noroviruses. *Appl. Environ. Microbiol. 2014•Am. Soc. Microbiol. [Internet]*, 80, 20, 6473–9, 2014. [cited 2025 Jan 14] Available from: https://journals.asm.org/doi/abs/10.1128/aem.02092-14.
108. Li, D., Baert, L., Coillie, E.V., Critical studies on binding-based RT-PCR detection of infectious Noroviruses. *Elsevier [Internet]*, 2011. [cited 2025 Jan 14]; Available from: https://www.sciencedirect.com/science/article/pii/S016609341100317X.
109. Tian, P., Yang, D., Jiang, X., Zhong, W., Cannon, J.L., Burkhardt Iii, W., Woods, J.W., Hartman, G., Specificity and kinetics of norovirus binding to magnetic bead-conjugated histo-blood group antigens. *J. Appl. Microbiol. 2010•academic.oup.com [Internet]*, 2010. [cited 2025 Jan 14]; Available from: https://academic.oup.com/jambio/article-abstract/109/5/1753/6721084.
110. Jiang, X., Wilton, N., Zhong, W., Farkas, T., Huang, P.W., Barrett, E., Guerrero, M., Ruiz-Palacios, G., Diagnosis of human caliciviruses by use of enzyme immunoassays. *J. Infect. Dis. 2000•academic.oup.com [Internet]*, 2000. [cited 2025 Jan 14]; Available from: https://academic.oup.com/jid/article-abstract/181/Supplement_2/S349/1025187.
111. Martino, B.D., Marsilio, F., Profio, F.D., Lorusso, E., Friedrich, K.G., Buonavoglia, C., *et al.*, Martella, V., Detection of antibodies against norovirus genogroup GIV in carnivores. *Clin. Vaccine Immunol. 2010•Am. Soc. Microbiol. [Internet]*, 17, 1, 180–2, 2010 Jan. [cited 2025 Jan 14] Available from: https://journals.asm.org/doi/abs/10.1128/cvi.00312-09.
112. Oliver, S.L., Wood, E., Asobayire, E., Wathes, D.C., Brickell, J.S., Elschner, M., *et al.*, Otto, P., Lambden, P.R., Serotype 1 and 2 bovine noroviruses are endemic in cattle in the United Kingdom and Germany. *J. Clin. Microbiol. 2007•Am. Soc. Microbiol. [Internet]*, 45, 9, 3050–2, 2007 Sep. [cited 2025 Jan 14] Available from: https://journals.asm.org/doi/abs/10.1128/jcm.02015-06.
113. Sugieda, M., Nagaoka, H., Kakishima, Y., Ohshita, T., Nakamura, S., Nakajima, S., Detection of Norwalk-like virus genes in the caecum contents of pigs. Brief report. *Arch. Virol*, 143, 6, 1215–21, 1998.
114. Wobus, C.E., Karst, S.M., Thackray, L.B., Chang, K.O., Sosnovtsev, S.V., Belliot, G., *et al.*, Karst, S.M., Thackray, L.B., Chang, K.O., Sosnovtsev, S.V., Belliot, G., Krug, A., Replication of Norovirus in Cell Culture Reveals a Tropism for Dendritic Cells and Macrophages. *PLoS Biol. 2004•journals.plos.org [Internet]*, 2, 12, 2004 Dec. [cited 2025 Jan 14] Available from: https://journals.plos.org/plosbiology/article?id=10.1371/journal.pbio.0020432.
115. Hsu, C.C., Wobus, C.E., Steffen, E.K., Riley, L.K., Livingston, R.S., Development of a microsphere-based serologic multiplexed fluorescent immunoassay and a reverse transcriptase PCR assay to detect murine norovirus 1 infection in. *Clin. Vaccine Immunol. 2005•Am. Soc. Microbiol. [Internet]*, 12, 10, 1145–51, 2005 Oct. [cited 2025 Jan 14] Available from: https://journals.asm.org/doi/abs/10.1128/cdli.12.10.1145-1151.2005.

116. Green, K.Y., Chanock, R.M., Kapikian, A.Z., *Google Scholar [Internet]*, [cited 2025 Jan 14]. Available from: https://scholar.google.com/scholar?hl=en&as_sdt=0%2C5&scioq=Alcala%2C+A.C.%2C+Hidalgo%2C+M.A.%2C+Obando%2C+C.%2C+Vizzi%2C+E.%2C+Liprandi%2C+F.%2C+Ludert%2C+J.E.%2C+2003.+Deteccion+molecular+de+calicivirus+entericos+de+bovinos+en+Venezuela.+Acta+Cientifica+Venezolana+54%2C+148%E2%80%93152.+&q=Green%2C+K.+Y.%2C+R.+M.+Chanock%2C+and+A.+Z.+Kapikian.+2001.+Human+Caliciviruses%2Cp.+841%E2%80%93874.+In+D.M.+Knipe+and+P.+M.+Howley+%28ed.%29%2C+Fields+Virology%2Cvol.+1.+Lippincott+Williams+%26Wilkins%2C+Philadelphia%2CPa.&btnG=.
117. Meyers, G., Wirblich, C., Thiel, H.J., Genomic and subgenomic RNAs of rabbit hemorrhagic disease virus are both protein-linked and packaged into particles. *Virol. 1991•Elsevier [Internet]*, 1991. [cited 2025 Jan 15]; Available from: https://www.sciencedirect.com/science/article/pii/004268229190437G.
118. Ehresmann, D. and Schaffer, F.L., Calicivirus intracellular RNA: Fractionation of 18–22 S RNA and lack of typical 5′-methylated caps on 36 S and 22 S San Miguel sea lion virus RNAs. *Virol. 1979•Elsevier [Internet]*, 1979. [cited 2025 Jan 14]; Available from: https://www.sciencedirect.com/science/article/pii/0042682279904264.
119. Katayama, K., Hansman, G., Oka, T., Ogawa, S., Takeda, N., Investigation of norovirus replication in a human cell line. *Arch. Virol. 2006•Springer [Internet]*, 2006. [cited 2025 Jan 15]; Available from: https://link.springer.com/article/10.1007/s00705-005-0720-9.
120. Povey, R., Wardley, R., Jessen, H., Feline picornavirus infection: the *in vivo* carrier state, 1973. [cited 2025 Jan 15]; Available from: https://www.cabidigitallibrary.org/doi/full/10.5555/19732205856.
121. Wardley, R., The clinical disease and patterns of excretion associated with three different strains of feline caliciviruses. *Elsevier [Internet]*, 1977. [cited 2025 Jan 15]; Available from: https://www.sciencedirect.com/science/article/pii/S0034528818332144.
122. Casanova, J.L. and Abel, L., Inborn errors of immunity to infection. *J. Exp. Med.*, 202, 2, 197–201, 2005 Jul 18.
123. Widdowson, M., Cramer, E., Hadley, L., Bresee, J.S., Beard, R.S., Bulens, S.N., Charles, M., Outbreaks of acute gastroenteritis on cruise ships and on land: identification of a predominant circulating strain of norovirus—United States, 2002. *J. Infect. Dis. 2004•academic.oup.com [Internet]*, 2004. [cited 2025 Jan 15]; Available from: https://academic.oup.com/jid/article-abstract/190/1/27/2191803.
124. Lopman, B., Reacher, M., Van Duijnhoven, Y., Hanon, F.X., Brown, D., Koopmans, M., Viral gastroenteritis outbreaks in Europe, 1995–2000. *Emerging Infect. Dis. 2003•ncbi.nlm.nih.gov [Internet]*, 2003. [cited 2025 Jan 15]; Available from: https://www.ncbi.nlm.nih.gov/pmc/articles/PMC2873740/.

125. Gallimore, C.I., Green, J., Lewis, D., Richards, A.F., Lopman, B.A., Hale, A.D., *et al.*, Eglin, R., Gray, J.J., Diversity of noroviruses cocirculating in the north of England from 1998 to 2001. *J. Clin. Microbiol. 2004•Am. Soc. Microbiol. [Internet]*, 42, 4, 1396–401, 2004 Apr. [cited 2025 Jan 15] Available from: https://journals.asm.org/doi/abs/10.1128/jcm.42.4.1396-1401.2004.
126. Lochridge, V.P. and Hardy, M.E., Snow Mountain virus genome sequence and virus-like particle assembly. *Virus Genes.*, 26, 1, 71–82, 2003.
127. Deng Y, Batten CA, Liu BL, Lambden PR, Elschner M, Gunther H, Otto P, Schnurch P, Eichhorn W, Herbst W, Clarke IN. Studies of epidemiology and seroprevalence of bovine noroviruses in Germany. *J. Clin. Microbiol.*, 41, 6, 2300–5, 2003 Jun.
128. Vinjé, J., Green, J., Lewis, D.C., Gallimore, C.I., Brown, D.W.G., Koopmans, M.P.G., Genetic polymorphism across regions of the three open reading frames of "Norwalk-like viruses.". *Arch. Virol. [Internet]*, 145, 2, 223–41, 2000. [cited 2025 Jan 15] Available from: https://link.springer.com/article/10.1007/s007050050020.
129. Jiang, X., Espul, C., Zhong, W.M., Cuello, H., Matson, D.O., Characterization of a novel human calicivirus that may be a naturally occurring recombinant. *Arch. Virol*, 144, 12, 2377–87, 1999.
130. Packer, C., Altizer, S., Appel, M., Brown, E., Martenson, J., O'brien, S.J., *et al.*, Roelke-Parker, M., Viruses of the Serengeti: patterns of infection and mortality in African lions. *J. Anim. Ecol. 1999• Wiley Online Lib. [Internet]*, 68, 6, 1161–78, 1999 Nov. [cited 2025 Jan 15] Available from: https://besjournals.onlinelibrary.wiley.com/doi/abs/10.1046/j.1365-2656.1999.00360.x.
131. Kadoi, K., Kamata, H., Ohba, S., Satoh, T., Inaba, Y., Experimental infection in SPF kittens with a particular calicivirus strain originally isolated from lion. *New Microbiol. 1998•europepmc.org [Internet]*, 1998. [cited 2025 Jan 15]; Available from: https://europepmc.org/article/med/9699205.
132. Endo, Y., Uema, M., Miura, R., Tsukiyama-Kohara, K., Tsujimoto, H., Yoneda, K., *et al.*, Kai, C., Prevalence of canine distemper virus, feline immunodeficiency virus and feline leukemia virus in captive African lions (Panthera leo) in Japan. *J. Vet. Med. Sci. 2004•jstage.jst.go.jp [Internet]*, 66, 12, 1587–9, 2004. [cited 2025 Jan 15] Available from: https://www.jstage.jst.go.jp/article/jvms/66/12/66_12_1587/_article/-char/ja/.
133. Spencer, J.A., Survey of antibodies to feline viruses in free-ranging lions. *South Afr. J. Wildl. Research-24-month Delayed Open Access 1991•journals.co.za [Internet]*, 1991. [cited 2025 Jan 15]; Available from: https://journals.co.za/doi/abs/10.10520/EJC116866.
134. Jiang, X., Huang, P., Zhong, W., Design and evaluation of a primer pair that detects both Norwalk-and Sapporo-like caliciviruses by RT-PCR. *Elsevier [Internet]*, 1999. [cited 2025 Jan 15]; Available from: https://www.sciencedirect.com/science/article/pii/S0166093499001147.

135. Zheng, D., Ando, T., Fankhauser, R., Beard, R.S., Glass, R.I., Monroe, S.S., Norovirus classification and proposed strain nomenclature. *Virol. 2006•Elsevier [Internet]*, 2006. [cited 2025 Jan 15]; Available from: https://www.sciencedirect.com/science/article/pii/S0042682205007658.
136. Reuter, G., Krisztalovics, K., Vennema, H., Koopmans, M., Szucs, G., Evidence of the etiological predominance of norovirus in gastroenteritis outbreaks - Emerging new-variant and recombinant noroviruses in Hungary. *J. Med. Virol.*, 76, 4, 598–607, 2005 Aug.

13

The Challenges of Developing Norovirus Vaccines and Therapies

Gita Chawla*, Tathagata Pradhan and Ojasvi Gupta

Department of Pharmaceutical Chemistry, School of Pharmaceutical Education and Research, Jamia Hamdard, Hamdard Nagar, New Delhi, India

Abstract

Norovirus is a primary contributor to acute gastroenteritis worldwide, posing unique challenges in vaccine and therapeutic development due to its genetic diversity, rapid mutation rates, and complex immune evasion mechanisms. This chapter explores the multifaceted hurdles in creating effective norovirus vaccines and therapies. It delves into the virus' epidemiology, the limitations of current treatment approaches, and the obstacles in vaccine development. These challenges include the absence of long-lasting immunity and difficulty achieving broad cross-protection against various strains. The chapter also examines the ongoing research efforts, novel strategies, and potential future breakthroughs that may pave the way for successful norovirus control measures.

Keywords: Norovirus, vaccines, challenges, epidemiology, genomics

13.1 Overview of Norovirus Epidemiology

Norovirus (NoV) ranks as one of the top global causes of viral gastroenteritis, accounting for more than 50% of outbreaks worldwide. Approximately, 58% of foodborne illnesses acquired in the US are caused by NoV as per data from the Centers for Disease Control and Prevention (CDC) [1]. The CDC identifies norovirus (NoV) as the leading cause of acute gastroenteritis across all age groups, frequently necessitating medical care in emergency departments, outpatient clinics, and community settings. As per WHO reports,

**Corresponding author*: gchawla@jamiahamdard.ac.in

Sachin Kumar, Shriyansh Srivastava, Rishabha Malviya and Ranjit Sah (eds.) Emerging Infectious Disease: Unravelling Norovirus and Pandemic Dynamics, (359–378) © 2026 Scrivener Publishing LLC

NoV is responsible for nearly 685 million cases of acute gastroenteritis each year, including 200 million infections in children under 5 years, making it a critical public health issue worldwide [2–4].

Norovirus (NoV) spreads mainly *via* the fecal–oral route, often through contaminated food, water, or surfaces. Direct transmission is also frequent, especially in confined spaces such as schools, nursing homes, and cruise ships, where outbreaks can escalate quickly. The virus is also resilient, surviving on surfaces for extended periods and resisting many standard cleaning and disinfecting methods. Norovirus typically incubates for 24 to 48 h, leading to symptoms within 12 to 48 h [1, 4]. These symptoms often include nausea, vomiting (which can be forceful and frequent), watery diarrhea, and abdominal cramps. Significant fluid loss from vomiting and diarrhea increases the risk of dehydration. Individuals may also experience mild flu-like symptoms, including low-grade fever, headache, muscle aches, and fatigue, which are generally less intense than the gastrointestinal distress [5]. Abdominal pain or cramping is common, sometimes accompanied by bloating. The body's fight against the virus can cause fatigue and weakness, which are exacerbated by dehydration. While most norovirus infections resolve within 1–3 days [6], the illness can last longer in vulnerable groups like teenagers, the elderly, or immunocompromised individuals, potentially requiring medical attention to prevent severe dehydration [4, 7]. Asymptomatic infections with norovirus have been documented, with individuals shedding the virus without showing any symptoms. Most notably, the virus has an extremely low infectious dose, with as few as 18–100 viral particles capable of leading to infection. Additionally, high levels of viral shedding occur, exceeding 10^9 particles/mL of stool during the early days of infection. This shedding can precede the onset of symptoms and be prolonged in immunocompromised individuals [6]. The virus is also remarkably stable, surviving in a temperature range from 0°C to 60°C [8]. Furthermore, norovirus undergoes frequent mutation and recombination, resulting in significant antigenic diversity, which complicates efforts to control its spread and develop lasting immunity [9, 10].

Norovirus infections typically follow a seasonal trend, with cases peaking during the colder months, especially in winter. However, the virus can cause outbreaks year-round, especially in densely populated areas and settings with compromised sanitation. This seasonality is more pronounced in temperate regions, though developing countries experience a high burden of norovirus cases regardless of the season due to environmental factors such as poor water.

Norovirus infections can be especially devastating in healthcare settings, leading to high rates of hospitalization and, in some cases, death. Moreover, malnutrition and co-infections in resource-limited regions exacerbate the impact of norovirus, leading to more severe outcomes and prolonged outbreaks [11].

13.2 Classification and Molecular Biology of Norovirus

Noroviruses (NoVs) are a type of non-enveloped virus possessing a positive-sense single-stranded RNA genome belonging to the *Caliciviridae* family under genus *Norovirus.* The *Caliciviridae* family consists of five other genera, including *Vesivirus, Lagovirus, Becovirus, Recovirus,* and *Sapovirus.* Caliciviruses are defined by their non-enveloped structure and single-stranded, positive-sense RNA genomes. These viruses are encased by a highly stable protein capsid with icosahedral symmetry. The human norovirus (NoV) genome spans approximately 7.5 to 7.7 kilobases (kb) long and consists of three open reading frames (ORFs) [12, 13]. The structure of norovirus is illustrated in Figure 13.1.

NoVs are classified into seven genogroups (GI–GVII), with GI and GII genogroups being the most relevant for human infections. However, substantial antigenic variations result in the absence of shared neutralization epitopes. Human noroviruses (HuNoVs) are a major cause of nonbacterial acute gastroenteritis globally, with GII. Since the discovery of norovirus, thousands of strains have been identified worldwide, displaying remarkable genetic and antigenic diversity. A phylogenetic analysis of the viral capsid genes, particularly the VP1 protein, has led to the classification of noroviruses into five distinct genogroups: GI, GII, GIII, GIV, and GV. Within each genogroup, the viruses are further subdivided into genotypes based on genetic clusters [13, 14].

Approximately 33 known genotypes of norovirus have been identified to date. Genogroup II (GII) contains the majority of human noroviruses besides GI and GIV which are also known to infect humans. GIV has also been detected in humans. The prototype strain of norovirus, i.e., Norwalk virus, falls under GI.1 (genogroup I and genotype 1). Contrastingly, GIII noroviruses are found in cattle, and GV strains infect mice, such as the murine norovirus (MNV). Additionally, porcine noroviruses are classified under GII, and feline and canine strains are categorized in GIV [15–17].

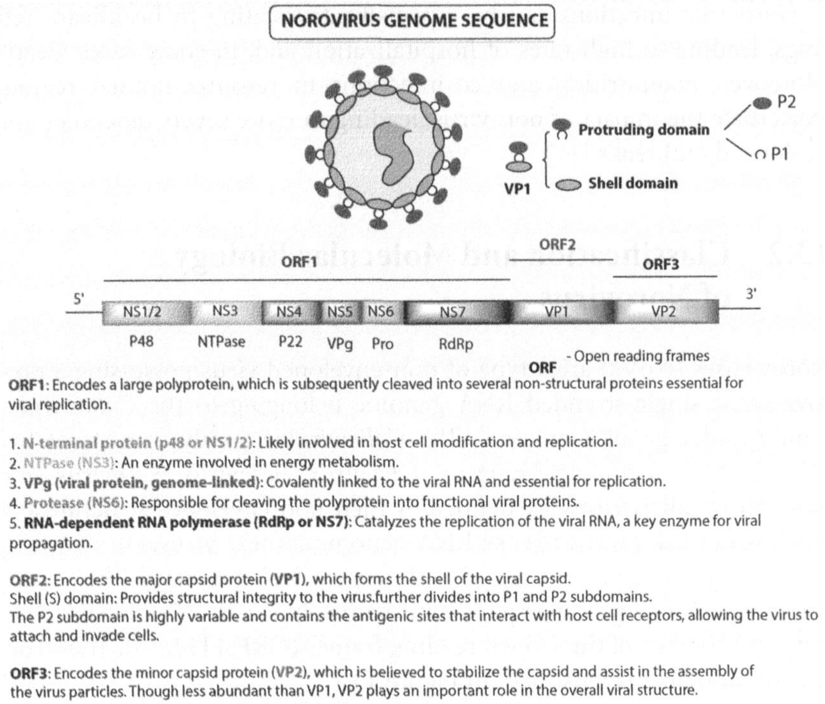

Figure 13.1 Structure of norovirus genome sequence.

Natural immunity following exposure to norovirus (NoV) appears to be largely specific to the strain or genotype, offering limited or no cross-protection between genogroups. Immunity acquired from a natural norovirus infection is estimated to last between 4 and 9 years. Genogroup II, genotype 4 (GII.4) is currently the most common strain in outbreaks, accounting for around 60%–90% of cases globally [18]. This genotype frequently undergoes genetic changes with new strains coming every 2 to 4 years due to RNA recombination [18, 19]. RNA recombination is observed in multiple genotypes, particularly GII.4, leading to the periodic appearance of new variants and subtypes with minimal cross-protection across genogroups. In Sydney, a new GII.4 strain emerged in 2012, and GII.17 cases have increased in Asia in recent years [15, 20].

Norovirus is highly contagious, lacks effective antiviral treatment, and is difficult to avoid, resulting in a strong candidate for vaccine development [21]. However, identifying a serological protection against infection has been challenging, as norovirus lacks an *in vitro* culture system for reliable study of neutralizing activity. The virus is believed to initiate infection by binding to human histo-blood group antigens (HBGAs). Blocking this binding interaction is being studied as a potential surrogate for neutralizing activity and is under evaluation in human challenge studies as a correlate of protection [22].

Efforts to develop a norovirus vaccine have faced obstacles due to limited methods for efficient virus propagation and the absence of small animal models for investigating immune responses. Consequently, current approaches often focus on non-replicating recombinant proteins, which are expressed as virus-like particles (VLPs) or P particles [23]. The major capsid protein, VP1, spontaneously assembles into 180-unit VLPs in cell culture, closely resembling the structure and antigenicity of natural norovirus capsids [7]. These VLPs can be produced in various systems, including recombinant viral vectors (such as baculovirus, Venezuelan equine encephalitis virus, vesicular stomatitis virus), insect and mammalian cells, plasmids, yeast, and even transgenic plants like tomatoes, potatoes, and tobacco [7]. Meanwhile, P particles—formed by 24 copies of the capsid's protrusion (P) domain—are highly stable and immunogenic and can be produced efficiently in *Escherichia coli*, offering another promising avenue for vaccine development [24].

13.3 Current Approaches to Norovirus Therapy

Current therapeutic approaches for norovirus (NoV) focus primarily on symptom management, experimental antiviral agents, and innovative passive immunization methods due to the absence of a widely available vaccine or approved antiviral therapy [12].

13.3.1 Symptomatic Treatment

Since norovirus infections are self-limiting in most cases, the primary focus of treatment is alleviating symptoms, especially those leading to dehydration. Typical symptoms include nausea, vomiting, diarrhea, and stomach cramps, which contribute to significant fluid and electrolyte loss. Symptomatic treatments could be done using oral rehydration solutions (ORS), intravenous fluids (lactated Ringer's solution or normal saline), antiemetic medications (ondansetron and metoclopramide), and antidiarrheal agents (loperamide) [25].

13.3.2 Antiviral Drug Development

Intending to directly target the virus, researchers have investigated several antiviral agents. Developing effective antivirals is challenging for norovirus due to the virus' high mutation rate and lack of a stable *in vitro* culture model for testing.

Examples of experimental antivirals are as follows:

- Nitazoxanide: Originally used as a broad-spectrum antiparasitic, nitazoxanide has demonstrated antiviral effects *in vitro* against various RNA viruses, including norovirus. Its mechanism is thought to involve interference with host factors that viruses rely on for replication. Clinical trials have shown some benefit in reducing symptoms' duration, but further studies are needed to establish its efficacy specifically for norovirus [26, 27].
- Favipiravir: It is an RNA-dependent RNA polymerase inhibitor that is approved in Japan for influenza. Favipiravir has shown potential *in vitro* for inhibiting norovirus replication by targeting the viral RNA polymerase. Though promising, it requires further testing in clinical trials for norovirus [28, 29].
- 2′-C-Methylcytidine: This nucleoside analog interferes with the viral RNA polymerase, showing effectiveness against norovirus in laboratory settings. By inhibiting replication, it may shorten the duration of symptoms. Studies on animal models suggest potential, but human trials are needed for confirmation [29, 30].
- Protease inhibitors: Since norovirus uses specific proteases to cleave polyproteins during its replication, protease inhibitors offer a possible target. Drugs like rupintrivir have been explored for their ability to inhibit viral proteases *in vitro*, although they remain in the early stages of research [31].

13.3.3 Passive Immunization Strategies

Passive immunization refers to the administration of pre-formed antibodies to provide immediate, short-term immunity or help reduce viral load. These approaches are beneficial for immunocompromised individuals or those at a high risk, such as the elderly, who may not respond effectively to a vaccine or antiviral drugs [32].

Examples of passive immunization approaches are as follows:

- Monoclonal antibodies (mAbs): Specific monoclonal antibodies have been developed to neutralize norovirus. For example, HuNoV mAb 5B18 targets the capsid protein of the virus. It has been effective in animal studies in blocking norovirus binding to HBGAs, the receptors that allow virus' entry into human cells [33].
- Immunoglobulin therapy: Studies are investigating the use of pooled human immunoglobulin from recovered or vaccinated individuals, which contains a broad array of antibodies against various norovirus strains. This approach could provide short-term immunity, helping reduce symptoms in exposed individuals [34, 35].
- Plant-produced antibodies: Research has explored using plants, such as genetically engineered tobacco, to produce antibodies that neutralize norovirus. These antibodies could be taken orally and are particularly useful in settings where traditional antibody production is costly or challenging [36].

13.3.4 Vaccine Candidates

Research into norovirus (NoV) vaccines has advanced considerably in recent years, with several promising candidates undergoing rigorous testing in both preclinical and clinical settings. Vaccine formulations are designed to target the two primary genogroups responsible for most human infections, GI and GII, to induce broad immune responses. Multiple candidate vaccines, including bivalent and trivalent virus-like particles (VLPs) and P particles, have demonstrated efficacy in animal models and early human trials. These VLPs, which mimic the virus structure without causing infection, stimulate strong immune responses by resembling the capsid proteins of NoV [12, 37]. Table 13.1 includes the various vaccines for noroviruses.

13.3.4.1 Virus-Like Particle (VLP) Vaccines

Virus-like particles (VLPs) mimic the structure of norovirus but lack the genetic material needed to cause infection, making them safe yet immunogenic candidates [38].

Table 13.1 List of different vaccine candidates against NoVs.

Vaccine candidate	Company	Antigens present in vaccine	Stage (preclinical/ clinical)	References
TAK-214	Takeda	GI.1/GII.4	Phase 2b	[7, 39, 41]
VXA-NVV-104	Vaxart	GI.1/GII.4	Phase 1	[41, 52]
Hansenulapolymorpha	NVSI	GI.1/GII.4	Phase 1	[41, 53]
Longkoma	IPS/Zhifei	GI.1/GII.3/ GII.4/ GII.17	Phase 2a	[41]
Norovirus VLP	Takeda	GI.1/GII.4	Preclinical studies	[21]

➢ Takeda's VLP vaccine (TAK-214): One of the most advanced norovirus vaccine candidates, TAK-214 targets both GI and GII norovirus strains. This intramuscular vaccine comprises of two VLPs: GI.1 (based on the Norwalk virus) and a consensus GII.4 VLP designed to cover multiple GII.4 variants. The GII.4 VLP component is based on sequences from three strains that circulated over several years, thus aiming to produce a cross-reactive immune response across different GII.4 subtypes [39]. Preclinical studies demonstrated that this formulation elicited broad antibody responses, showing the potential to counteract various GII.4 strains. In clinical trials, Takeda's candidate vaccine has been tested with different formulations and dosages to find an optimal balance of efficacy and tolerability. In phase 1 and 2 studies, doses ranging from 5 to 150 μg of each VLP, with or without the adjuvant MPL (3-O-desacyl-4′-monophosphoryl lipid A) alongside aluminum hydroxide, were assessed. The vaccine was generally well tolerated, although adjuvanted formulations did result in some injection site reactions, such as mild pain and tenderness. Nonetheless, these reactions were temporary, and systemic adverse effects were similar to those observed in placebo groups, indicating a favorable safety profile [7, 39–41].

- Moderna's mRNA-based VLP vaccine: Moderna is exploring an mRNA approach to produce norovirus VLPs within the body. By delivering mRNA encoding norovirus VLPs, the body generates the VLPs itself, prompting an immune response. This approach is still in preclinical development but holds promise due to the rapid adaptability of mRNA platforms [41].

13.3.4.2 P Particle-Based Vaccines

The P particle vaccine approach uses the P domain of the norovirus capsid protein, which forms small sub-virus-sized particles. These particles are highly immunogenic and can bind to host cell receptors, mimicking the initial steps of norovirus infection [41].

- P particle vaccines: Research has shown that P particles are stable and strongly immunogenic, inducing both humoral and cellular immune responses. A notable P particle-based vaccine has demonstrated efficacy in preclinical studies and is believed to be a potential oral vaccine candidate due to the particles' stability [38].

13.3.4.3 Recombinant Protein Vaccines

Recombinant protein vaccines include producing parts of the norovirus in a laboratory and then using these proteins to stimulate an immune response in vaccinated individuals [42].

- NVX-GII.4 by Novavax: This recombinant vaccine targets the GII.4 norovirus strain, responsible for most human infections. Produced in insect cell cultures, the vaccine was shown to produce strong immune responses in preclinical studies, and ongoing trials aim to establish its efficacy in humans.

13.3.4.4 Bivalent and Multivalent VLP Vaccines

To address the diversity of norovirus strains, several companies and research groups are working on bivalent (targeting two strains) or multivalent vaccines that target multiple genogroups and genotypes. The human papillomavirus (HPV) vaccines, Gardasil and Cervarix, are multivalent vaccines that utilize virus-like particles (VLPs) and have demonstrated strong effectiveness in preventing infections and precancerous lesions for the HPV strains that they target [43, 44]. The success of these VLP-based HPV vaccines offers a promising model for developing similar approaches for other viruses, such as norovirus [41, 45].

- Bivalent GI and GII VLP vaccines: Bivalent vaccines, which target both GI.1 and GII.4 genotypes, are under study. Clinical trials have shown that bivalent vaccines provide broad immune responses and are well tolerated by recipients [46].
- Valneva's multivalent VLP vaccine: Valneva, a European biotech company, is developing a multivalent norovirus vaccine with VLPs representing multiple genogroups. Early trials have indicated safety and immunogenicity, and further studies are underway to confirm efficacy across various norovirus strains [47].

13.3.4.5 Oral Vaccines

Oral vaccines are ideal for norovirus, as they can directly stimulate immunity in the gut, where the virus infects.

- Vaxart's oral norovirus vaccine: Vaxart is developing an oral tablet vaccine using recombinant adenoviral vectors encoding norovirus antigens. Preliminary studies have shown promising results, and ongoing clinical trials aim to test the vaccine's efficacy in providing local immunity in the gastrointestinal tract [41, 48].

13.3.4.6 Plant-Derived Vaccines

Plant-based vaccines involve using genetically modified plants, like potatoes or tobacco, to express norovirus proteins. These proteins are harvested and used in vaccine formulations or, in some cases, can be ingested directly as food-based vaccines.

- Medicago's plant-based VLPs: Medicago, known for its work on plant-derived vaccines, has explored using tobacco plants to produce norovirus VLPs. These plant-derived VLPs are currently in the early stages of development, but they show potential as a cost-effective and scalable option for norovirus vaccination [49].

13.3.5 Monoclonal Antibody-Based Passive Immunization

Passive immunization with monoclonal antibodies can provide immediate, though short-term, immunity. This approach is particularly valuable for immunocompromised patients or during outbreaks. Monoclonal antibodies targeting norovirus capsid proteins have shown efficacy in blocking the virus' entry into cells [50]. For example, antibodies specific to GII.4 have been successful in reducing infection rates in experimental models. These antibodies could potentially be administered as prophylaxis for high-risk individuals [51].

13.4 Challenges in Norovirus Vaccine Development

Developing an effective vaccine against norovirus presents a unique set of challenges, which range from the virus' complex biology to limitations in laboratory models [12]. These challenges have hindered the production of a widely effective vaccine despite ongoing research and some promising vaccine candidates [54] as displayed in Table 13.2.

13.4.1 Genetic Diversity and Rapid Evolution

- ➢ High mutation rate: NoVs are highly diverse and rapidly evolving, especially strains within the GII.4 genotype, which is responsible for most global outbreaks. This high mutation rate results in the frequent emergence of new variants with distinct antigenic profiles, often every 2–4 years, similar to the way influenza changes [54].
- ➢ Antigenic variation: Norovirus strains lack common neutralization epitopes across genogroups, and the antigenic diversity means that immunity developed against one strain or genotype often does not protect against others. Consequently, designing a single vaccine that effectively targets all strains is extremely challenging, as it must address a wide array of antigens [41, 54].

Table 13.2 Key challenges faced in norovirus vaccine development [5].

Key area	Challenges	Potential solution	References
Vaccine effectiveness	Strain selection: Numerous norovirus genotypes contribute to cases of acute gastroenteritis in the US, with emerging strains (e.g., GII.4, GII.17, GII.2) posing additional challenges.	Development of multivalent vaccines to cover various key genotypes. Periodic updates to vaccines may be necessary to address newly emerging genotypes and maintain effectiveness.	[58, 59]
	Defining and assessing norovirus illness: Sero-response and viral shedding levels differ significantly across age groups, immune status, and prior exposure history. Various studies define severity inconsistently, complicating comparisons.	Conduct studies to better understand sero-response patterns post-infection and clarify the impact of viral shedding after symptoms resolve. Establish a standardized illness definition and a clinical severity scale to facilitate consistent assessments across studies.	[60]
	Duration of protection: It remains unclear how long immunity from natural infection or vaccination persists. Additionally, no established correlate of protection exists for norovirus.	Challenge and vaccine studies should be utilized to assess long-term immunity, while the newly developed *in vitro* culture system can help validate current markers and identify reliable indicators of immunity.	[60]

(*Continued*)

Table 13.2 Key challenges faced in norovirus vaccine development [5]. (*Continued*)

Key area	Challenges	Potential solution	References
Defining the target population	High-income countries: Data for vaccine efficacy for children and older adults is limited. Little research exists on the cost-effectiveness of vaccinating specific groups, such as travelers, military personnel, healthcare workers, or food handlers.	Conduct efficacy studies within these key populations to assess protection levels, considering factors like previous exposure and immune health. Surveillance and modeling studies can provide insights into transmission patterns and help develop cost-effectiveness models for various groups.	[54, 59]
	Low and middle-income countries: In regions with high transmission rates, the full impact of norovirus on acute gastroenteritis remains unclear, and genotype data is sparse.	Longitudinal cohort studies can help quantify norovirus incidence and role in acute gastroenteritis, offering a clearer picture of its overall burden. Establishing a global norovirus genotype surveillance network will enable standard data practices and better tracking of circulating strains.	[54, 59]

13.4.2 Lack of Long-Lasting Immunity

- Short duration of natural immunity: Immunity following natural infection is often short-lived, estimated to last between a few months to several years. This limited immunity does not reliably protect against reinfection, particularly by different strains. Since vaccines ideally confer long-lasting immunity, this presents a significant barrier.
- Lack of cross-protection: Natural immunity tends to be strain- or genotype-specific, with little cross-protection between genogroups. Thus, developing a vaccine with broad coverage against norovirus remains challenging, as it requires inducing immunity that is effective across different strains and genotypes.

13.4.3 Absence of Reliable *In Vitro* Culture System

- Challenges in virus cultivation: NoVs do not grow easily in standard cell culture systems, making it difficult to understand the virus' replication and test vaccine efficacy in a controlled environment. Though recent advances have enabled some success in culturing norovirus in specific cell types, these methods are still limited and not universally applicable, restricting the scope for vaccine development [41].
- Model limitations for immunogenicity testing: Without a reliable *in vitro* culture model or stable small-animal models, researchers lack effective ways to test vaccine efficacy against norovirus. Human challenge studies, where volunteers are deliberately infected under controlled conditions, are one of the few ways to test vaccines, but they are expensive and ethically complex [41].

13.4.4 Unclear Correlates of Protection

- Uncertain immune markers: Unlike many other viruses, there are no well-established correlates of protection (i.e., immune markers) for norovirus, which would indicate successful immunity against infection or disease. For example, while antibodies that block norovirus binding to human histo-blood group antigens (HBGAs) are a potential marker, this has not yet been definitively proven as an indicator of immunity [21, 55].

- Challenges in evaluating neutralizing antibodies: Neutralizing antibodies are typically used to assess immunity for other viruses, but because norovirus lacks an effective *in vitro* system, measuring neutralization accurately is difficult [21, 55].

13.4.5 Variation in Host Susceptibility

- Genetic differences in susceptibility: Susceptibility to norovirus infection can depend on genetic factors, particularly the expression of specific HBGAs, which serve as receptors for the virus. Individuals who do not express these antigens (known as "non-secretors") are often naturally resistant to certain strains. This genetic variation complicates vaccine development, as it is challenging to design a vaccine that is universally effective across different human populations with varying HBGA profiles [56].
- Impact of age and immunity: The immune response to norovirus differs across age groups, including young children, the elderly, and immunocompromised individuals. This variability adds to the difficulty of designing a vaccine that offers effective protection for diverse populations with varying immune strengths [56].

13.4.6 High Cost and Complexity of Clinical Trials

- Complexity of conducting human challenge studies: Since animal models are not fully effective for norovirus vaccine studies, human challenge studies are often necessary to assess vaccine efficacy. These studies, where volunteers are intentionally exposed to norovirus, require rigorous ethical, safety, and regulatory oversight. They are also costly and logistically complex, limiting their feasibility on a large scale [57].
- Need for multivalent vaccines: Given the diversity of norovirus strains, an effective vaccine likely needs to be multivalent (i.e., targeting multiple strains). Designing, manufacturing, and testing multivalent vaccines significantly increase both development time and costs [57].

13.5 Conclusion and Future Perspectives

Norovirus continues to pose a significant global health challenge, particularly as there are no approved vaccines or antiviral treatments available to prevent or control infection. Considerable advancements have been made in research tools and vaccine development strategies, especially with recombinant technologies that produce virus-like particles (VLPs) and P particle complexes. These approaches have resulted in promising vaccine candidates, some of which have progressed to advanced clinical trials with encouraging safety and immunogenicity profiles. However, norovirus diversity and frequent emergence of new strains underscore the need for multivalent vaccine formulations that can provide broad protection across genotypes.

Further efforts are necessary to refine the models for norovirus infection, as current cell culture and small animal systems are limited in consistency and replication efficiency. The development of robust and cost-effective *in vitro* and *in vivo* models would support both vaccine and antiviral research by enabling the production of inactivated or live-attenuated vaccine candidates and facilitating antiviral screening. The recent establishment of *in vitro* culture models, though still in early stages, holds promise for validating immune response markers, understanding viral replication, and optimizing drug development.

Moreover, vaccine strategies that combine multiple genotypes or use chimeric and multivalent formulations appear to be the most promising for achieving broad-spectrum protection, akin to influenza vaccine strategies. In parallel, antiviral therapies, including nucleoside analogs and immune-stimulatory compounds, should continue to be explored. Developing an effective norovirus culture system would enhance these efforts by improving the evaluation of drug–virus interactions and supporting the preclinical testing of both vaccine and antiviral candidates.

By addressing these key areas, norovirus vaccine and antiviral research is poised to significantly impact public health, potentially reducing the economic and healthcare burden of norovirus-associated illness. With continued focus on robust vaccine formulations, antiviral drug development, and improved diagnostic models, there is hope for the introduction of effective preventive and therapeutic solutions against norovirus soon.

AI disclosure statement: The author acknowledges the use of AI-based tools for minor language editing or grammatical correction. However, all intellectual contributions, analyses, and conclusions presented in this chapter are entirely the author's own.

References

1. Robilotti, E., Deresinski, S., Pinsky, B.A., Norovirus. *Clin. Microbiol. Rev.*, 28, 1, 134–64, 2015.
2. Omatola, C.A., Mshelbwala, P.P., Okolo, M.-L.O., Onoja, A.B., Abraham, J.O., Adaji, D.M., *et al.*, Noroviruses: Evolutionary Dynamics, Epidemiology, Pathogenesis, and Vaccine Advances—A Comprehensive Review. *Vaccines*, 12, 6, 590, 2024.
3. Winder, N., Gohar, S., Muthana, M., Norovirus: an overview of virology and preventative measures. *Viruses*, 14, 12, 2811, 2022.
4. Carlson, K.B., Dilley, A., O'Grady, T., Johnson, J.A., Lopman, B., Viscidi, E., A narrative review of norovirus epidemiology, biology, and challenges to vaccine development. *npj Vaccines*, 9, 1, 94, 2024.
5. Hallowell, B.D., Parashar, U.D., Hall, A.J., Epidemiologic challenges in norovirus vaccine development. *Hum. Vaccin. Immunother.*, 15, 6, 1279–83, 2019.
6. Woodward, J., Gkrania-Klotsas, E., Kumararatne, D., Chronic norovirus infection and common variable immunodeficiency. *Clin. Exp. Immunol.*, 188, 3, 363–70, 2017.
7. Baehner, F., Bogaerts, H., Goodwin, R., Vaccines against norovirus: state of the art trials in children and adults. *Clin. Microbiol. Infect.*, 22, S136–S9, 2016.
8. Pogan, R., Dülfer, J., Uetrecht, C., Norovirus assembly and stability. *Curr. Opin. Virol.*, 31, 59–65, 2018.
9. Ushijima, H., Fujimoto, T., Müller, W.E., Hayakawa, S., Norovirus and foodborne disease: a review. *Food Safety*, 2, 3, 37–54, 2014.
10. Patel, M.M., Hall, A.J., Vinjé, J., Parashar, U.D., Noroviruses: a comprehensive review. *J. Clin. Virol.*, 44, 1, 1–8, 2009.
11. Tsai, H., Yune, P., Rao, M., Norovirus disease among older adults. *Ther. Adv. Infect. Dis.*, 9, 20499361221136760, 2022.
12. Li, J., Predmore, A., Divers, E., Lou, F., New interventions against human norovirus: progress, opportunities, and challenges. *Annu. Rev. Food Sci. Technol.*, 3, 1, 331–52, 2012.
13. Rocha-Pereira, J., Neyts, J., Jochmans, D., Norovirus: targets and tools in antiviral drug discovery. *Biochem. Pharmacol.*, 91, 1, 1–11, 2014.
14. Kroneman, A., Vega, E., Vennema, H., Vinjé, J., White, P.A., Hansman, G., *et al.*, Proposal for a unified norovirus nomenclature and genotyping. *Arch. Virol.*, 158, 2059–68, 2013.
15. Parra, G.I., Emergence of norovirus strains: A tale of two genes. *Virus Evol.*, 5, 2, vez048, 2019.
16. White, P., Evolution of norovirus. *Clin. Microbiol. Infect.*, 20, 8, 741–5, 2014.
17. Peña-Gil, N., Santiso-Bellón, C., Gozalbo-Rovira, R., Buesa, J., Monedero, V., Rodríguez-Díaz, J., The role of host glycobiology and gut microbiota in rotavirus and norovirus infection, an update. *Int. J. Mol. Sci.*, 22, 24, 13473, 2021.

18. Tran, T.H., Trainor, E., Nakagomi, T., Cunliffe, N.A., Nakagomi, O., Molecular epidemiology of noroviruses associated with acute sporadic gastroenteritis in children: global distribution of genogroups, genotypes and GII. 4 variants. *J. Clin. Virol.*, 56, 3, 269–77, 2013.
19. Desai, R., Hembree, C.D., Handel, A., Matthews, J.E., Dickey, B.W., McDonald, S., *et al.*, Severe outcomes are associated with genogroup 2 genotype 4 norovirus outbreaks: a systematic literature review. *Clin. Infect. Dis.*, 55, 2, 189–93, 2012.
20. Zhang, P., Hao, C., Di, X., Chuizhao, X., Jinsong, L., Guisen, Z., *et al.*, Global prevalence of norovirus gastroenteritis after emergence of the GII. 4 Sydney 2012 variant: a systematic review and meta-analysis. *Front. Public Health*, 12, 1373322, 2024.
21. Esposito, S. and Principi, N., Norovirus vaccine: priorities for future research and development. *Front. Immunol.*, 11, 1383, 2020.
22. Prasad, B.V., Shanker, S., Muhaxhiri, Z., Deng, L., Choi, J.-M., Estes, M.K., *et al.*, Antiviral targets of human noroviruses. *Curr. Opin. Virol.*, 18, 117–25, 2016.
23. Shirbaghaee, Z. and Bolhassani, A., Different applications of virus-like particles in biology and medicine: vaccination and delivery systems. *Biopolymers*, 105, 3, 113–32, 2016.
24. Tan, M. and Jiang, X., Subviral particle as vaccine and vaccine platform. *Curr. Opin. Virol.*, 6, 24–33, 2014.
25. Faure, C., Role of antidiarrhoeal drugs as adjunctive therapies for acute diarrhoea in children. *Int. J. Pediatr.*, 2013, 1, 612403, 2013.
26. Siddiq, D.M., Koo, H.L., Adachi, J.A., Viola, G.M., Norovirus gastroenteritis successfully treated with nitazoxanide. *J. Infect.*, 63, 5, 394–7, 2011.
27. Rossignol, J.F. and El-Gohary, Y.M., Nitazoxanide in the treatment of viral gastroenteritis: a randomized double-blind placebo-controlled clinical trial. *Aliment. Pharmacol. Ther.*, 24, 10, 1423–30, 2006.
28. Rocha-Pereira, J., Jochmans, D., Dallmeier, K., Leyssen, P., Nascimento, M., Neyts, J., Favipiravir (T-705) inhibits *in vitro* norovirus replication. *Biochem. Biophys. Res. Commun.*, 424, 4, 777–80, 2012.
29. Jin, Z., Tucker, K., Lin, X., Kao, C.C., Shaw, K., Tan, H., *et al.*, Biochemical evaluation of the inhibition properties of favipiravir and 2′-C-methylcytidine triphosphates against human and mouse norovirus RNA polymerases. *Antimicrob. Agents Chemother.*, 59, 12, 7504–16, 2015.
30. Rocha-Pereira, J., Jochmans, D., Dallmeier, K., Leyssen, P., Cunha, R., Costa, I., *et al.*, Inhibition of norovirus replication by the nucleoside analogue 2′-C-methylcytidine. *Biochem. Biophys. Res. Commun.*, 427, 4, 796–800, 2012.
31. Chang, K.-O., Kim, Y., Lovell, S., Rathnayake, A.D., Groutas, W.C., Antiviral drug discovery: norovirus proteases and development of inhibitors. *Viruses*, 11, 2, 197, 2019.

32. Ghosh, S., Malik, Y.S., Kobayashi, N., Therapeutics and immunoprophylaxis against noroviruses and rotaviruses: the past, present, and future. *Curr. Drug Metab.*, 19, 3, 170–91, 2018.
33. Shanker, S., Czakó, R., Sapparapu, G., Alvarado, G., Viskovska, M., Sankaran, B., *et al.*, Structural basis for norovirus neutralization by an HBGA blocking human IgA antibody. *Proc. Natl. Acad. Sci.*, 113, 40, E5830–E7, 2016.
34. Vega, C.G., Bok, M., Ebinger, M., Rocha, L.A., Rivolta, A.A., González Thomas, V., *et al.*, A new passive immune strategy based on IgY antibodies as a key element to control neonatal calf diarrhea in dairy farms. *BMC Vet. Res.*, 16, 1–9, 2020.
35. Dai, Y.-C., Zhang, X.-F., Tan, M., Huang, P., Lei, W., Fang, H., *et al.*, A dual chicken IgY against rotavirus and norovirus. *Antiviral Res.*, 97, 3, 293–300, 2013.
36. Stander, J., Mbewana, S., Meyers, A.E., Plant-derived human vaccines: Recent developments. *BioDrugs.*, 36, 5, 573–89, 2022.
37. El-Kamary, S.S., Pasetti, M.F., Mendelman, P.M., Frey, S.E., Bernstein, D.I., Treanor, J.J., *et al.*, Adjuvanted intranasal Norwalk virus-like particle vaccine elicits antibodies and antibody-secreting cells that express homing receptors for mucosal and peripheral lymphoid tissues. *J. Infect. Dis.*, 202, 11, 1649–58, 2010.
38. Tan, M., Huang, P., Xia, M., Fang, P.-A., Zhong, W., McNeal, M., *et al.*, Norovirus P particle, a novel platform for vaccine development and antibody production. *J. Virol.*, 85, 2, 753–64, 2011.
39. Sherwood, J., Mendelman, P.M., Lloyd, E., Liu, M., Boslego, J., Borkowski, A., *et al.*, Efficacy of an intramuscular bivalent norovirus GI. 1/GII. 4 virus-like particle vaccine candidate in healthy US adults. *Vaccine*, 38, 41, 6442–9, 2020.
40. Atmar, R.L., Bernstein, D.I., Harro, C.D., Al-Ibrahim, M.S., Chen, W.H., Ferreira, J., *et al.*, Norovirus vaccine against experimental human Norwalk Virus illness. *N. Engl. J. Med.*, 365, 23, 2178–87, 2011.
41. Tan, M., Norovirus vaccines: current clinical development and challenges. *Pathogens*, 10, 12, 1641, 2021.
42. Philip, A.A. and Patton, J.T., Generation of recombinant rotaviruses expressing human norovirus capsid proteins. *J. Virol.*, 96, 22, e01262–22, 2022.
43. Yadav, R., Zhai, L., Tumban, E., Virus-like particle-based L2 vaccines against HPVs: where are we today? *Viruses*, 12, 1, 18, 2019.
44. Uddin, M.N., Kouzi, S.A., Hussain, M.D., Strategies for developing oral vaccines for human papillomavirus (HPV) induced cancer using nanoparticle mediated delivery system. *J. Pharm. Pharm. Sci.*, 18, 2, 220–34, 2015.
45. Debbink, K., Lindesmith, L.C., Baric, R.S., The state of norovirus vaccines. *Clin. Infect. Dis.*, 58, 12, 1746–52, 2014.
46. Parra, G.I., Bok, K., Taylor, R., Haynes, J.R., Sosnovtsev, S.V., Richardson, C., *et al.*, Immunogenicity and specificity of norovirus Consensus GII.

4 virus-like particles in monovalent and bivalent vaccine formulations. *Vaccine*, 30, 24, 3580–6, 2012.
47. Yang, J.-X., Tseng, J.-C., Yu, G.-Y., Luo, Y., Huang, C.-Y.F., Hong, Y.-R., *et al.*, Recent advances in the development of toll-like receptor agonist-based vaccine adjuvants for infectious diseases. *Pharmaceutics*, 14, 2, 423, 2022.
48. Zhang, M., Fu, M., Hu, Q., Advances in human norovirus vaccine research. *Vaccines*, 9, 7, 732, 2021.
49. Monreal-Escalante, E., Ramos-Vega, A., Angulo, C., Bañuelos-Hernández, B., Plant-based vaccines: antigen design, diversity, and strategies for high level production. *Vaccines*, 10, 1, 100, 2022.
50. Smith, H.Q. and Smith, T.J., The dynamic capsid structures of the noroviruses. *Viruses.*, 11, 3, 235, 2019.
51. Alvarado, G., Ettayebi, K., Atmar, R.L., Bombardi, R.G., Kose, N., Estes, M.K., *et al.*, Human monoclonal antibodies that neutralize pandemic GII. 4 noroviruses. *Gastroenterol.*, 155, 6, 1898–907, 2018.
52. Cummings, J.F., Tucker, S. (eds), Potent Immune Responses to Norovirus G1. 1 Evaluated in Elderly Subjects following Oral Tablet Delivery in a Phase 1 Placebo-Controlled Study, in: *Proceedings of the World Vaccine Congress*, 2022.
53. Zhang, J., Tang, F., Zhang, X., Liu, Z., Liang, Y., *et al.*, Physicochemical and immunological characterization of two forms of recombinant norovirus GII. 4 virus-like particles assembled in Hansenula polymorpha. *Chin. J. Microbiol. Immunol.*, 37, 12, 927–32, 2017.
54. Cortes-Penfield, N.W., Ramani, S., Estes, M.K., Atmar, R.L., Prospects and challenges in the development of a norovirus vaccine. *Clin. Ther.*, 39, 8, 1537–49, 2017.
55. Melhem, N.M. and Abou Hassan, F.F., Norovirus Correlates of Protection. *Norovirus*, 157–87, 2019.
56. Lucero, Y. and Vidal, R., Norovirus vaccines under development. *Vaccine*, 36, 36, 5435–41, 2018.
57. Mattison, C.P., Cardemil, C.V., Hall, A.J., Progress on norovirus vaccine research: public health considerations and future directions. *Expert Rev. Vaccin.*, 17, 9, 773–84, 2018.
58. Melhem, N.M., Norovirus vaccines: correlates of protection, challenges and limitations. *Hum. Vaccin. Immunother.*, 12, 7, 1653–69, 2016.
59. Seo, H., Duan, Q., Zhang, W., Vaccines against gastroenteritis, current progress and challenges. *Gut Microbes*, 11, 6, 1486–517, 2020.
60. Riddle, M.S. and Walker, R.I., Status of vaccine research and development for norovirus. *Vaccine*, 34, 26, 2895–9, 2016.

14

The Importance of Animal Models and *In Vitro* Cultivation for Norovirus Research

Prarambh S.R. Dwivedi[1], Masmarika Mohan[2], Pukar Khanal[3], Sanjit Sah[4,5,6] and C.S. Shastry[1*]

[1]*Department of Pharmacology, NGSM Institute of Pharmaceutical Sciences (NGSMIPS), Nitte (Deemed to be University), Mangalore, Karnataka, India*
[2]*Department of Molecular Genetics and Cancer, Nitte University Centre for Science Education and Research (NUCSER), Nitte (Deemed to be University), Mangalore, Karnataka, India*
[3]*Silicon Script Sciences Private Limited, Bharatpur, Ghorahi, Dang, Nepal*
[4]*Department of Pediatrics, Dr. D. Y. Patil Medical College, Hospital and Research Centre, Dr. D. Y. Patil Vidyapeeth (Deemed-to-be-University), Pimpri, Pune, Maharashtra, India*
[5]*Department of Public Health Dentistry, Dr. D.Y. Patil Dental College and Hospital, Dr. D.Y. Patil Vidyapeeth (Deemed-to-be-University), Pune, Maharashtra, India*
[6]*Department of Medicine, Korea Universtiy, Seoul, South Korea*

Abstract

Globally, human noroviruses (HuNoVs) are a major source of acute gastroenteritis which significantly elevates the societal and financial costs for treatment. As of right now, there are no approved medications or vaccinations for the treatment or prevention of HuNoVs. *In vitro* and *in vivo* infection models that are well characterized are lacking; this has hindered the development of HuNoV defences. Despite being a notably difficult pathogen to study, recent advancements in animal models for norovirus and *in vitro* cultivation systems have significantly enhanced our knowledge of norovirus biology, replication, cell tropism, pathogenesis, and immune responses. Hence, this chapter aims to explore various *in vitro* and *in vivo* models to date that can be used for research advancement.

**Corresponding author*: csshastry@nitte.edu.in, drcshastry@gmail.com;
ORCID: https://orcid.org/0000-0002-2697-0449

Sachin Kumar, Shriyansh Srivastava, Rishabha Malviya and Ranjit Sah (eds.) Emerging Infectious Disease: Unravelling Norovirus and Pandemic Dynamics, (379–412) © 2026 Scrivener Publishing LLC

Even though these viruses are ubiquitous, they only infect certain species, and until recently, *in vitro* technologies have made it possible to cultivate noroviruses. The present HuNoV culture methods have been developed with the use of insights into norovirus replication offered by murine norovirus and other caliciviruses. Similarly, an optimal animal model for human HuNoV research should employ HuNoV itself or a closely related virus with high genetic homology, accurately replicating the biological processes and clinical manifestations of HuNoV infection in the chosen animal host. HuNoVs are members of the Caliciviridae family, a genetically heterogeneous norovirus genus, which has extremely limited species specificity. Therefore, with differing degrees of clinical symptoms, HuNoV infection in a non-native host is restricted to closely related primates (chimpanzees and neonatal pig-tail macaques), immunocompromised animals (mice), and gnotobiotic pigs and calves.

Keywords: Acute gastroenteritis, animal models, cell culture, human norovirus, murine norovirus, human intestinal enteroids, virus like particles, 3D organoid

14.1 Introduction

Human norovirus (HuNoV) is one of the major virologic outbursts for acute gastroenteritis (AGE), affecting around 685 million people worldwide, making it a major cause of AGE globally. Roughly 1.5 million fatalities worldwide are attributed to AGE, of which 136,000 to 278,000 are caused by norovirus infection [1]. Outbreaks of norovirus usually happen in environments where individuals live in close quarters and may readily spread the infection to one another. Norovirus is one of the most often reported causes of diarrhea among travelers in small settings (https://wwwnc.cdc.gov/travel/yellowbook/2024/infections-diseases/norovirus). For the management of AGE associated with norovirus, there are currently no treatments authorized by the US Food and Drug Administration [1]. The majority of norovirus-induced AGE episodes are self-limiting, last 2 to 3 days, and may be treated with supportive care and fluids. In severe situations, hospitalization may be necessary along with medical intervention to prevent the loss of fluid [2]. Without suitable models, our knowledge of viral replication, transmission, and pathogenesis is still lacking, which impedes the development of efficient treatment plans. The advances in understanding HuNoVs have been difficult since, although common, these viruses cause diseases peculiar to a particular species which has not been cultivated successfully *in vitro* until recently [3].

In the present scenario for the treatment of HuNoV infections, there is no specific antiviral drug [4]. The treatment for HuNoV-associated AGE majorly involves nurtured care, like fluid and electrolyte replacement, to prevent dehydration and alleviate symptoms [5]. Vulnerable populations like young children, elderly, and immunocompromised individuals are at a significantly higher risk due to the absence of specific antiviral therapies for HuNoVs, posing a significant risk to the public health [6]. In addition, there is an increase in economic burden corresponding to the absence of targeted pharmacotherapy, leading to increased healthcare costs and productivity losses resulting from outbreaks of HuNoV infections [7]. Due to these challenges, there is a significant progress in developing an effective *in vitro* and *in vivo* model to study HuNoVs. Innovative research focuses on establishing robust cell culture systems to understand the full viral life cycle, encompassing viral entry, replication, and release. To understand the viral pathogenesis and test for possible antiviral medications, experiments are done by stimulating the complex interactions between the virus and the host cells [8]. On the other hand, novel *in vivo* models are under investigation to imitate HuNoV infection in animals, addressing the challenge against the absence of small animal models which accurately represent HuNoV in humans. Genetically modified mice and gnotobiotic piglets have been used in recent studies to replicate specific aspects of HuNoV infection, which has been shown to be promising [9]. In this chapter, different animal models for HuNoVs with their advantages and disadvantages will be discussed along with the advancements of various *in vitro* techniques to cultivate HuNoVs.

14.2 Importance of Animal and Cell Culture Models for Human Norovirus

Research using various animal and cell culture procedures for HuNoV has made substantial progress in the last two decades. However, challenges remain in developing effective systems replicating HuNoV in laboratory conditions [10].

14.2.1 Importance of Cell Culture Models

In vitro methods for HuNoV cultivation are crucial for expanding our knowledge of the virus and creating efficient countermeasures. Traditional methods encountered significant obstacles due to the inability to replicate HuNoV effectively in conventional cell lines, which hindered research on the virus' biology and host interactions [11]. Cell culture models play a pivotal role in studying HuNoV at the cellular level. Historically, cultivating

HuNoV *in vitro* has been difficult, but recent advancements have led to more effective cell culture systems. Human gut organoids and 3D cell cultures are useful tools for studying how the virus interacts with host cells, i.e., understanding how it enters these cells and the type of cellular response that may occur. These models have potential in high-throughput screening of antiviral compounds and vaccinable candidates which ultimately accelerates drug discovery by enabling researchers to assess the effectiveness of treatment under controlled conditions. Notably, 3D cell culture models, such as the INT-407 small intestinal epithelium model, have shown promise by permitting the natural growth of HuNoV, which is crucial for understanding its pathogenesis [12]. There is a raise in the replicability and reproducibility of HuNoV infection after the development of *in vitro* culture systems using human intestinal organoids. This advancement in technology allows researchers to investigate HuNoV–host interactions in a physiologically appropriate environment that mimics the human intestinal tract, thereby providing critical insights into viral behavior and pathogenicity [13]. The use of human intestinal enteroids (HIEs) is a potential breakthrough for studying HuNoV using human organoids [14]. HIEs are stem cell-derived cultures mimicking the human intestinal epithelial cells, which have successfully cultivated multiple norovirus genotypes, particularly the prevalent GII.4 strains, facilitating detailed studies of viral pathogenesis and immune responses in a controlled setting [15]. These models display the natural architecture of the human gut, thus enhancing the physiological relevance of studies and subduing the limitations of 2D cultures.

14.2.2 Importance of Animal Models

Numerous animal models have been developed to study norovirus, each displaying a unique feature in viral pathogenesis and immune response. Among all, *rhesus macaques* have emerged as a favorable non-human primate model, displaying the ability to endure HuNoV and duplicating human infection dynamics, which includes a viral replication model in the jejunum and immune responses [16]. Additionally, Gnostic pigs have been used to examine the pathophysiology and immunity of HuNoV, offering a useful system for comprehending host–pathogen interactions [17]. However, the use of large animals makes it difficult to screen multiple drug candidates. Neonatal mice infected with the murine norovirus strain WU23 displayed symptoms similar to those of human gastroenteritis, enabling in-depth research on the immune system and norovirus-induced diarrhea [18]. Moreover, the intestinal villi isolated from mice have been displayed to mimic the infection and replication of HuNoV; this helps to facilitate research on the virus and potential therapeutic interventions [19]. Among the surrogate models, the feline norovirus (FNoV) cat model is a potential surrogate model, valuable for studying HuNoV and beneficial for comparative research examining the structural determinant of HuNoV alongside those of FNoV. These studies enhance the understanding of immune responses and help in identifying vaccine candidates for both human and feline noroviruses [20]. Lastly, environmental DNA analysis suggests that avian species may serve as reservoirs for HuNoVs, indicating a broader ecological context for studying these pathogens [21].

These models supply a method to study initiation, progression, and immune interactions associated to HuNoV in a similar environment as human pathology. Scientists can use animal models to perform experiments which cannot be done on humans directly, due to either impracticality or ethical reasons, such as testing the safety and efficacy of potential vaccines and antiviral agents before they are applied in clinical trials. The complete life cycle of HuNoV has to be characterized, including viral replication/transmission/immune response, so that workable vaccines/drugs can be developed. Moreover, advancements in genetically modified mouse models that support HuNoV replication facilitate targeted research into specific genes involved in the viral life cycle and host interactions, contributing to a deeper interpretation of viral biology and potential interventions [22].

The development of animal models for HuNoV has faced limitations, but several promising models have been explored. Non-human primates are considered the most relevant for studying HuNoV due to their physiological similarities to humans, making them ideal for understanding the virus' behavior in a comparable biological context. Non-human primates, such as chimpanzees, *rhesus macaques*, and pigtail macaques, offer insights into viral behavior and immune responses, although they may not always exhibit gastrointestinal symptoms.

14.3 Animal Models of HuNoV Infection

In the quest to understand HuNoV, researchers have investigated a variety of animal models. These models span both small and large animal species, which provide various insights into the virus' behavior and response mechanisms in a living host. There is an ongoing development of murine models that have shown promise in supporting HuNoV replication and studying its biology. However, challenges remain in recreating the full spectrum of human disease, given the differences in host responses between humans and other species. Ongoing efforts focus on improving these models to better mimic the human host environment and enhance our understanding of HuNoV infections. Thus far, it has been demonstrated that gnotobiotic calves, macaques, chimpanzees, pigs, and Rag- and γc-deficient mice may all contract a restricted variety of HuNoV strains [23]. All of the features of HuNoV infection, however, cannot be properly captured by a single animal model. Each model has unique properties that allow for the investigation of particular facets of HuNoV biology [24]. The different types of animal models used for screening drugs against HuNoV are given in Figure 14.1 and Table 14.1.

14.3.1 Mice Model

Murine models have advanced recently and are widely used in investigating HuNoV infections. This model has a genetically modified mice that has been found to be more favorable for HuNoV replication, such as BALB/c Rag-γc-deficient animals. [25]. Humanized and non-humanized forms of BALB/c Rag-γc-deficient mice lack certain immune components, increasing the vulnerability to HuNoV infections; both forms have displayed to reproduce GII.4 strains on increased viral load [26]. The name BALB/c Rag-γc arises from development of a "humanized" mice model using a BALB/c-deficient mouse with recombination activation genes (Rag) 1 or 2 and the traditional gamma chain (γc), known as Rag-γc [25].

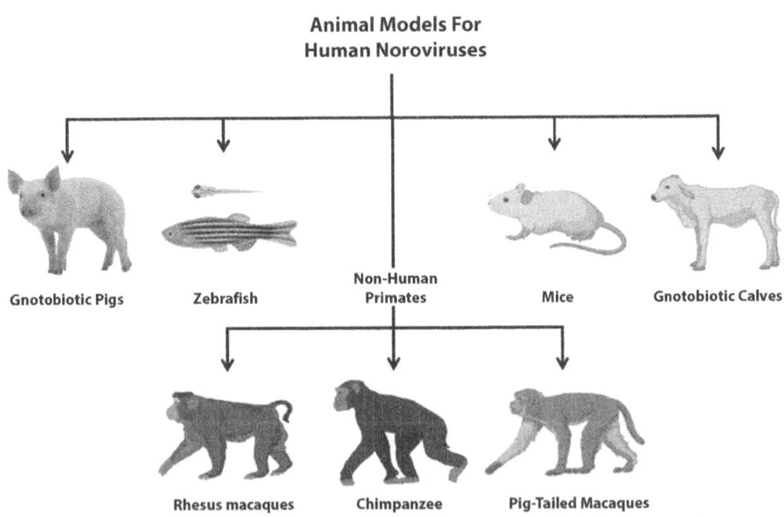

Figure 14.1 Animal models used for screening drugs against human norovirus (Source of images: https://www.pngwing.com/; https://www.biorender.com/).

Table 14.1 Different types of models available for the screening of human norovirus, along with their details.

Model type	Description	Advantages	Limitations	Complexity	Cost level	Genetic editing capacity	Physiological replication	References
In vitro models								
Cell culture models	Culturing HuNoV in human B cells, notably the GII.4-Sydney strain	Allows for detailed analysis of viral attachment and replication	Limited ability to fully replicate human conditions	Moderate	Low	Limited to inherent cell capabilities	Limited replication capabilities	[55]
3D models (organoids)	Three-dimensional organoid cultures derived from human intestinal cells	Mimics the human gut environment, allowing for realistic viral interaction studies	Still in development; may not fully replicate all aspects of human biology	Moderate to high	Moderate	Can utilize CRISPR and other techniques for editing	Significant replication mimicking human physiology	[11]

(*Continued*)

Table 14.1 Different types of models available for the screening of human norovirus, along with their details. (*Continued*)

Model type	Description	Advantages	Limitations	Complexity	Cost level	Genetic editing capacity	Physiological replication	References
In vivo models								
Zebrafish model	Utilizing zebrafish larvae (*Danio rerio*) infected with HuNoV	High viral replication; suitable for antiviral studies	May not fully mimic complex human immune responses	Moderate	Moderate	Genetic manipulation of zebrafish is feasible.	Moderate replication with viral dynamics	[50]
Gnotobiotic pigs and calves	Using pigs and calves raised in germ-free conditions for HuNoV infection	Physiological similarity to humans; valuable for studying infectivity	Limited availability and higher maintenance costs	High	High	Advanced genetic editing is possible.	Strong physiological replication similar to humans	[17]
Mice model	Using genetically modified mice, particularly BALB/c Rag-γc-deficient mice	First genetically manipulable small-animal model for HuNoV; supports viral replication	Not all wild-type mice are susceptible; high-maintenance	High	Moderate	Advanced genetic editing capabilities are available.	Limited replication capacity	[56]

(*Continued*)

Table 14.1 Different types of models available for the screening of human norovirus, along with their details. (*Continued*)

Model type	Description	Advantages	Limitations	Complexity	Cost level	Genetic editing capacity	Physiological replication	References
In vivo models								
Recovirus model	Utilizing caliciviruses within the recovirus genus to study HuNoV antigenic diversity	Offers insights into HuNoV pathogenesis, immune escape mechanisms, and vaccine development	Limited direct applicability to human responses	Moderate	Moderate	Potential for genetic manipulation exists.	Minimal physiological relevance	[51]
Non-human primates as *in vivo* models								
Rhesus macaque Model	Non-human primate model for studying HuNoV infection	Useful for exploring viral pathogenicity and natural immunity	Ethical considerations and cost of maintenance	High	High	Advanced genetic editing is possible.	High physiological relevance	[33, 57]
Chimpanzee Model	Chimpanzees are used for studying HuNoV infections and vaccine development	Close genetic resemblance to humans; suitable for immune response studies	Ethical considerations and high cost	High	High	Advanced genetic editing is possible.	Excellent physiological replication	[58]

The BALB/c Rag-γc-deficient mice when embedded with human CD34 hemopoietic stem cells were examined along with non-engrafted siblings and immunocompetent wild-type controls using pooled stool isolates from HuNoV-positive patients [27]. Both the humanized and non-humanized BALB/c Ragγc-deficient mice displayed to support the replication of a GII.4 strain of HuNoV, as indicated by increased viral loads relative to the initial input. In contrast, immunocompetent wild-type BALB/c mice displayed no signs of infection [28]. Similarly, induction *via* intraperitoneal route in BALB/c strain, a combination with specific genetic background, is essential to facilitate subclinical HuNoV infections in Rag-γc-deficient mice [3]. In addition, the presence of macrophage-like cells in the spleen and liver of these mice indicated the ability of HuNoV to replicate, assuring the presence of a humanized form for NoV infection [29]. The murine strain, CR6 is known to cause sustained NoV infection in adult mice by targeting the intestinal tuft cells. In a study conducted by Kennedy *et al.*, they observed that CR6 was naturally transmitted from contaminated dams exclusively in juvenile mice. When wild-type neonatal mice were directly inoculated with CR6 *via* oral route, independent of active replication, there was a buildup of viral RNA in the ileum and a protracted virus shedding in the feces. This viral exposure triggered both innate and adaptive immune responses, including the expression of interferon-stimulated genes and the production of MuNoV-specific antibodies [30]. Notably, viral uptake was facilitated by passive absorption of luminal virus in the ileum, a process that was inhibited by cortisone acetate administration, preventing the accumulation of viral RNA in the ileum. Neonates deficient in interferon signaling within hematopoietic cells were vulnerable to productive infection, viral dissemination, and lethality, all of which were dependent on the canonical MuNoV receptor CD300LF [31].

14.3.2 Chimpanzee Model

Chimpanzees are acknowledged as a vital animal model for investigating HuNoV infections due to their genetic and physiological similarities to humans. These similarities enhance our understanding of the disease processes induced by HuNoV in humans, thereby improving the applicability and translatability of research findings to human health outcomes [32]. Chimpanzee research has made significant progress in our knowledge of HuNoV pathogenesis. According to Wyatt *et al.*, the first HuNoV infections in chimpanzees occurred in 1978 [33]. Despite the fact that none of the chimpanzees given filtered fecal samples containing HuNoV orally showed symptoms typically linked to HuNoV infection, these animals did develop norovirus-specific antibody responses, and their feces contained viral antigens [3]. In a study, researchers assessed viral shedding, replication, and immunological response followed through a controlled infection in chimpanzees. These findings are essential for figuring out possible intervention routes and creating strong vaccinations. Seronegative chimpanzees inoculated intravenously with the HuNoV strain Norwalk virus (NorV) did not exhibit clinical signs of gastroenteritis [34]. However, the timing and duration of viral shedding in their stool, as well as the serum antibody responses, were comparable to those observed in humans. During the period of viral shedding, NorV RNA was detected in both intestinal and liver biopsies, with NorV antigen expression noted in cells within the small intestinal lamina propria [34]. In two chimpanzees that were rechallenged with NorV at intervals of 4, 10, and 24 months after initial infection, resistance to reinfection was observed, with NV-specific serum antibodies positively correlating with this protection. Additionally, the immunogenicity and efficacy of virus-like particles (VLPs) derived from NorV (genogroup I, GI) and MD145 (genogroup II, GII) noroviruses were evaluated as potential vaccines. The chimpanzees who were vaccinated intramuscularly using GI VLPs displayed protection against NorV; however, when vaccinated with the GII VLP vaccine challenged at 2 and 18 months post-vaccination, they did not display such protection [35]. Hence, this establishes chimpanzee as an effective animal model for studying norovirus replication and immune responses and demonstrates that NorV VLP vaccines can induce protective homologous immunity, even after extended periods.

14.3.3 Macaques Model

Rhesus macaques have been established as a promising animal model for studying HuNoV infections. This model is valuable due to the susceptibility of these macaques to oral infection with HuNoV from various genogroups, allowing researchers to observe relevant viral dynamics and immune responses that closely resemble those in humans [36]. Infected *rhesus macaques* exhibited patterns of viral shedding and immune responses comparable to those seen in human infections. The research demonstrated variability in the duration of virus shedding among individual animals, similar to what is observed in humans, as well as the development of adaptive immune responses against the virus. Macaques displayed only minor pathological signs during infection, including alterations in epithelial cell surface glycosylation patterns in the small intestine [37]. Additionally, the detection of viral proteins and RNA in intestinal biopsies confirmed the presence of the virus in chromogranin A expressing epithelial cells, indicating that the infection mirrors some aspects of human pathology [38].

Research states a greater susceptibility of several non-human primates to norovirus infection, including *rhesus macaques*, cotton-top tamarins, *cynomolgus macaques*, and common marmosets. Common marmosets and cotton-top tamarins showed minimal viral multiplication after oral injection, as seen by short viral shedding episodes; these species did not show any clinical signs or antibody responses. Throughout the investigation, the *cynomolgus macaques* exhibited no signs of viral replication [39]. *Rhesus macaques*, on the other hand, were shown to be vulnerable to infection by the NV since one of them showed evidence of sustained viral shedding and generated particular IgM and IgG antibody responses to the virus [40]. A more positive result was observed in research employing young pigtail macaques (*Macaca nemestrina*) exhibiting symptoms of infection with the Toronto virus P2-A, a HuNoV strain classified as GII.3. Two of the three monkeys had diarrhea and were dehydrated; the other had no symptoms at all. The animals all produced IgG antibodies and the virus was excreted in their feces for at least 3 weeks. Furthermore, when the virus infected more baby macaques, the same illness progression and viral shedding patterns were seen. Interestingly, the mothers of the affected babies also showed signs of natural infection, suggesting that HuNoV can spread across this species of monkey [41]. These findings indicate a need to further develop a model for studying the pathogenesis and transmission of HuNoV using *Macaca nemestrina*, *cynomolgus macaques*, and *rhesus macaques*.

14.3.4 Gnotobiotic Calves and Pigs as Model

Gnotobiotic calves have been shown to possess consistent susceptibility to HuNoV infections with indications similar to that of human NoV infections; the calves were seen to display symptoms like diarrhea and viral shedding *via* feces, confirming replication of the virus in the calves [42]. Studies have reported that all calves inoculated with NoV infection shed the virus with a significant increase in immune and gastrointestinal responses. Calves were observed to have increased production of antibodies; 67% of the calves developed antibody responses specific to HuNoV infection. This indicates the extent of immunogenic activity [43]. Pro-inflammatory cytokines are critical components indicating the immune response against viral infections; the presence of pro-inflammatory cytokines was observed. This also demonstrates how HuNoV infection elicits immune reactions in calves and can be applicable to human infections.

In addition, the calves displayed gastrointestinal lesions (villous atrophy and loss of epithelial cells) leading to a significant damage to the intestine, resembling the pathogenesis of HuNoV in humans [42].

Gnotobiotic pigs have proven to be a relevant model for carrying out research on HuNoV due to their physiological similarities in gastrointestinal structure and immunology with humans, making them a potential model for understanding the pathogenesis of HuNoV, evaluating potential therapeutic strategies and developing vaccines [43]. Traditional cell culture methods for HuNoV have been largely ineffective, making the gnotobiotic pig model a valuable alternative [44]. The pigs are raised in a controlled environment, in the absence of conventional microbiota allowing researchers to evaluate the intensity of interactions with HuNoV and other microbial communities. This controlled environment allows researchers to resemble human-like conditions to study the virus' infectivity and immune response [45]. In contrast to the findings of the mouse model, the use of gnotobiotic pigs has empowered researchers to assess HuNoV tropism and identify precise cell types targeted during infection. The results displayed that the main target of HuNoV is enterocytes, which are the intestinal lining's absorptive cells [11]. This authenticates previous observations in gnotobiotic pigs, where the virus was primarily detected in the intestinal epithelium (duodenum and jejunum). Gnotobiotic pigs have shown promise in studying HuNoV-induced changes in the immune response. A study has demonstrated that RAG2/IL2RG-deficient gnotobiotic pigs with a severe combined immunodeficiency (SCID) exhibited prolonged HuNoV infections due to a critical role of functional lymphocytes in eliminating the virus [46]. The ability to study HuNoV in a gnotobiotic pig model that closely resembles human physiology creates opportunities for identifying treatment plans and preventative measures against HuNoV infections and provides further insights into the disease.

14.3.5 Zebrafish Model

The zebrafish (*Danio rerio*) model has emerged to be a robust model for studying HuNoV infections as studies have displayed that upon injection of HuNoV directly into the yolk at 3 days post-fertilization (dpf), zebrafish larvae display a significant increase in viral replication, making it a suitable model for investigating the viral pathogenesis of HuNoV infection [4]. Studies have reported that HuNoV replicates to high titers in zebrafish larvae, with the highest replication observed on the second day [47]. The zebrafish embryo model presents to be an improved model as compared to zebrafish larvae as embryos facilitate continuous viral passaging and a prolonged viral behavior is observed in the embryos. In addition, the zebrafish embryos allow the evaluation of HuNoV infectivity and its interactions with histo-blood group antigens (HBGAs) [48]. A study was conducted using zebrafish to assess the efficacy of disinfection against HuNoV using UV254 light, which demonstrated a significant reduction in HuNoV infectivity [49]. However, the zebrafish larvae is a highly effective model for studying HuNoV replication; studies have reported the ability of larvae to replicate both HuNoV genotype I (GI) and genotype II (GII) at high titers, which peaked around 2 days post-infection and was detectible until the 6th day [47]. The genetic similarity between zebrafish and humans is around 70% and 84% for genes linked to illness. Furthermore, zebrafish larvae constitute a reliable model for screening HuNoV because of their optical transparency, which makes it simpler to monitor viral replication and related biological processes using real-time imaging and whole-mount labeling [50].

14.3.6 Recovirus Model

HuNoVs utilize a recovirus model to replicate and express viral components effectively by utilizing its unique properties. Genetically engineered recoviruses are used for enabling viral replication, pathogenesis, and host interactions for the effective development of therapeutic strategies [51]. Recoviruses or recombinant virus are engineered to induce specific characteristics for the study and manipulation of viral interactions with host cells [52]. Researchers use recoviruses that carry HuNoV genetic material to examine aspects of the virus, including replication, tropism, and immune response. Due to the lack of robust culture systems for HuNoV, the recoviruses model is more effective to understand the regulatory mechanisms of viral replication, which are often difficult to study in other models [53].

Although recovirus presents a promising framework, it does not fully replicate the complexities of HuNoV infections observed in humans [54]. Traditional animal models have their own limitations, such as the expense and ethical concerns of using larger animals. One significant line of research has investigated the relationship between different genotypes and serotypes of recoviruses to model antigenic diversity in HuNoV. Recoviruses are also used to study viral receptors and the mechanisms of HuNoV entry into host cells. Research involving recoviruses has enabled the exploration of virus–host interactions at the cellular level.

14.4 Advancement of *In Vitro* Models for Norovirus

Research on HuNoVs has faced significant challenges, particularly due to the difficulty in cultivating the virus *in vitro*. This obstacle has hindered a comprehensive understanding of the virus, its pathogenesis, and the development of effective treatments and vaccines. This topic aims to thoroughly examine the advancements in *in vitro* models for HuNoVs, covering early cultivation attempts, significant breakthroughs in human epithelial systems, the development of reverse genetic systems, and the characterization of HuNoV strains.

14.4.1 Early Attempts to Cultivate HuNoVs

Challenges in Cultivation

After the identification of HuNoVs as a source of AGE in 1972, numerous efforts to grow these viruses in traditional primary and immortalized cell cultures were unsuccessful [59]. Unlike many other viruses, HuNoVs have unique replication requirements and complex interactions with host cells that were not well comprehended in the early phases of research. Numerous cell lines, including primary human cells and established epithelial cell lines, were tested; however, these efforts generally failed to yield consistent results [24].

Initial Cell Culture Models

Initial efforts to utilize cell cultures, including human embryonic kidney (HEK) cells, human lung fibroblasts, and various epithelial cell lines, did not result in notable replication of HuNoV or observable cytopathic effects. Despite facing early challenges, some progress was achieved by developing differentiated intestinal epithelial cells cultivated as monolayers or within three-dimensional (3D) cultures [60, 61].

A significant portion of the initial research on cultivating HuNoVs *in vitro* centered on examining the capacity of intestinal epithelial cells to reproduce the virus [62]. An infected individual experiences temporary inflammation of the stomach and intestines, leading to pain, diarrhea, vomiting, and fatigue. Thus, it was hypothesized that norovirus infects intestinal epithelial cells and releases pro-inflammatory cytokines. In fact, *in vivo* data from intestinal sections of acutely infected animals also supported HuNoV's preference for epithelial cells [63]. These initial cultivation studies provided numerous valuable insights. For instance, when inoculating cultured cells with stool filtrates, it is essential to ensure that any observed cytopathic effects (CPE) are due to viral replication and not through other toxic components. This process includes the application of controls such as the treatment of filtrates with polymyxin B to mitigate lipopolysaccharides, the utilization of irradiated or heat-inactivated stools, and the demonstration that CPE or virus replication can be neutralized by virus capsid-specific antibodies, while pre-immune sera or isotype control antibodies do not exhibit this effect [62].

Initially, research was extensively performed in classical cell lines to determine their ability to cultivate a wide range of viruses. Different approaches were experimented by the researchers such as incorporating proteolytic enzymes that facilitate the growth of other mucosal viruses and by using stool samples with high viral genomic loads and rigorously optimizing the culture conditions, which proved unsuccessful as these efforts consistently failed to replicate HuNoVs *in vitro* [64]. Significant challenges were faced due to the inability of the HuNoV viruses to replicate in traditional cell lines which stalls the process to understand the virus' biology, pathogenesis, and potential treatments [65]. To overcome these challenges, there should be more sophisticated and physiologically suitable models to study HuNoVs and the development of alternative systems that mimic human intestinal epithelium such as human intestinal enteroids [62].

A breakthrough after four decades in the field caused the creation of the first effective cultivation system for HuNoVs [66]. The three culture systems currently used for studying the norovirus (NoVs) are a B-cell-based system, a human intestinal enterocyte (HIE) system, and a Vero-cell-based system [11]. The B-cell-based culture system utilizes human B cells. Specifically, the BJAB cell line is derived from human B lymphoblastoid cell line for studying the immune response to noroviruses [67]. BJAB cell lines can be used to analyze the affect of noroviruses on B-cell functions such as immune signaling, antibody production, and studying the mode of evasion of the virus by modulating the immune response. Even though BJAB cell lines are used in studying certain aspects of the noroviurs biology, it cannot mimic the human intestinal environment conditions or aspects of norovirus pathogenesis involving the gut epithelium. The BJAB cells are usually used alongside other models for a more thorough comprehension study of the virus [68]. The human intestinal enterocyte (HIE) derived from human intestinal enteroids are a 3D organoid culture system created from adult intestinal stem cells [69]. HIE mimics the architecture and functionality of the human intestine more than that of traditional cell cultures. This model is more physiologically relevant for studying the replication of the noroviruses by maintaining the complex cellular structures and interactions similar to the gut conditions, which allows the researcher to study the viral

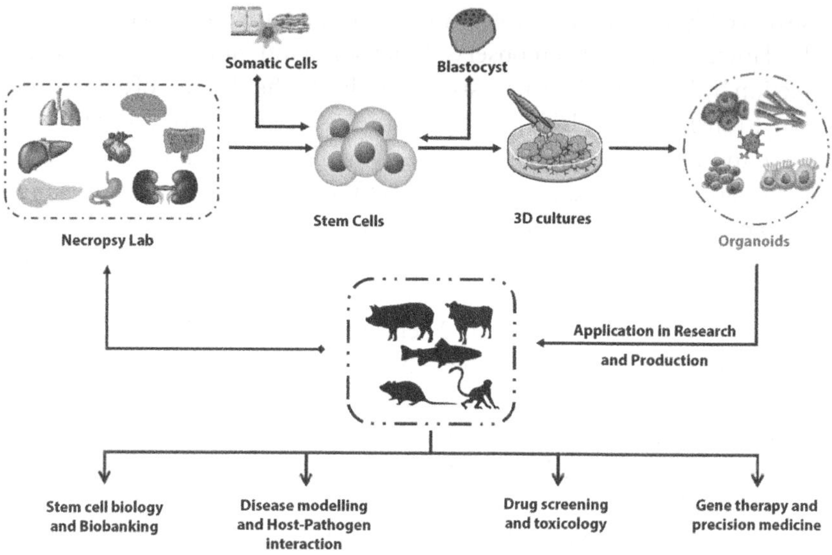

Figure 14.2 A schematic representation of organoid cultures from humans and mice into various animals to promote the screening of HuNoVs (Source of images: https://www.pngwing.com/).

behavior and virus–host interactions like the conditions of the human intestinal environment [11]. Vero cells originating from the kidneys of African green monkey are used for Vero-cell-based system. Compared to the B-cell and HIE culture methods, the yield of virus from Vero-cells is only a minimal amount. Each of these culture systems provides different insights into norovirus biology [2]. The initial infection mechanism is studied using the B-cell system. The detailed replication of norovirus is studied using the physiologically relevant HIE model. To study the viral assay, the Vero-cell-based system is used as a general tool. A combination of all the three systems contributes to a more comprehensive study of understanding the pathogenesis and potential therapeutic approaches [66]. Various organoid cultures from humans and mice into various animals to promote the screening of HuNoVs are depicted in Figure 14.2.

14.4.2 Key Discoveries That Helped to Cultivate the Human Epithelial System

Research conducted between 1972 to 1990 identified some individuals who possess a genetic resistance to NV infection. It was proposed that a specific host receptor might be absent in these individuals, accounting for their resistance [70]. Norovirus typically binds to specific receptors on intestinal cells. However, it has been observed that some individuals are resistant to infection by NV [71, 72]. This resistance is due to the absence of a functional fucosyltransferase 2 (FUT2) enzyme, which is necessary for the expression of certain fucosylated glycans or HBGAs on the intestinal villi [73]. These individuals are referred to as "non-secretors" because they do not express these antigens [66]. The absence of these receptors means that Norwalk virus-like particles cannot bind to the intestinal cells of non-secretors, providing them with natural resistance to the virus [74]. Therefore, among the population with a functional FUT2 gene, some individuals remained resistant to NV infection, implying that either a memory immune response or another unknown factor provides protection from NV infection [74]. This hypothesis was further validated when independent controlled human infection studies revealed that only secretor-positive volunteers, those with a functional FUT2 enzyme, were susceptible to NV infection [75]. While the presence of external HBGAs enhances HuNoV attachment to the surface of B cells and infection, merely expressing HBGAs is not enough for *in vitro* permissiveness. This is evident as HBGAs on intestinal epithelial cells (IECs) do not make these cells vulnerable to viral infection [76].

Furthermore, bile acids were found to enhance HuNoV infection, underscoring their importance in refining *in vitro* culture conditions [11]. Interestingly, bile played a role in HuNoV replication in enteroids and intestinal organoids in a strain-specific manner. Its effects were observed at the initial stages of infection, impacting the host cells rather than the virus directly [14]. For instance, replication of specific HuNoV strains, including GI.1, GII.3, and GII.17, is either reliant on or improved by the presence of bile acids. On the other hand, pandemic-causing strains such as GII.4 do not need bile acids for replication, though their replication might be enhanced when bile acids are present. These findings emphasize that bile acids can trigger various cellular responses like raising the levels of ceramide on the apical membrane *via* the sphingosine-1-phosphate receptor 2 (S1PR2) by boosting endosomal uptake and acidification which aids in viral replication in HIEs [77]. Stains that depend on bile acid only require SIPR2, whereas GIL.4 does not require bile acid as it does not depend on it [78]. Hence, it shows that various cellular responses might be triggered by bile acids that aid in viral replication.

In 1977, Fogh and colleagues first established a human epithelial cell line, namely, Caco-2, originating from colon carcinoma. When cultured in appropriate conditions, these cells are extensively used in gastrointestinal research due to their ability to differentiate into cells resembling enterocytes [79]. Caco-2 cells, upon reaching confluency, naturally differentiate themselves by adopting the features akin to mature intestinal enterocytes, development of microvilli, formation of tight junctions, and expression of different enzyme and transporters. These characteristics make the Caco-2 cells an important *in vitro* model for studying intestinal absorption, drug transport, and the biology of enteric pathogens [80]. Caco-2 cells have been known to be effective in evaluating enteric viruses. However, they have been unsuccessful in facilitating the replication of HuNoV. This has been observed to be due to the inability of the virus to enter the cells, and even if the virus enters the cell, it has suppressed the replication of the virus. This has impacted the ability of Caco-2 cells to be used for studying the life cycle, pathogenesis, and immune evasion mechanisms of HuNoV [81]. Due to these challenges with cell lines, it has become necessary to find newer models that physiologically replicate the human gut environment. This realization drove the development of alternative *in vitro* models, such as HIEs, which have since become instrumental in HuNoV research.

Enterocytes extracted from human tissue would quickly undergo apoptosis, making it difficult to assess viral replication. However, a groundbreaking discovery changed this landscape: HIEs derived from stem cells harvested from human intestinal crypts provided a viable solution [82]. These HIEs closely mimic the natural intestinal epithelium, offering a supportive environment for HuNoV growth and thus enabling comprehensive studies of viral replication and pathogenesis [62]. These diverse, specialized HIEs are unaltered and functionally active cultures that include multiple intestinal cell types (enterocytes, goblet, enteroendocrine, and Paneth cells) and react to different agonists, regardless of being grown in three-dimensional structures or as flat monolayers [83]. Once HuNoV were successfully cultured in human intestinal epithelial cells, derived from various small intestinal segments of immunocompetent individuals, the replication was confirmed through the expression of viral proteins, production and release of virus particles into the culture medium, and successful virus passaging [83]. The cultivation method reflects biological relevance, with replication patterns aligning with known host restrictions based on human blood group antigen (HBGA) expression.

For years, cultivating primary human intestinal epithelial cells *in vitro* remained an elusive goal. Enterocytes extracted directly from human tissue would quickly undergo apoptosis, thwarting any attempts to study viral replication effectively. The challenge faced by using 2D cell cultures was finally overcome by the development of HIEs, which are derived from stem cells located in human intestinal crypts. The HIEs mimic the human intestinal epithelium by incorporating growth factors and cytokines, manufacturing a viable environment facilitating HuNoV growth [11].

14.4.3 Reverse Genetic Systems

The origination of the reverse genetic system was a breakthrough for studying norovirus research as this model allows scientists to mold genetic material as to attain a similar pathophysiology of HuNoV [15]. The inability of HuNoV to replicate in tissue cultures has hindered the progress of identifying novel drugs and vaccines for its treatment; this has been seen to be reversed with research on reversed genetic systems [84]. Reverse genetics provides a valuable framework for examining the role of specific proteins of HuNoV throughout the course of an infection. By utilizing this approach, researchers can gain deeper insights into how each protein contributes to the virus' behavior and pathogenicity. This understanding could be instrumental in developing attenuated vaccines, which are designed to stimulate an immune response by using a weakened form of the virus. Thereby, it holds the potential to significantly advance vaccine development against HuNoV by elucidating the functions of its individual proteins [36].

Reverse and forward genetics are two main approaches in functional genomics, where reverse genetics deals with alteration of the genetic material to study the resultant phenotype such as ectopic expression and mutagenesis [85]. In the field of virology, reverse genetics is an advanced and an impactful approach to study the meticulous gene alterations to generate specific phenotypes that help the researchers to understand the functions of each gene and to study their regulatory mechanisms. This helps in exploring the observable traits after the change that occurs at the genetic level and an in-depth understanding of the gene function [86]. A genetically modified virus can be created from starch by using the cloned cDNA. This will significantly expand the view on virus replication and pathogenicity and understanding of numerous RNA virus families [87].

Noroviruses are single-stranded RNA viruses, which lack the outer envelope with an intricate structural characteristic that makes them difficult to study [68]. The genomic size of RNA is approximately 7.3 to 8.3 kb in length and sub-genomic RNA measures about 2.4 kb in length. Along with the three open reading frames (ORFs) of the normal virus' genome, murine norovirus features an additional fourth ORF [88]. Reverse genetics plays a major role in creating helper plasmids, cloning cDNA, modifying full-length genome fragments, co-transfecting, identifying virus offspring, and determining sequences. Thus, it is possible to examine host pathways, pathogen relationships, and viral protein activities by using reverse genetic systems to restore infectious norovirus *in vitro* [15].

The development of HuNoV replicon-bearing cells in a human hepatoma cell line, the B-cell culture system, the stem-cell-derived intestinal organoid system, and zebrafish larvae infection models have all provided valuable tools to dissect the norovirus life cycle. Replicative, transmissive, competitive, and epidemiological fitness are key elements of the overall viral fitness, which conceptually determines how well a virus "fits" into its environment [89]. Now, researchers can make specific mutations or deletions in the viral and host genomes using CRISPR/Cas9 and advanced genomic tools for precision gene editing. As a result, infection-related host variables and crucial viral genes can be more easily identified. New developments in reverse genetic systems are expected to shed more light on the biology of HuNoV and aid in the creation of innovative vaccinations and antiviral treatments [90]. The model systems that can be used for different genogroups of human norovirus are presented in Table 14.2.

14.4.4 Human Norovirus Strains

Noroviruses belong to the Caliciviridae family of viruses, which has a diverse genetic makeup. They include three ORFs that make up their 7.6-kb single-stranded positive-sense RNA genome [96]. A big polyprotein, ORF1, is cleaved into six mature non-structural proteins. The primary capsid protein VP1, which can self-assemble into virus-like particles (VLPs), is encoded by ORF2, and the minor structural protein VP2 is encoded by ORF3 [97]. The shell (S) and protruding (P) domains are the two primary domains that make up the norovirus capsid. The P domain holds the primary antigenic diversity determinants, while the S domain acts as a protective scaffold for the RNA. HBGAs of the host interact with the norovirus capsid. For the entry and replication of viruses, this interaction with HBGAs is known to be crucial, and intraspecies diversity can be achieved by genetic polymorphism in the genes that regulate HBGA production [98].

Genetic Diversity of HuNoVs

In the middle of the 1990s, NoV strains began to be classified according to their genogroups and genotypes as an outcome of RNA-dependent RNA polymerase partial nucleotide sequencing [99]. Noroviruses are classified into different genetic groups based on the amino acid sequence of their primary capsid protein [100]. They are classified into 10 distinct genogroups, labeled GI through GX, based on the differences in the amino acid sequences of their viral proteins. Among these genogroups, over 36 unique

Table 14.2 Different types of models that can be used for various genogroups of human norovirus.

Geno group/ strain	Substrains	Characteristics	Models used	References
GI	Norwalk virus, Desert Shield virus, Southampton virus	Causes gastroenteritis outbreaks; less prevalent today compared to GII strains	Cell culture, murine models	[91]
GII	GII.4 (various variants), GII.17	Dominant strain globally; GII.4 is associated with significant outbreaks and high transmission rates.	Gnotobiotic pigs, rhesus macaque models, 3D organoids	[92]
GIII	HuGov strain	Primarily infects humans but less prevalent than GII; contributes to genetic diversity	Zebrafish models, murine models	[48]
GIV	Various animal-associated strains	Rarely reported in human outbreaks; primarily associated with animal infections	Cell culture models, murine models	[21]
GV	Limited specific substrains	Considered less common among predominant strains; specific substrains are not well-documented	Gnotobiotic pigs, 3D models	[17]
GVI	Emerging strains	Part of the GII category, associated with potential outbreaks but less documented prevalence	Zebrafish models, murine models	[47]
GVII	GVII.1, GVII.2	Emerging variants associated with increasing outbreak trends; geographical variability in presence	Rhesus macaque models, murine models	[33, 93]
GVIII	Limited identification	Less documented; primarily associated with specific strains linked to animal infections	Cell culture, animal models	[94, 95]
GIX	Various electropherotypes	Characterized by distinct electropherotypes with specific outbreak associations noted	Gnotobiotic pigs, 3D organoids	[11]

genotypes have been identified, with representatives from five particular genogroups—GI, GII, GIV, GVIII, and GIX—that are known to be capable of causing infections [96]. Humans are known to be infected by noroviruses from the genogroups GI, GII, and GIV. Within these genogroups, GI encompasses eight distinct genotypes, while GII includes at least 19 different genotypes [101]. GIV.1 and GIV.NA1 are two of the genotypes from GIV genogroups that are found to be infecting humans [102]. Among the various norovirus genogroups, GII viruses are the most frequently detected in human cases [103]. Notably, patients with AGE are more likely to display GII viruses, with GII.3, GII.4, GII.6, and GII.7 as the common genogroups. This indicates the probability of GII viruses being linked with AGE and its various genotypes; this allows the researchers to perform targeted studies to inhibit gastroenteritis. The genotype GII.3 was initially observed in children and newborn [104]. Noroviruses of genotype GII.4 have been the predominant cause of gastroenteritis outbreaks in Europe and the United States since the mid-1990s. There is evidence to suggest that the introduction of new phylogenetically linked clusters or variants of GII.4 noroviruses is correlated with an increase in the global frequency of large-scale epidemics [105]. Six GII.4 variants have surfaced and produced global epidemics since the mid-1990s: Grimsby_1996, Farmington_Hills_2002, Hunter_2004, Den_Haag_2006, New Orleans_2009, and Sydney_2012. The Sydney_2012 and Grimsby_1996 variants each dominated for more than 5 years in the human population despite the fact that the transition of the prominent variants can happen quickly (2 to 3 years) [102]. The novel strain known as GII.4 Sydney was first discovered in Australia in March 2012 and has emerged as the most common strain worldwide [103].

Research indicates that approximately 5% of GII.4 noroviruses undergo continuous genetic variant evolution. It is proposed that these viruses possess a mechanism that enables them to avoid the immune system or modify their receptor binding patterns. Additionally, the GII.4 genotype frequently experiences genetic recombination, which broadens their diversity [98]. According to several recent publications, the main capsid proteins of GII.4 strains are quickly changing, leading to the emergence of new epidemic strains with changed antigenicity. Most of these modifications are taking place in the P2 subdomain that is exposed to the surface [100].

While the recently identified GII.P17-GII.17 genotype has taken over as the main strain in several regions of Asia [106], HuNoV exhibits significant antigenic diversity, making it challenging to develop antibodies that are effective against a broad range of strains. As a result, enzyme immunoassays often have limited sensitivity when it comes to detecting these diverse viral types [107].

Even though phylogenetic distances are used for classification, genotypes and genogroups typically differ by 20%–40% and 40%–60% of their amino acid sequence, respectively. Human virus strains only replicate a limited amount in animals despite some research indicating their prevalence in a variety of animal species. The creation of a reliable animal model for HuNoV has been hampered by this notable species' specificity [108]. There are currently no approved vaccines or antivirals for HuNoV on the market despite their discovery more than 40 years ago. Nevertheless, it has recently been demonstrated that human milk oligosaccharides (HMOs), which resemble HBGA structures, can prevent norovirus VLPs from attaching to HBGAs, making them promising candidates for potentially managing HuNoV infection [98]. Thus, achieving a better understanding of why these strains have not managed to spread globally could provide valuable insights for effectively managing potential norovirus pandemics [102].

14.5 Conclusion

Our investigations into human viruses significantly depend on animal models. Small animal models, especially knock-out and transgenic mouse models, serve as crucial tools for deepening our mechanistic understanding of the interactions between viruses and their hosts on both organismal and cellular levels. In contrast, large animal models play an essential role in translational studies and are particularly advantageous for preclinical efficacy testing of antiviral therapies and vaccines. Although significant progress has been made in understanding HuNoVs over the past 20 years, our knowledge of these viruses remains less comprehensive compared to other human viruses.

The field continues to grapple with the challenge of identifying a model that accurately replicates HuNoV disease. Currently, no "ideal" model exists; however, various animal models have been developed to study HuNoV by either overcoming the virus' species barrier or utilizing surrogate viruses. Each of these models offers unique advantages and limitations, and addressing these challenges remains a key objective for future research. A promising direction for future advancements could be the development of host-adapted HuNoVs, a milestone that has not yet been achieved in any existing HuNoV animal model. Additionally, progress may come from exploring less-studied species, such as newborn pigtail macaques, or investigating new animal hosts. Furthermore, the discovery of new surrogate viruses or the identification of additional HuNoV strains that can more easily adapt to or transmit among different species could significantly enhance our understanding of these viruses. While the data generated from both current and future models are crucial for advancing our knowledge of HuNoVs, the ultimate validation of these findings depends on their correlation with human hosts. This correlation will be the determining factor in assessing the suitability of each model for studying specific aspects of HuNoV biology.

Abbreviations

3D	three-dimensional
2D	two-dimensional
AGE	acute gastroenteritis
CPE	cytopathic effects
CRISPR	clustered regularly interspaced short palindromic repeats
DNA	deoxyribonucleic acid
FNoV	feline norovirus
FUT2	fucosyltransferase 2
HBGA	histo-blood group antigens
HEK	human embryonic kidney
HIE	human intestinal enteroids
HMO	human milk oligosaccharides
HuNoV	human norovirus
IEC	intestinal epithelial cells
MuNoV	murine norovirus
NoV	norovirus
NorV	Norwalk virus
ORF	open reading frames
RAG	recombination activating genes
RNA	ribonucleic acid
S1PR2	sphingosine-1-phosphate receptor 2
SCID	severe combined immunodeficiency
VLP	virus-like particles

AI disclosure statement: The author acknowledges the use of AI-based tools for minor language editing or grammatical correction. However, all intellectual contributions, analyses, and conclusions presented in this chapter are entirely the author's own.

References

1. Carlson, K.B., Dilley, A., O'Grady, T., Johnson, J.A., Lopman, B., Viscidi, E., A narrative review of norovirus epidemiology, biology, and challenges to vaccine development. *npj Vaccines*, 9, 94, 2024. https://doi.org/10.1038/s41541-024-00884-2.
2. Robilotti, E., Deresinski, S., Pinsky, B.A., Norovirus. *Clin. Microbiol. Rev.*, 28, 134–64, 2015. https://doi.org/10.1128/cmr.00075-14.
3. Karst, S.M., Wobus, C.E., Goodfellow, I.G., Green, K.Y., Virgin, H.W., Advances in Norovirus Biology. *Cell Host Microbe*, 15, 668–80, 2014. https://doi.org/10.1016/j.chom.2014.05.015.
4. Kreins, A.Y., Roux, E., Pang, J., Cheng, I., Charles, O., Roy, S., *et al.*, Favipiravir induces HuNoV viral mutagenesis and infectivity loss with clinical improvement in immunocompromised patients. *Clin. Immunol.*, 259, 109901, 2024. https://doi.org/10.1016/j.clim.2024.109901.
5. Graves, N.S., Acute Gastroenteritis. *Prim. Care Clin. Off. Pract.*, 40, 727–41, 2013. https://doi.org/10.1016/j.pop.2013.05.006.
6. Lewis, M.A., Cortés-Penfield, N.W., Ettayebi, K., Patil, K., Kaur, G., Neill, F.H., *et al.*, Standardization of an antiviral pipeline for human norovirus in human intestinal enteroids demonstrates nitazoxanide has no to weak antiviral activity. *Antimicrob. Agents Chemother.*, 67, e00636–23, 2023. https://doi.org/10.1128/aac.00636-23.
7. Gidey, K., Gidey, M.T., Hailu, B.Y., Gebreamlak, Z.B., Niriayo, Y.L., Clinical and economic burden of healthcare-associated infections: A prospective cohort study. *PLoS One*, 18, e0282141, 2023. https://doi.org/10.1371/journal.pone.028214.
8. Papafragkou, E., Hewitt, J., Park, G.W., Greening, G., Vinjé, J., Challenges of Culturing Human Norovirus in Three-Dimensional Organoid Intestinal Cell Culture Models. *PLoS One*, 8, e63485, 2013. https://doi.org/10.1371/journal.pone.0063485.
9. Ruiz, S.I., Zumbrun, E.E., Nalca, A., Animal Models of Human Viral Diseases. *Anim. Model Study Hum. Dis.*, 2, 853–901, 2017. https://doi.org/10.1016/B978-0-12-809468-6.00033-4.
10. Kocher, J. and Yuan, L., Norovirus Vaccines and Potential Antinorovirus Drugs: Recent Advances and Future Perspectives. *Future Virol.*, 10, 899–913, 2015. https://doi.org/10.2217/fvl.15.57.
11. Ettayebi, K., Crawford, S.E., Murakami, K., Broughman, J.R., Karandikar, U., Tenge, V.R., *et al.*, Replication of human noroviruses in stem cell–derived human enteroids. *Science*, 353, 1387–93, 2016. https://doi.org/10.1126/science.aaf5211.
12. Gardner, J. and Herbst-Kralovetz, M., Three-Dimensional Rotating Wall Vessel-Derived Cell Culture Models for Studying Virus-Host Interactions. *Viruses*, 8, 304, 2016. https://doi.org/10.3390/v8110304.

13. Chia, S.P.S., Kong, S.L.Y., Pang, J.K.S., Soh, B.S., 3D Human Organoids: The Next "Viral" Model for the Molecular Basis of Infectious Diseases. *Biomedicines*, 10, 1541, 2022. https://doi.org/10.3390/biomedicines10071541.
14. Ettayebi, K., Tenge, V.R., Cortes-Penfield, N.W., Crawford, S.E., Neill, F.H., Zeng, X.L., *et al.*, New Insights and Enhanced Human Norovirus Cultivation in Human Intestinal Enteroids. *mSphere*, 6, 1, 2021. https://doi.org/10.1128/msphere.01136-20.
15. Zhang, Q., Zhu, S., Zhang, X., Su, L., Ni, J., Zhang, Y., *et al.*, Recent insights into reverse genetics of norovirus. *Virus Res.*, 325, 199046, 2023. https://doi.org/10.1016/j.virusres.2023.199046.
16. Rimkute, I., Chaimongkol, N., Woods, K.D., Nagata, B.M., Darko, S., Gudbole, S., *et al.*, A non-human primate model for human norovirus infection. *Nat. Microbiol.*, 9, 776–86, 2024. 10.1038/s41564-023-01585-7.
17. Cheetham, S., Souza, M., Meulia, T., Grimes, S., Han, M.G., Saif, L.J., Pathogenesis of a Genogroup II Human Norovirus in Gnotobiotic Pigs. *J. Virol.*, 80, 10372–81, 2006. Available from: https://journals.asm.org/doi/10.1128/JVI.00809-06.
18. Peiper, A.M., Helm, E.W., Nguyen, Q., Phillips, M., Williams, C.G., Shah, D., *et al.*, Infection of neonatal mice with the murine norovirus strain WU23 is a robust model to study norovirus pathogenesis. *Lab Anim.*, 52, 119–129, 2023 https://doi.org/10.1038/s41684-023-01166-5.
19. Seir, J.J., José, M.C., Pierre, R., Arturo, G.C., Mouse intestinal villi as a model system for studies of norovirus infection. *Acta Virol.*, 67, 24–41, 2023. Available from: http://www.elis.sk/index.php?page=shop.product_details&flypage=flypage.tpl&product_id=8028&category_id=190&option=com_virtuemart.
20. Takano, T., Ryu, M., Doki, T., Kusuhara, H., Immunodominant B-Cell Linear Epitope on the VP1 P Domain of a Feline Norovirus Cat Model. *Pathogens*, 11, 731, 2022. Available from: https://www.mdpi.com/2076-0817/11/7/731.
21. Sato, Y., Yasuda, J., Sakurai, M., Animal-sourced model of human norovirus infection predicted using environmental DNA metabarcoding analysis. *J. Freshw. Ecol.*, 39, 1–12, 2024. Available from: https://www.tandfonline.com/doi/full/10.1080/02705060.2023.2293171.
22. Zhao, B., Hu, L., Song, Y., Patil, K., Ramani, S., Atmar, R.L., *et al.*, Norovirus Protease Structure and Antivirals Development. *Viruses*, 13, 1–14, 2021. Available from: https://doi.org/10.3390/v13102069.
23. Ludwig-Begall, L.F., Mauroy, A., Thiry, E., Noroviruses—The State of the Art, Nearly Fifty Years after Their Initial Discovery. *Viruses*, 13, 1541, 2021. Available from: https://www.mdpi.com/1999-4915/13/8/1541.
24. Chandran, S. and Gibson, K.E., Improving the Detection and Understanding of Infectious Human Norovirus in Food and Water Matrices: A Review of Methods and Emerging Models. *Viruses*, 16, 776, 1–24, 2024. Available from: https://www.mdpi.com/1999-4915/16/5/776.

25. Taube, S., Kolawole, A.O., Höhne, M., Wilkinson, J.E., Handley, S.A., Perry, J.W., *et al.*, A Mouse Model for Human Norovirus. *MBio*, 4, 2013. Available from: https://journals.asm.org/doi/10.1128/mBio.00450-13.
26. Kolawole, A.O., Gonzalez-Hernandez, M.B., Turula, H., Yu, C., Elftman, M.D., Wobus, C.E., Oral Norovirus Infection Is Blocked in Mice Lacking Peyer's Patches and Mature M Cells. *J. Virol.*, 90, 1499–506, 2016. Available from: https://journals.asm.org/doi/10.1128/JVI.02872-15.
27. Blümich, S., Zdimerova, H., Münz, C., Kipar, A., Pellegrini, G., Human CD34 + Hematopoietic Stem Cell–Engrafted NSG Mice: Morphological and Immunophenotypic Features. *Vet. Pathol.*, 58, 161–80, 2021. Available from: http://journals.sagepub.com/doi/10.1177/0300985820948822.
28. Santiso-Bellón, C., Gozalbo-Rovira, R., Buesa, J., Rubio-del-Campo, A., Peña-Gil, N., Navarro-Lleó, N., *et al.*, Replication of Human Norovirus in Mice after Antibiotic-Mediated Intestinal Bacteria Depletion. *Int. J. Mol. Sci.*, 23, 10643, 2022. Available from: https://www.mdpi.com/1422-0067/23/18/10643.
29. Thorne, L., Lu, J., Chaudhry, Y., Bailey, D., Goodfellow, I., Targeting macrophage- and intestinal epithelial cell-specific microRNAs against norovirus restricts replication *in vivo*. *J. Gen. Virol.*, 99, 1621–32, 2018. Available from: https://www.microbiologyresearch.org/content/journal/jgv/10.1099/jgv.0.001065.
30. Kennedy, E.A., Aggarwal, S., Dhar, A., Karst, S.M., Wilen, C.B., Baldridge, M.T., Age-associated features of norovirus infection analysed in mice. *Nat. Microbiol.*, 8, 1095–107, 2023. https://doi.org/10.1038/s41564-023-01383-1.
31. Compton, S.R., Prevention of murine norovirus infection in neonatal mice by fostering. *J. Am. Assoc. Lab. Anim. Sci.*, 47, 25–30, 2008. http://www.ncbi.nlm.nih.gov/pubmed/18459709.
32. Bok, K., Parra, G.I., Mitra, T., Abente, E., Shaver, C.K., Boon, D., *et al.*, Chimpanzees as an animal model for human norovirus infection and vaccine development. *Proc. Natl. Acad. Sci.*, 108, 325–30, 2011. https://doi.org/10.1073/pnas.1014577107.
33. Rockx, B.H.G., Bogers, W.M.J.M., Heeney, J.L., van Amerongen, G., Koopmans, M.P.G., Experimental norovirus infections in non-human primates. *J. Med. Virol.*, 75, 313–20, 2005. https://doi.org/10.1002/jmv.20273.
34. Omatola, C.A., Mshelbwala, P.P., Okolo, L.O., Onoja, A.B., Abraham, J.O., Adaji, D.M., *et al.*, Noroviruses: Evolutionary Dynamics, Epidemiology, Pathogenesis, and Vaccine Advances—A Comprehensive Review. *Vaccines*, 12, 590, 2024. https://doi.org/10.3390/vaccines12060590.
35. Wyatt, C.A., Andrus, L., Brotman, B., Huang, F., Lee, D.H., Prince, A.M., Immunity in Chimpanzees Chronically Infected with Hepatitis C Virus: Role of Minor Quasispecies in Reinfection. *J. Virol.*, 72, 1725–30, 1998. https://doi.org/10.1128/jvi.72.3.1725-1730.1998.
36. Todd, K.V. and Tripp, R.A., Human Norovirus: Experimental Models of Infection. *Viruses*, 11, 151, 2019. https://doi.org/10.3390/v11020151.

37. Wachtman, L. and Mansfield, K., Viral Diseases of Nonhuman Primates. *Nonhum. Primates Biomed. Res.*, 2, 1–104, 2012. https://doi.org/10.1016/B978-0-12-381366-4.00001-8.
38. Karandikar, U.C., Crawford, S.E., Ajami, N.J., Murakami, K., Kou, B., Ettayebi, K., *et al.*, Detection of human norovirus in intestinal biopsies from immunocompromised transplant patients. *J. Gen. Virol.*, 97, 2291–300, 2016. https://doi.org/10.1099/jgv.0.000545.
39. Mätz-Rensing, K. and Bleyer, M., Viral Diseases of Common Marmosets. *Common Marmoset. Captiv. Biomed. Res.*, 251–64, 2019. https://doi.org/10.1016/B978-0-12-811829-0.00015-7.
40. Gardner, M.B. and Luciw, P.A., Macaque Models of Human Infectious Disease. *ILAR J.*, 49, 220–55, 2008. https://doi.org/10.1093/ilar.49.2.220.
41. Subekti, D.S., Tjaniadi, P., Lesmana, M., McArdle, J., Iskandriati, D., Budiarsa, I.N., *et al.*, Experimental infection of Macaca nemestrina with a Toronto Norwalk-like virus of epidemic viral gastroenteritis. *J. Med. Virol.*, 66, 400–6, 2002. https://doi.org/10.1002/jmv.2159.
42. Souza, M., Azevedo, M.S.P., Jung, K., Cheetham, S., Saif, L.J., Pathogenesis and Immune Responses in Gnotobiotic Calves after Infection with the Genogroup II.4-HS66 Strain of Human Norovirus. *J. Virol.*, 82, 1777–86, 2008. https://doi.org/10.1128/JVI.01347-07.
43. Yuan, L., Jobst, P.M., Weiss, M., Gnotobiotic Pigs. *Gnotobiotics*, 349–68, 2017. https://doi.org/10.1016/B978-0-12-804561-9.00005-0.
44. Lei, S., Twitchell, E.L., Ramesh, A.K., Bui, T., Majette, E., Tin, C.M., *et al.*, Enhanced GII.4 human norovirus infection in gnotobiotic pigs transplanted with a human gut microbiota. *J. Gen. Virol.*, 100, 1530–40, 2019. https://doi.org/10.1099/jgv.0.001336.
45. Yuan, L., Gnotobiotic pig models: Illuminating the enigma of human norovirus infection and immunity. *Open Access Gov.*, 40, 34–5, 2023. https://doi.org/10.56367/OAG-040-10359.
46. Boettcher, A.N., Loving, C.L., Cunnick, J.E., Tuggle, C.K., Development of Severe Combined Immunodeficient (SCID) Pig Models for Translational Cancer Modeling: Future Insights on How Humanized SCID Pigs Can Improve Preclinical Cancer Research. *Front. Oncol.*, 8, 2018. https://doi.org/10.3389/fonc.2018.00559.
47. Van Dycke, J., Ny, A., Conceição-Neto, N., Maes, J., Hosmillo, M., Cuvry, A., *et al.*, A robust human norovirus replication model in zebrafish larvae. Suthar, M., (ed), *PLoS Pathog.*, 15, e1008009, 2019. https://doi.org/10.1371/journal.ppat.1008009.
48. Tan, M.T.H., Gong, Z., Li, D., Use of Zebrafish Embryos To Reproduce Human Norovirus and To Evaluate Human Norovirus Infectivity Decay after UV Treatment. *Appl. Environ. Microbiol.*, 89, e00115-23, 2023. https://doi.org/10.1128/aem.00115-23.

49. Yoon, S.R., Ha, S., Park, B., Yang, J.S., Dang, Y.M., Ha, J.H., Effect of Ultraviolet-C Light-Emitting Diode Treatment on Disinfection of Norovirus in Processing Water for Reuse of Brine Water. *Front. Microbiol.*, 13, 1–10, 2022. https://doi.org/10.3389/fmicb.2022.885413.
50. Liu, J., Yuan, X., Fan, C., Ma, G., Application of the zebrafish model in human viral research. *Virus Res.*, 341, 199327, 2024. https://doi.org/10.1016/j.virusres.2024.199327.
51. Gan, T., Droit, L., Vernon, S., Barouch, D.H., Wang, D., Isolation of a rhesus calicivirus that can replicate in human cells. *Virol.*, 582, 83–9, 2023. https://doi.org/10.1016/j.virol.2023.03.012.
52. Bulcha, J.T., Wang, Y., Ma, H., Tai, P.W.L., Gao, G., Viral vector platforms within the gene therapy landscape. *Signal Transduct. Target Ther.*, 6, 53, 2021. https://doi.org/10.1038/s41392-021-00487-6.
53. Strtak, A.C., Perry, J.L., Sharp, M.N., Chang-Graham, A.L., Farkas, T., Hyser, J.M., Recovirus NS1-2 Has Viroporin Activity That Induces Aberrant Cellular Calcium Signaling To Facilitate Virus Replication. *mSphere*, 4, e00506–19, 2019. https://doi.org/10.1128/mSphere.00506-19.
54. Thiede, L., Pogan, R., Uetrecht, C., Norovirus—A Viral Capsid in Perpetual Flux, in: *Physical Virology*, vol. 24, Springer Series in Biophysics, Springer, Cham, 2023, https://doi.org/10.1007/978-3-031-36815-8_8.
55. Jones, M.K., Grau, K.R., Costantini, V., Kolawole, A.O., de Graaf, M., Freiden, P., Graves, C.L., Koopmans, M., Wallet, S.M., Tibbetts, S.A., Schultz-Cherry, S., Wobus, C.E., Vinjé, J., Karst, S.M., Human norovirus culture in B cells. *Nat. Protoc.*, 10, 1939–1947, 2015. https://doi.org/10.1038/nprot.2015.121.
56. Arias, A., Ureña, L., Thorne, L., Yunus, M.A., Goodfellow, I., Reverse Genetics Mediated Recovery of Infectious Murine Norovirus. *J. Vis. Exp.*, 2012. https://doi.org/10.3791/4145.
57. Le Bras, A., *Rhesus macaques* for human norovirus studies. *Lab Anim. (NY)*, 53, 87–87, 2024. https://doi.org/10.1038/s41684-024-01356-9.
58. Wyatt, R.G., Greenberg, H.B., Dalgard, D.W., Allen, W.P., Sly, D.L., Thornhill, T.S., Chanock, R.M., Kapikian, A.Z., Experimental infection of chimpanzees with the Norwalk agent of epidemic viral gastroenteritis. *J. Med. Virol.*, 2, 89–96, 1978. https://doi.org/10.1002/jmv.1890020203.
59. Lay, M.K., Atmar, R.L., Guix, S., Bharadwaj, U., He, H., Neill, F.H., *et al.*, Norwalk virus does not replicate in human macrophages or dendritic cells derived from the peripheral blood of susceptible humans. *Virol.*, 406, 1–11, 2010. https://doi.org/10.1016/j.virol.2010.07.001.
60. Duizer, E., van Duynhoven, Y., Vennema, H., Koopmans, M., Letter to the Editor. *Public Health*, 118, 455–6, 2004. https://doi.org/10.1016/j.puhe.2004.07.001.
61. Duizer, E., Schwab, K.J., Neill, F.H., Atmar, R.L., Koopmans, M.P.G., Estes, M.K., Laboratory efforts to cultivate noroviruses. *J. Gen. Virol.*, 85, 79–87, 2004. https://doi.org/10.1099/vir.0.19478-0.

62. Estes, M.K., Ettayebi, K., Tenge, V.R., Murakami, K., Karandikar, U., Lin, S.C., *et al.*, Human Norovirus Cultivation in Nontransformed Stem Cell-Derived Human Intestinal Enteroid Cultures: Success and Challenges. *Viruses*, 11, 638, 2019. https://doi.org/10.3390/v11070638.
63. Villabruna, N., Schapendonk, C.M.E., Aron, G.I., Koopmans, M.P.G., de Graaf, M., Human Noroviruses Attach to Intestinal Tissues of a Broad Range of Animal Species. *J. Virol.*, 95, 1–18, 2021. https://doi.org/10.1128/jvi.01492-20.
64. Bhar, S. and Jones, M.K., *In Vitro* Replication of Human Norovirus. *Viruses*, 11, 547, 2019. https://doi.org/10.3390/v11060547.
65. Love, M.S. and Choy, R.K.M., Emerging treatment options for cryptosporidiosis. *Curr. Opin. Infect. Dis.*, 34, 455–462, 2021. https://doi.org/10.1097/QCO.0000000000000761.
66. Hasing, M.E. and Pang, X.L., Norovirus: Molecular Epidemiology, Viral Culture, Immunity, and Vaccines. *Clin. Microbiol. Newsl.*, 43, 33–43, 2021. https://doi.org/10.1016/j.clinmicnews.2021.02.002.
67. Jones, M.K., Watanabe, M., Zhu, S., Graves, C.L., Keyes, L.R., Grau, K.R., *et al.*, Enteric bacteria promote human and mouse norovirus infection of B cells. *Science*, 346, 755–9, 2014. https://doi.org/10.1126/science.1257147.
68. Bartnicki, E., Cunha, J.B., Kolawole, A.O., Wobus, C.E., Recent advances in understanding noroviruses. *F1000Res.*, 6, 79, 2017. https://doi.org/10.12688/f1000research.10081.1.
69. Zou, W.Y., Blutt, S.E., Crawford, S.E., Ettayebi, K., Zeng, X.L., Saxena, K., *et al.*, Human Intestinal Enteroids: New Models to Study Gastrointestinal Virus Infections. *Methods Mol. Biol.*, 1576, 229–47, 2017. https://doi.org/10.1007/7651_2017_1.
70. Le Pendu, J., Ruvoën-Clouet, N., Kindberg, E., Svensson, L., Mendelian resistance to human norovirus infections. *Semin. Immunol.*, 18, 375–86, 2006. https://doi.org/10.1016/j.smim.2006.07.009.
71. Lindesmith, L., Moe, C., Marionneau, S., Ruvoen, N., Jiang, X., Lindblad, L., *et al.*, Human susceptibility and resistance to Norwalk virus infection. *Nat. Med.*, 9, 548–53, 2003. https://doi.org/10.1038/nm860.
72. Marionneau, S., Ruvoën, N., Le Moullac–Vaidye, B., Clement, M., Cailleau–Thomas, A., Ruiz–Palacois, G., *et al.*, Norwalk virus binds to histo-blood group antigens present on gastroduodenal epithelial cells of secretor individuals. *Gastroenterol.*, 122, 1967–77, 2002. https://doi.org/10.1053/gast.2002.33661.
73. Haga, K., Ettayebi, K., Tenge, V.R., Karandikar, U.C., Lewis, M.A., Lin, S.C., *et al.*, Genetic Manipulation of Human Intestinal Enteroids Demonstrates the Necessity of a Functional Fucosyltransferase 2 Gene for Secretor-Dependent Human Norovirus Infection. *MBio*, 11, 1–10, 2020. https://doi.org/10.1128/mbio.00251-20.
74. Lindesmith, L.C., Brewer-Jensen, P.D., Mallory, M.L., Jensen, K., Yount, B.L., Costantini, V., *et al.*, Virus–Host Interactions Between Nonsecretors

and Human Norovirus. *Cell Mol. Gastroenterol. Hepatol.*, 10, 245–67, 2020. https://doi.org/10.1016/j.jcmgh.2020.03.006.

75. Nordgren, J. and Svensson, L., Genetic Susceptibility to Human Norovirus Infection: An Update. *Viruses*, 11, 226, 2019. https://doi.org/10.3390/v11030226.
76. Esseili, M.A., Gao, X., Boley, P., Hou, Y., Saif, L.J., Brewer-Jensen, P., *et al.*, Human Norovirus Histo-Blood Group Antigen (HBGA) Binding Sites Mediate the Virus Specific Interactions with Lettuce Carbohydrates. *Viruses*, 11, 833, 2019. https://doi.org/10.3390/v11090833.
77. Kong, F., Saif, L.J., Wang, Q., Roles of bile acids in enteric virus replication. *Anim. Dis.*, 1, 2, 2021. https://doi.org/10.1186/s44149-021-00003-x.
78. Tenge, V., Ayyar, B.V., Ettayebi, K., Crawford, S.E., Hayes, N.M., Shen, Y.T., *et al.*, Bile acid-sensitive human norovirus strains are susceptible to sphingosine-1-phosphate receptor 2 inhibition. Wobus, C.E., (ed), *J. Virol.*, 98, e02020–23, 2024. https://doi.org/10.1128/jvi.02020-23.
79. van Breemen, R.B. and Li, Y., Caco-2 cell permeability assays to measure drug absorption. *Expert Opin. Drug Metab. Toxicol.*, 1, 175–85, 2005. https://doi.org/10.1517/17425255.1.2.175.
80. Angelis, I. and Turco, L., Caco-2 Cells as a Model for Intestinal Absorption. *Curr. Protoc. Toxicol.*, 47, 2011. https://doi.org/10.1002/0471140856.tx2006s47.
81. Pohl, C., Szczepankiewicz, G., Liebert, U.G., Analysis and optimization of a Caco-2 cell culture model for infection with human norovirus. *Arch. Virol.*, 167, 1421–143, 2022. https://doi.org/10.1007/s00705-022-05437-3.
82. Poole, N.M., Rajan, A., Maresso, A.W., Human Intestinal Enteroids for the Study of Bacterial Adherence, Invasion, and Translocation. *Curr. Protoc. Microbiol.*, 50, e55, 2018. https://doi.org/10.1002/cpmc.55.
83. Narwankar, R. and Esseili, M.A., Replication of Human Norovirus in Human Intestinal Enteroids Is Affected by Fecal Sample Processing. *Viruses*, 16, 241, 2024. https://doi.org/10.3390/v16020241.
84. Vashist, S., Bailey, D., Putics, A., Goodfellow, I., Model Systems for the Study of Human Norovirus Biology. *Future Virol.*, 4, 353–67, 2009. https://doi.org/10.2217/fvl.09.18.
85. Khan, A., Raess, M., de Angelis, M.H., Moving forward with forward genetics: A summary of the INFRAFRONTIER Forward Genetics Panel Discussion. *F1000Res.*, 10, 456, 2021. https://doi.org/10.12688/f1000research.25369.1.
86. Chen, H., Liu, H., Peng, X., Reverse genetics in virology: A double edged sword. *Biosaf. Heal.*, 4, 303–13, 2022. https://doi.org/10.1016/j.bsheal.2022.08.001.
87. Taube, S. and Wobus, C.E., A novel reverse genetics system for human norovirus. *Trends Microbiol.*, 22, 604–6, 2014. https://doi.org/10.1016/j.tim.2014.10.001.
88. Borg, C., Jahun, A.S., Thorne, L., Sorgeloos, F., Bailey, D., Goodfellow, I.G., Murine norovirus virulence factor 1 (VF1) protein contributes to viral fitness

during persistent infection. *J. Gen. Virol.*, 102, 001651, 2021. https://doi.org/10.1099/jgv.0.001651.
89. Ludwig-Begall, L.F., Lu, J., Hosmillo, M., de Oliveira-Filho, E.F., Mathijs, E., Goodfellow, I., *et al.*, Replicative fitness recuperation of a recombinant murine norovirus – *in vitro* reciprocity of genetic shift and drift. *J. Gen. Virol.*, 101, 510–22, 2020. https://doi.org/10.1099/jgv.0.001406.
90. Khoshandam, M., Soltaninejad, H., Mousazadeh, M., Hamidieh, A.A., Hosseinkhani, S., Clinical applications of the CRISPR/Cas9 genome-editing system: Delivery options and challenges in precision medicine. *Genes. Dis.*, 11, 268–82, 2024. https://doi.org/10.1016/j.gendis.2023.02.027.
91. Atmar, R.L. and Estes, M.K., The Epidemiologic and Clinical Importance of Norovirus Infection, Gastroenterol. *Clin. North Am.*, 35, 275–290, 2006. https://doi.org/10.1016/j.gtc.2006.03.001.
92. Vinjé, J., Advances in Laboratory Methods for Detection and Typing of Norovirus. *J. Clin. Microbiol.*, 53, 373–381, 2015. https://doi.org/10.1128/JCM.01535-14.
93. He, Z., Liu, B., Tao, Y., Li, C., Xia, M., Zhong, W., Jiang, X., Liu, H., Tan, M., Norovirus GII.17 Natural Infections in Rhesus Monkeys, China. *Emerg. Infect. Dis.*, 23, 316–319, 2017. https://doi.org/10.3201/eid2302.161077.
94. Bertolotti-Ciarlet, A., Chen, R., Estes, M.K., Venkataram Prasad, B.V., IV, 1. Structure of norwalk virus: the prototype human calicivirus. *Perspect. Med. Virol.*, 9, 455–466, 2003. https://doi.org/10.1016/S0168-7069(03)09027-X.
95. Hardy, M. and Estes, M., Completion of the Norwalk virus genome sequence. *Virus Genes.*, 12, 287–290, 1996. https://doi.org/10.1007/BF00284649.
96. Lo, M., Doan, Y.H., Mitra, S., Saha, R., Miyoshi, S., Kitahara, K., *et al.*, Comprehensive full genome analysis of norovirus strains from eastern India, 2017–2021. *Gut. Pathog.*, 16, 3, 2024. https://doi.org/10.1186/s13099-023-00594-5.
97. Karst, S.M. and Baric, R.S., What Is the Reservoir of Emergent Human Norovirus Strains? *J. Virol.*, 89, 5756–9, 2015. https://doi.org/10.1128/jvi.03063-14.
98. Doerflinger, S.Y., Weichert, S., Koromyslova, A., Chan, M., Schwerk, C., Adam, R., *et al.*, Human Norovirus Evolution in a Chronically Infected Host. *mSphere*, 2, e00489–20, 2017. https://doi.org/10.1128/msphere.00352-16.
99. Winder, N., Gohar, S., Muthana, M., Norovirus: An Overview of Virology and Preventative Measures. *Viruses*, 14, 2811, 2022. https://doi.org/10.3390/v14122811.
100. Lindesmith, L.C., Donaldson, E.F., Baric, R.S., Norovirus GII.4 Strain Antigenic Variation. *J. Virol.*, 85, 231–42, 2011. https://doi.org/10.1128/jvi.01364-10.
101. Zheng, D.P., Ando, T., Fankhauser, R.L., Beard, R.S., Glass, R.I., Monroe, S.S., Norovirus classification and proposed strain nomenclature. *Virol.*, 346, 312–23, 2006. https://doi.org/10.1016/j.virol.2005.11.015.

102. Parra, G.I., Emergence of norovirus strains: A tale of two genes. *Virus Evol.*, 5, vez048, 2019. https://doi.org/10.1093/ve/vez048.
103. Lee, B.E. and Pang, X.L., New strains of norovirus and the mystery of viral gastroenteritis epidemics. *Can. Med. Assoc. J.*, 185, 1381–2, 2013. https://doi.org/10.1503/cmaj.130426.
104. Lin, S.C., Bai, G.H., Lin, P.C., Chen, C.Y., Hsu, Y.H., Lee, Y.C., *et al.*, Molecular and Genetics-Based Systems for Tracing the Evolution and Exploring the Mechanisms of Human Norovirus Infections. *Int. J. Mol. Sci.*, 24, 9093, 2023. https://doi.org/10.3390/ijms24109093.
105. Parra, G.I., Abente, E.J., Sandoval-Jaime, C., Sosnovtsev, S.V., Bok, K., Green, K.Y., Multiple Antigenic Sites Are Involved in Blocking the Interaction of GII.4 Norovirus Capsid with ABH Histo-Blood Group Antigens. *J. Virol.*, 86, 7414–26, 2012. https://doi.org/10.1128/jvi.06729-11.
106. de Graaf, M., van Beek, J., Koopmans, M.P.G., Human norovirus transmission and evolution in a changing world. *Nat. Rev. Microbiol.*, 14, 421–33, 2016. https://doi.org/10.1038/nrmicro.2016.48.
107. Escudero-Abarca, B.I., Suh, S.H., Moore, M.D., Dwivedi, H.P., Jaykus, L.A., Selection, Characterization and Application of Nucleic Acid Aptamers for the Capture and Detection of Human Norovirus Strains. *PLoS One*, 9, e106805, 2014. https://doi.org/10.1371/journal.pone.0106805.
108. Nagasawa, K., Matsushima, Y., Motoya, T., Mizukoshi, F., Ueki, Y., Sakon, N., *et al.*, Genetic Analysis of Human Norovirus Strains in Japan in 2016–2017. *Front. Microbiol.*, 9, 2018. https://doi.org/10.3389/fmicb.2018.00001.

15

The Role of Human Intestinal Enteroids in Norovirus Research

Apeksha Shetty[1], Anushree Deshpande[2], Raushan Kumar Chaudhary[3] and Prakash Patil[4*]

[1]Department of Pharmacy Practice, NGSM Institute of Pharmaceutical Sciences (NGSMIPS), Nitte (Deemed to be University), Mangalore, Karnataka, India
[2]Department of Pharmacy Practice, KLE College of Pharmacy, Belagavi, KLE Academy of Higher Education and Research (KAHER), Belagavi, Karnataka, India
[3]Centre for Integrative Omics Data Science (CIODS), Yenepoya (Deemed to be University), Mangalore, Karnataka, India
[4]Central Research Laboratory (CRL), KS Hegde Medical Academy (KSHEMA), Nitte (Deemed to be University), Mangalore, Karnataka, India

Abstracts

Human intestinal enteroids (HIEs) are an extremely useful model for studies of small intestine pathology across all life stages. HIEs are three-dimensional tissue structures with stem cells obtained from human intestinal surgical specimens or biopsies, and they are the *in vivo* epithelial lining of the small intestine. Experimental studies demonstrated that gene expression profiles of HIEs in neonates and adults are different, in which genes related to bile acid binding and transport increase and genes related to biological adhesion and cell migration are decreased in infant HIEs. HIEs in infants may offer immense insights on the pathophysiological mechanisms of diseases specific to the neonatal and early childhood period. For instance, it has been demonstrated that the roles of HBGAs in producing HuNoVs as well as the role of bile acids in the producing HuSaVs are significant. HIEs are composed of absorptive cells, mucus-secreting cells, Paneth cells, and germ cells, which together maintain homeostasis and function of gastrointestinal epithelium. The paper emphasizes the role of ion transport associated in maintaining the physiological intestinal function, particularly the duodenal area. Studies of HIEs have also contributed to understanding the underlying ionic mechanisms of the human small intestine. Thus, the study results have important

**Corresponding author*: prakashpatil@nitte.edu.in

Sachin Kumar, Shriyansh Srivastava, Rishabha Malviya and Ranjit Sah (eds.) Emerging Infectious Disease: Unravelling Norovirus and Pandemic Dynamics, (413–446) © 2026 Scrivener Publishing LLC

inferences both for our understanding of small intestinal diseases in the lifespan and for the fact that the physiological state and developmental stage should be considered when modeling small intestinal diseases for their model construction. Compared with typical IHCs, HIEs are valuable resources to investigate small intestine diseases and unveil disease mechanisms.

Keywords: Intestinal enteroids, ligands, norovirus, protease, replication, tropism

15.1 Introduction to Human Intestinal Enteroids (HIEs)

15.1.1 Definition and Significance of HIEs

The human small digestive tract is a crucial organ for nutrient intake, digestion as well as eliminating waste products. Although it is vital, it is prone to a multitude of infectious and non-infectious diseases that may severely affect health at any age. Models that precisely model the postnatal morphological and functional development of human small intestine are of particular importance for furthering our knowledge of the pathophysiological basis of underlying illness and for evolution of targeted treatments [1]. Human intestinal enteroids (HIEs) are useful in the research of small intestine disease. These so-called three-dimensional tissue-engineered cultures recapitulate the *in vivo* epithelium heterogeneity and are generated from human intestine surgical tissues or biopsy stem cells. Since HIEs contain a heterogeneous mix of intestinal cell types, these are a perfect model system to investigate drug activity, host–virus interactions, cell–cell interactions, and gut physiology. However, there are scant studies employing HIEs of newborn and young children and few systematic comparisons of different models [2]. It has been demonstrated by researchers that HIE infants are distinct from HIE adults with respect to transcriptional, morphologic, and functional features, and thus this aspect is used to model diseases specific to the infant period. The analysis of this study revealed that HIEs in infants exhibit unique gene expression profiles, such as upregulated genes in bile acid binding and transport. By contrast, genes related to biological adhesion and cell migration were downregulated. Gene set enrichment analysis showed that the most overrepresented pathways were processing of RNA and ribonucleoprotein complex formation. In addition, transcript expression of secretory/absorptive cell markers was also investigated between infant and grown-up's HIEs. The findings demonstrated that in child HIEs compared with adult cultures, indicators for intestinal absorptive cells, intestinal mucosal epithelial cells, and enteroendocrine cells were shown at high levels. However, there were no evident discrepancies between the identification markers of Paneth and tuft cells. High MUC2 expression was also observed in baby samples, suggesting that the baby HIEs preserve the properties of the donor tissues [1, 2].

In conclusion, human intestinal enteroids are an important tool for studying small intestinal disorders across the lifespan. The results emphasize the need to consider the developmental age and physiological state in designing models to investigate small intestinal pathologies. Therefore, the application of infant HIEs can offer more information regarding the pathophysiological mechanisms of disease in pediatrics as well as early childhood years.

Foodborne gastroenteritis is a condition that affects almost all age groups and is mainly caused by the human noroviruses (HuNoVs). However, research for forming effective treatments was limited by the inability to develop a reproducible *in vitro* culture method. HIEs served as the main tool for studying HuNoV proliferation in human cells in recent years. HIEs are able to be grown to yield two-dimensional monolayers or multicellular tissue aggregates which completely recapitulate the small intestine tissue. Intestinal stem cells are extracted through an *ex-vivo* model of gastro-intestinal crypts, and this has enabled the study of how HuNoV interacts with the host and provided the evidence that some HuNoV strains need some necessary expression of histo-blood group antigen (HBGAs) for their successful initiation of an illness. Trans-replication of few enteric viruses, such as rotavirus, enterovirus, coronavirus, and adenovirus, has also been achieved using a similar HIE model [3]. Studies are being conducted to unmask the mysteries of human sapoviruses (HuSaVs) and the interaction between HuSaVs and human cells. Studying these viruses in a unique culture way of human intestinal epithelium, human intestinal enteroids (HIEs), has taken to important advances in knowledge regarding how HuSaVs can replicate and spread. The HIE model has been used to investigate the involvement of bile acids in HuSaV replication as well as the functional significance of certain cell surface determinants known as HBGAs in mediating virus entry into susceptible cells [1, 4]. The study found that HIEs support the growth of HuNoVs, with enterocytes being the main target for infection and replication. The findings suggest that HuNoVs can infect HIEs formed on all parts of the small intestine, which in return supports viral replication. The results also show that HuNoVs depend on HBGA expression for a productive infection [5].

In general, this shows that HIEs are a powerful tool for understanding HuNoV replication and thus help in designing effective interventions against these viruses. The findings also offered new insights into HuNoV–host interactions, thereby emphasizing the role of HBGA expression in an efficient infection.

15.1.2 The Components of HIEs: Enterocytes, Goblet Cells, Paneth Cells, and Germ Cells

The enterocytes, mucus-secreting cells, Paneth cells, and germ cells are some of the various cell types that make up the complex system known as the human intestine enteroids, refer Figure 15.1 to see various cell types present in the HIEs. In these cell types, they cooperate for the health and function of the gastrointestinal epithelium. The Lgr5+ crypt base columnar-stem cells (CBC-SC), seen at the base of the crypts, are in control of the continuous renewal of the epithelium through a tightly controlled cycle of cell division, growth, and death [6]. While CBC-SCs move toward the lumen of the intestine, the cells begin to differentiate into multiple epithelial cell subpopulations such as absorptive cells, secretory cells, enteroendocrine cells, Paneth cells, and tuft cells. In the lumen, epithelial absorptive cells take up ions, nutrients as well as water, whereas enteroendocrine cells, which control metabolism, gut motility, and satiety, secrete hormone peptides as response to the food and microbial stimuli [7]. Mucin proteins that are produced by goblet cells form the mucous barrier which keeps the

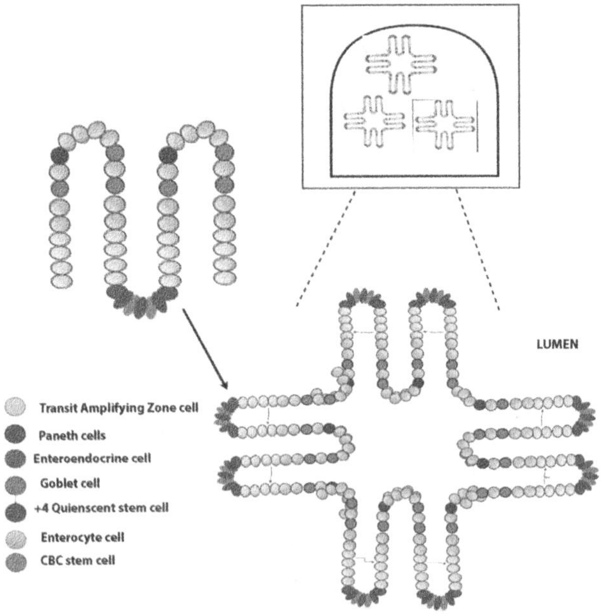

Figure 15.1 *Ex vivo* culture of gastro-intestinal stem cells from the base of crypts, which generate indefinitely propagating enteroid cultures. These cells are similar to those found in *in vivo* cultures.

mucosa away from a direct link with the intestinal microbiota. The Paneth cells are usually found in the small gastro-intestinal tract and secrete growth hormones and antimicrobial peptides that either regulate the composition of the microbiota or preserve the crypt. The tuft cells are a recently identified subtype of intestinal epithelial cells (IEC) and hypothesized to function in mucosal immunity and host–microbiome communication [2].

The wide variety of cell types which are present in human intestinal enteroids work together in order to maintain balance in the gastrointestinal tract as well as to ensure proper functioning of the enteric nervous system. The intestinal epithelium is known for its special morphology, comprising invaginations (crypts) that contain CBC-SCs. The epithelium is self-maintained by being regenerated through a well regulating cell cycle, cell growth, and death. Thus, the coordinated activity of the different IEC subtypes is very important not only to keep normal gut function but also to prevent diseases [1, 4, 6].

15.1.3 Activity of HIEs: Ion Transport and Immune Function

Absorption

The human small intestine is a very important organ for nutrient absorption and its epithelial layer is a main player in this process. HIEs is being a useful approach in studying the mechanisms of cell nutrient uptake and also in knowing how the dietary component alters as well as affects the intestinal function. These 3D cultures are morphologically similar to the small intestine with villi which rise the effective area for absorption. Absorption of dietary components completely depends on the interaction of several factors like the morphology of the tissue, gut motility and transit time, metabolic reactions, and also the presence of nutrient transporters in the intestinal membrane. In gastrointestinal diseases such as inflammatory bowel disease, disrupted and altered epithelial physiology was observed due to which the absorption processes are impaired. Studies have put forward various assays to investigate the effects of dietary components on human enteroids, thus focusing on the uptake, transport, and metabolism of labeled nutrients such as carnitine, palmitate, glucose, and fructose. These studies have shown the capacity of dietary components to modulate nutrient uptake as well as their metabolism on human intestinal enteroids (HIEs). For example, alcohol and its by-products have been shown to impair biotin absorption in enteroid monolayers.

There has also been research showing the influence of food ingredients on celiac disease, a disease characterized by insufficient intestinal absorption due to small intestinal villous atrophy caused by an immune response to gluten. In those studies, in which enteroid monolayers and enteroids derived from patient biopsies were taken, the epithelium showed little inflammation, decreased proliferation, and changed differentiation at baseline. These findings show the importance of epithelial-specific immune responses in celiac disease and thus suggest that defects in absorption may be initiated by these responses [6, 7]. Alterations in glucose homeostasis and the activities of carrier proteins such as sodium-glucose linked cotransporter-1 (SGLT1) are associated with obesity. Gastrointestinal immune cell infiltration and chronic low-grade systemic inflammation are additional hallmarks of this disease. Necrotizing enterocolitis and celiac disease are two small intestinal disorders that can cause short gut, a serious malabsorption condition. Recent evidence suggests that human organoids can serve as a helpful tool for tissue revival in short gut patients [8].

Thus, HIEs are useful for nutrient uptake and immune activity in the human small intestine. These self-assembling models maintain the morphology of the small intestine and can be used to investigate the influence of food ingredients on the absorption and metabolism of nutrients. The investigation of HIEs has further shown the pathogenesis of celiac disease and its related GI disorders, emphasizing the role of epithelial-selective immune responses in such conditions. Predicting the novel therapies for short bowel syndrome could be established through further research on application of HIEs to tissue regeneration of these patients [8].

Ion Transport

The human gastro-intestinal enteroid model has been extensively applied for understanding the intricate ion transport processes in the human small intestine. Recent works have emphasized the role of ion transport in physiological gastro-intestinal homeostasis especially in the duodenal area. The results suggest that ion transport is a multifaceted process involving various mechanisms, including Na+/H+ exchange, Cl^- secretion, and HCO_3^- secretion. In contrast to earlier reports, it has been found that both differentiated and undifferentiated enteroids are involved in fluid and anion secretion as well as Na+ absorption, proposing that, in fact, the crypt and villus compartments are not spatially separated. In addition, the information reveals that sodium–hydrogen exchanger 3 (NHE3) activity is maintained over a few passages of enteroids and that knockdown of NHE3 by small hairpin RNA (shRNA) inhibits acid resistance. The function of HCO_3^- and selective transporters in maintaining the pHi balance of specialized intestinal enteroids has been less investigated. The results indicate that while sodium potassium chloride cotransporter 1 (NKCC1) function is not essential for HCO_3^- transport in the presence of increased secretory conditions, cystic fibrosis transmembrane conductance regulator (CFTR) and electrogenic sodium bicarbonate cotransporter 1 (NBCe1) are essential [8, 9]. The significance of these findings lies in their ability to provide mechanistic insight into the pathophysiology of secretory diarrhea as well as in the evolution of new therapeutic interventions for dysregulated ion transport disorders. Targeting ion transport mediators pharmacologically or genetically using human intestinal enteroids is a powerful tool to investigate the impacts of ion transport on intracellular pH (pHi) homeostasis, abnormally rapid anion/fluid secretion, and neutral Na+ uptake. The present study also shows the role of ion transport to maintain the physiological integrity of the intestinal tract, thus providing a better understanding of the mechanisms of ion transport in the small intestine. The implications of these findings are significant, as they may give in to the evolution of new treatments as well as provide a greater understanding of the pathophysiology of gastrointestinal disorders. In addition, the study's results have inference for the evolution of new treatments for gastrointestinal diseases, such as cholera, which are characterized by dysregulation of ion transport mechanisms, and also the study offered novel understanding of ion transport in the human small intestinal tract and underline the role of ion transport for normal intestinal functions [8, 9].

Mostly, the results obtained from the study will provide important outcome in understanding about the movement of ions and its impact on maintaining the normal function of the human small intestine. Targeting ion transport machineries pharmacologically or genetically by human intestinal enteroids offers a significant resource to explore the role of ion transport on neutral sodium absorption and on intracellular pH balancing and on triggered anion/fluid secretion.

15.2 Components Required for Norovirus Activation

15.2.1 Receptors and Ligands Involved in Norovirus Binding and Entry

Human norovirus (HNV) genome comprises three open reading frames (ORFs) which encode to the main and smaller capsid proteins as well as unformed proteins. The VP1 protein's P domain has an extreme P2 subdomain that plays a critical role in binding to human HBGAs. The most variable part of the genome is the P2 subdomain, and this variation within the site shows the distinct binding patterns of HNV. Specifically, site A and site B are the two recognized sites in the P2 subdomain that facilitate the interaction between the virus and HBGAs; they ascertain the specificity and strength of the binding between the virus and HBGAs. The HBGA receptors that HNV binds to is basically a complex group of carbohydrate receptors, including A/B and Lewis residues. Different HNV strains bind to specific HBGAs, whereas some strains do not bind to any HBGAs at any point. The binding patterns between HNV and HBGAs are specific and vary across different strains [10]. The binding of HNV to HBGAs is thought to occur through a combination of hydrogen-bonding arrays and longer-range interactions between the P2 subdomain of the virus and the HBGAs. While site B is in charge of maintaining the interactions, site A is in charge of identifying the fucose categories or A/B ending sugar of the HBGAs. Changes in these locations affect the intensity and binding specificity of HNV–HBGA interactions. HNV binding sites on HBGAs exhibit strong intra-genogroup conservation, but not inter-genogroup conservation. For instance, although the precise binding locations are different for each genogroup, the GI and GII genogroups connect to HBGAs in the same region of the P2 domain. The GI.1 Norwalk virus binds to a single monomer of the P2 subdomain, while the GII.4 VA387 virus binds across the dimer. In addition to binding to HBGAs, HNV also interacts with other receptors such as heparan sulfate, a cell surface proteoglycan which is present on epithelial cells. This binding is mediated by the P domain and is involved in the initial attachment of the virus to the cell surface. Other ligands like sialic acid, glycerol, and tannic acid also have been shown to interact with HNV. Sialic acid is a general ligand for Caliciviridae family that blocks the HBGA binding of HNV. Glycerol and tannic acid have been shown to combine with the same HBGA binding site as fucose and has been identified to block HBGA interaction by HNV [10–26]. The genome of the human norovirus encodes a complex protein that assembles the viral capsid, which interacts with human carbohydrate receptors (so-called HBGAs). Two positions (site A and site B) within the virus' P2 subdomain, site A and site B, are known to be different in terms of the binding specificity and efficacy of norovirus–HBGA interaction. These sites have a key function in identifying the fucose-bearing HBGAs on the layer of intestinal epithelial cells. Veroviruses of various strains possess distinct binding profiles, with high conservation of the binding sites within genogroups, but not across genogroups. Understanding these interactions is very important for the evolution of effective treatments and vaccines against norovirus infections. The discovery of these ligands and their interactions with HNV has an important inference for the evolution of antiviral therapies against norovirus infections. Studying how these ligands bind to HNV helps for the invention of new therapeutic approaches by aiming at blocking HNV interaction with its receptor and also has implications on understanding the norovirus pathogenesis. Since HNV is known to bind multiple drugs and ligands, it is plausible to hypothesize that the virus has evolved to make use of these interactions to gain access to host cells [1, 10–26]. The complete process has been explained in Figure 15.2 which shows the Norovirus entry, attachment along with receptors and ligands which facilitate it.

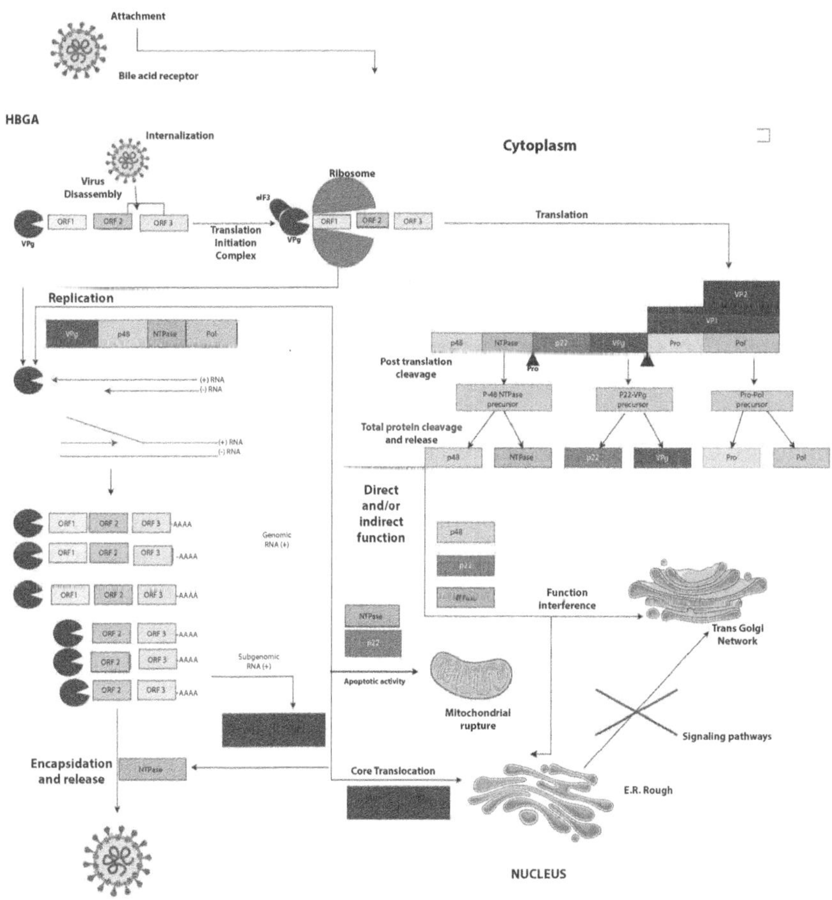

Figure 15.2 Diagram showing the pattern of norovirus entry and attachment, which is facilitated by host glycans, soluble cofactors, and divalent cations.

In conclusion, the human norovirus attachment to its receptors is a complex process where the interaction of multiple ligands and receptors is involved. The identification of these ligands and their binding to HNV is of great significance in developing antiviral therapy as well as to know more about the pathogenesis of norovirus.

In addition, studies have seen that HNV can adhere to certain foodstuffs, including leafy greens and molluscan shellfish, which are known to have frequent implication in norovirus outbreaks. The carbohydrate content of these foods is of greatest importance in determining the HNV binding selectivity. It was also shown that older leaves bind greater amounts to the norovirus GII.4 VLPs, whereas younger leaves bind less strongly when exposed to lectins that target sialic acid, galNAc, and GlcNAc. Few study evidences for the norovirus binding compounds are mentioned in the Table 15.1. In molluscan shellfish, HBGA residues have been identified in gastrointestinal tracts of oyster, whereas type A-like HBGAs have been identified in oysters and other shellfish. This tells that these shellfish may be able to bind to norovirus genogroups I and II. HNV binding again to same shellfish is further influenced by the specific carbohydrates like sialic acid-like structures. Additionally, HBGA or HBGA-like moieties have been seen in bacteria and studies have reported that some of the bacteria, e.g., Enterobacter cloacae, mainly associate with human norovirus in suspension. This interaction depends on HBGA-like residues on the bacterial surface, particularly type H-like moieties. Other bacteria, such as lactic acid bacteria, have also identified binding to human norovirus. In norovirus binding mechanism, ligand and receptor identification is being the important consequence in further explaining the norovirus pathogenies. The ability of HNV in binding to multiple receptors and ligands shows that the virus has evolved to exploit these interactions in order to get in the host cells. The knowledge of such interactions can be useful in understanding the pathogenesis of norovirus and their prevention or treatment [10–26].

Thus, a virus that binds to the human histo-blood group antigens most actively was also, in addition, found to attach to other receptors like heparan sulfate and to food products such as leafy greens and molluscan shellfish. The carbohydrate content of these food products determines the binding affinity specificity of HNV mainly. In addition, human norovirus binding by some bacteria has the potential such that these interactions may modulate the control of norovirus spread *via* food. These study on interactions also helps in the evolution of effective prevention and cure strategies for norovirus infection.

Table 15.1 Few binding compounds of norovirus with evidence.

Study reference	Binding compound	Evidence
Tamura *et al.* (2004) [14]	Heparan sulfate	Binding to VLPs, reduction of binding by heparinase I
Hansman *et al.* (2012) [17]	Citrate	Binding to VLPs, structural mimicry of HBGA binding, competition for binding site
Rydell *et al.* (2009) [19]	Sialic acid	Binding to VLPs, inhibition of HBGA binding, strain variation in binding
Bally *et al.* (2012) [27]	Monoglycosylceramides	Binding to VLPs, multivalent interaction required for binding
Zhang *et al.* (2012) [22]	Tannic acid	Inhibition of norovirus attachment to HBGAs, common in Chinese medicinal herbs
Liu *et al.* (2015) [21]	Glycerol	Binding to HBGA binding pocket, outcompeting Lewis antigens
Feng and Jiang (2007) [28]	Panel of 5,000 compounds	Identification of 14 potential inhibitors of norovirus binding to HBGAs
Rademacher *et al.* (2011) [29]	Compound 160 (inhibitory polymer)	Inhibition of HBGA binding to VLPs, identified in nuclear magnetic resonance-based screening

15.2.2 Enzymes that are Required for Norovirus Replication

Norovirus (HuNoV) protease, or 3CL Pro (NS6), is an important protease enzyme that cleaves the polyprotein, thus creating individual proteins for the replication of the virus. The protease has a molecular size which is approximately 19.4kDa of the Picornavirus 3C protease. The cleavage process includes early cleavage sites and post-cleavage sites, with the later ones cleaved at a much faster rate than the former. Optimal pH and the temperature of the HuNoV protease's proteolytic activity is about 8.6 and 37°C, respectively. The S2 substrate-adhering pocket as well as the active domain of GII.4 protease is different from those of the GI homologue protein, according to the protein's crystal structure. Together with its catalytic histidine, the conserved arginine in the GII.4 protease's active site limits the entry of inhibitors and substrates into the S2 pocket, increasing the enzyme's sensitivity to pH variations [30, 31].

ProPol, an intracellular precursor version of HuNoV protease, possesses both polymerase protease capabilities. ProPol functions best at a pH of 8.6 to 9.0, which is higher than the pH at which the viral protease functions on its own. Additionally, ProPol is further susceptible to $CaCl_2$ exposure than the protease itself. In addition to its protease activity, When Mn^{+2} is present, ProPol can nucleotidylate a VPg protein. It has also been shown that ProPol has protein breakdown ability that is equivalent or better than the viral protease at all polyprotein breakdown sites when the activity of protease was compared to that of ProPol utilizing peptides, indicating the five cleavage regions of the HuNoV polyprotein. ProPol can therefore act as the HuNoV protease's dominant form [30, 31]. The primary role of the HuNoV polymerase protein, sometimes referred to as Pol (NS7), is to promote the replication of the genome of the viral gene. HuNoV protease cleaves the polymerase at residues 1,190 as well as 1,699 of the polyproteins, yielding a matured protein having molecular weight of 56.8 kDa and 509 amino acids. Resolutions of 2.17 and 2.95 Å have been reported for the three-dimensional configuration of the NV Pol protein in both the absence and presence of metals, accordingly. P kinds of HuNoVs have been identified using the Pol region; they have been shown to be made up of 60 P types. The Pol area of GI, GII, GIII, and GIV diverged approximately 1,150 years ago, according to a molecular development analysis of the domain coding for the Pol protein. About 570 years ago, the Pol area of GII separated from GIV, giving rise to three major lineages of HuNoV GII. Furthermore, an analysis of various HuNoV GII gene sequences indicates that the HuNoV GII's development at $4.33 \times 10-3$ to $8.98 \times 10-3$ substitutions/site/year is present in the four Pol proteins. The GII Pol protein was shown to have multiple amino acid substitutions (39–107 sites) across various genotypes, and each genotype had a different set of negative selection sites. These suggests that the HuNoV GII Pol region is susceptible to rapid evolution involving multiple amino acid substitution events and that the rate of evolution may depend on the gene type [30, 31].

In conclusion, both the HuNoV protease and polymerase enzymes play important roles in the viral replication and in the pathological process by cleaving the polyprotein into individual proteins and thus replicating the viral genome.

15.2.3 Cellular Factors Which Help in Regulating Norovirus Replication and Tropism

Norovirus (HuNoV) binding to HBGAs is a critical tread in the life cycle of the virus. Polymorphic glucosaccharides (i.e., HBGAs) are presented on the lining of epithelial and red blood cells, and HuNoV possess specific binding sites on its VP1 protein that bind to these antigens. The binding site has shown high diversity among the norovirus genogroups, with each genogroup/genotype having its own unique HBGAs recognition pattern, whereas the ability of individuals to be infected with HuNoV is related to the expression of functional fucosyltransferase 2 (FUT2) enzyme which catalyzes the formation of HBGAs. Individuals who carry the FUT2 enzyme, secretors, have a higher rate of infecting with HuNoV compared to non-secretors. This is due to the absence in non-secretors of HBGAs that HuNoV is known to bind to [32–35]. Studies have shown that HBGA variability is linked to HuNoV infection, and different HuNoV strains have definite binding order to HBGAs. These patterns can be divided into two main groups: (a) those recognizing A/B and H antigens and (b) those recognizing Lewis antigens. On the other hand, the binding patterns of HuNoV strains with regard to their ability to target individuals are also implicated. The potential of the virus to bind to the HBGA is a very crucial part in the life cycle of the virus since it allows the entry of the virus into the host cell and its multiplication. Variation in the HBGA binding sites on HuNoV may help these viruses overcome the host immune system and to adapt to diverse surroundings. Therefore, HBGA acts as a regulator of HuNoV for its replication and tropism, as it presents a specific binding site for the virus. The heterogeneity of HBGA binding sites in general norovirus genogroups allows the virus to infect various hosts and circumvent host immunity [32–35].

15.3 Pathophysiology of Norovirus Infection

15.3.1 Mechanism of Infection of Norovirus Into the Gut: Attachment, Internalization, and Replication

Norovirus (NoV) infection leads to acute gastrointestinal syndrome including stomach flu or winter vomiting disease. The pathophysiology of NoV infection is a complex cascade of events that ultimately culminates in symptom onset. The primary step in the pathogenesis of NoV infection is

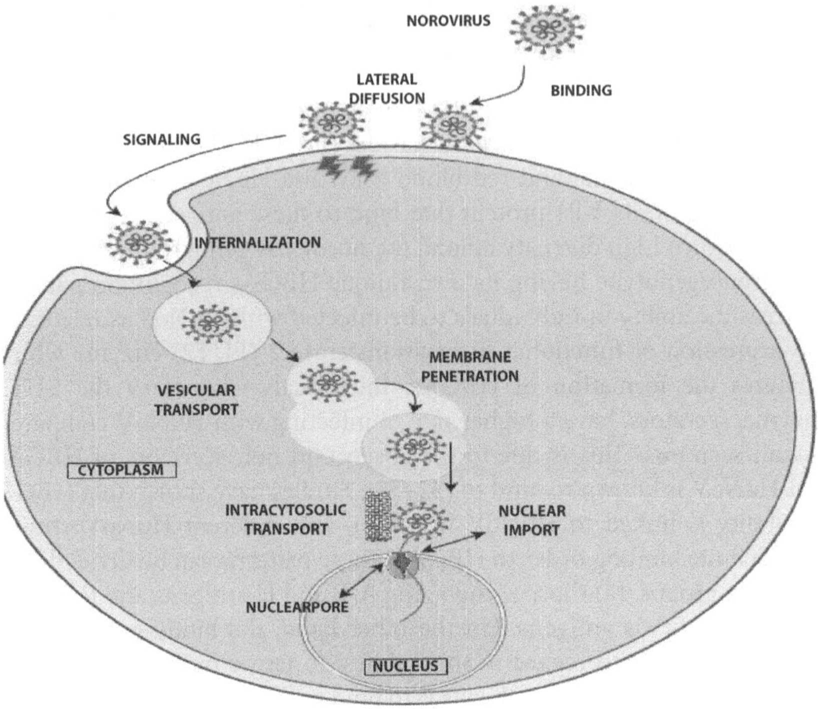

Figure 15.3 Endocytic entry of virus: A complex process which involves virus binding to the binding factors and receptors, followed by endocytic internalization of virus, continued by vesicular trafficking, penetration of the virus or capsids into the cytosolic compartment through the membrane, and finally transport into the nuclear pore complex and release of the viral gene into the nucleus of the cells.

viral binding to epithelial cells of the human gut. It is due to the expression on the surface of these cells of HBGAs, with which this VP1 binds [31, 36]. Viral attachment to the host cells is a crucial point in the process of infection because this step allows the virus in entering the host cells and beginning their replication. After the virus has attached to host cells, it is taken inside the vacuole through clathrin-mediated endocytosis. It is an intracellular process where the virus enters the vesicle which later combine with a lysosome, thus releasing the virus in the cytoplasm [37, 38]. The process of endocytic internalization of virus is simply explained through diagram in Figure 15.3. Then, the virus further uses its RNA-dependent RNA polymerase (RdRp) in viral gene replication and produces a series of viral mRNAs. A newly replicating viral RNA afterward gets encapsidated into newly formed viral particles, which are then released out by

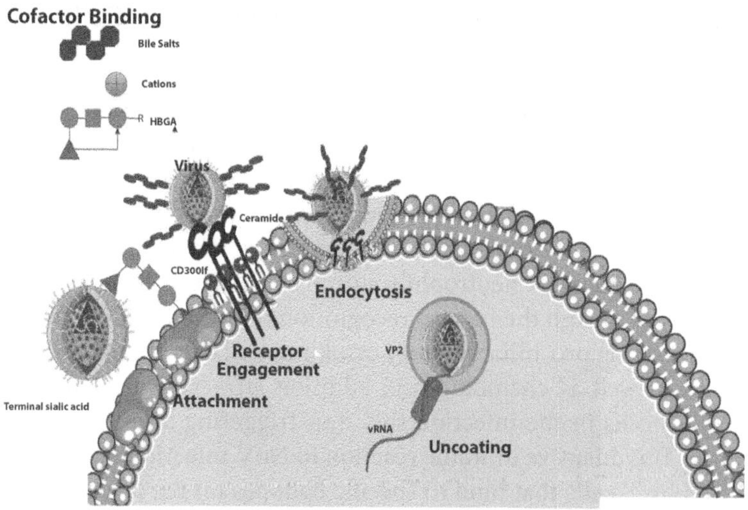

Figure 15.4 Diagram showing the replicative cycle of HuNoV and its protein function. The viral replication begins with the process of attachment of the P2 region to a receptor and host co-receptors, followed by internalization, disassembly, and, lastly, releasing the RNA into the cytoplasm. The viral proteins, including VP1, VP2, Pro, Pol, P48, NTPase, and VPg, play various roles in translation, polyprotein processing, protein migration, and replication.

budding process. During budding, the viral envelope is acquired from the host cell membrane, which contains HBGAs that were recognized by the virus during its initial binding step [37, 39]. The protein function during replicative cycle of HuNoV is shown in Figure 15.4. The newly formed viral particles are then free to infect other cells of the gut, which results in a rapid increase in viral load and symptom onset (diarrhea, vomiting, and abdominal cramp). The clinical picture of NoV infection is expected to result from the failure of the whole normal gut epithelial function and, as a consequence, from the dyshomeostasis of electrolyte and fluid balance [37, 40]. The clinical course of NoV infection can be varied as to age, immune status, and comorbid conditions. Generally, children <5 years old and older people are at increased risks for serious disease because of the immature or decrepit immune system [31, 36]. Apart from direct action on the intestinal epithelium, NoV infection is known to induce host immunological responses, including inflammation and tissue injury. Immune response is specified by the activating few immune cells, such as T-cells and macrophage, releasing the cytokines and chemokines for the viral clearance of the body [15].

15.3.2 Immune Response to Norovirus Infection: Cellular and Humoral Immunity

The immunological response to norovirus (NoV) infection is a complex immune process including cellular as well as humoral immunity. The main aim of the immunological response is to eliminate the virus, thus preventing its replication as well as subsequent disease development [38, 41, 42]. The early response to NoV infection is marked by innate immune cell activation, which includes neutrophils and macrophages, thus encountering these viruses through the pattern recognition receptors (PRRs) and subsequently secreting pro-inflammatory cytokines and chemokines [43]. These cytokines as well as chemokines in addition attracts T-cells and natural killer (NK) cells in the infection site, thus triggering an intense immune response. This adaptive immune reaction to NoV infection is largely intermediated by T-cells that bind to specific epitopes on the layer of the virus by major histocompatibility complex (MHC) molecules [44]. CD4+ T-cells play a main role in the immunological response coordination by triggering other immune cells, such as CD8+ T-cells and B-cells through cytokine production. CD8+ T-cells directly recognize and eliminate cells that are infected, whereas B-cells act in neutralizing the virus by producing antibodies [45]. These antibodies neutralize the virus load by preventing it from binding to host cells or by eliminating them through phagocytosis or complement-mediated lysis. The levels of IgA as well as IgG antibodies against norovirus is significantly seen as elevated in patients with acute gastroenteritis [46]. The immunological response to norovirus infection is also influenced by one's genetic background, as certain genetic variants can affect the responsiveness to infection and the disease severity [47]. For example, an individual with a specific allele of the HLA-DQ2 gene is seen as more sensitive to severe norovirus infection due to impaired T-cell recognition of the virus [48].

Thus, in brief, the immunological response to NoV infection is a multistore process that involves both cellular as well as humoral immunity. Innate immune cell activation and, in turn, adaptive immune cell activation, T-cells and B-cells, are very critical for the resolution of infection from the host's body including its replication. Antibody production against a specific antibody determinant on the virus surface is also important for viral inactivation and the spread of infection. Hence, more research is required to explain the mechanisms of immunological response to NoV infection and also to build a potent therapeutic strategy to treat NoV-associated disease.

15.3.3 Few Clinical Symptoms of Norovirus Infection: Diarrhea, Vomiting, and Dehydration

Norovirus infection is among the most endemic contagious diseases affecting millions throughout world. The clinical features of norovirus infection are seen in gastrointestinal symptomatic forms such as diarrhea and vomiting to dehydration as the most serious manifestation. According to WHO, NoV infection is responsible for around 267,000 hospitalizations and 180,000 deaths yearly in children aged <5 years [49]. The most frequently seen manifestations of NoV infection are diarrhea, vomiting, and abdominal pain and these may continue from several days to several weeks. The frequency and intensity of symptoms can be variable, but the disease is started typically with a fulminant diarrhea which may be watery or bloody and is often associated with vomiting. The extremity of the symptoms can also be influenced by the individual's age, with older adults and young children being more susceptible to severe illness due to their compromised immune systems [50]. Dehydration is a common complication of NoV infection, particularly in children <5 years and older adults [51]. Dehydration can occur rapidly due to the rapid loss of fluids and electrolytes due to diarrhea and vomiting. If untreated, it can progress to serious complications such as hypovolemia, multiorgan failure, and even death. The aggressiveness of a NoV infection may also be dependent on a few pre-existing medical conditions like chronic diseases like diabetes, renal cell disease, and heart disease. In addition, one with a compromised immune system, like those undergoing chemotherapy or taking immunosuppressive medications, are also at an increased risk of severe illness. The diagnosis of norovirus infection is mainly made based on clinical symptoms, laboratory tests, and epidemiological investigations. Laboratory tests such as stool testing for viral particles or antibodies as well as serological tests to detect IgM and IgG antibodies against NoV are used [49, 52].

Therefore, it is concluded that NoV infection is an extremely contagious and propagating disease that is able to produce severe gastrointestinal manifestations such as diarrhea and vomiting and dehydration. The extremity of the disease can be significantly different based on age, pre-existing medical conditions, and immune status. Dehydration is a very frequent complication of NoV infection which can lead to a fatal outcome if untreated. Early detection and management of NoV infection is very important to prevent the complications and to decrease the associated morbidity and mortality.

15.4 Norovirus Cell Tropism

15.4.1 Viral Tropism: Process of Norovirus Targeting Specific Cell Types in the Gut

Norovirus is a highly transmissible and widespread cause of acute gastrointestinal illness with various symptoms such as diarrhea, vomiting, and dehydration. The tropism of NoV is also required in studying the pathogenesis of NoV and also to design efficacious therapeutic interventions. Recent studies have illuminated the specific cell types targeted by NoV in the gut, which are required for its replication and dissemination. NoV has also been seen mainly to infect enterocytes of the small intestine where absorption of nutrients and electrolytes occur [34]. The virus uses particular receptors, such as HBGA and sulfated glycans, for their binding into the surface of the enterocyte and penetration into the cell [35]. Binding of NoV to these receptors activates a signaling transduction chain that further results in the activation of viral replication and transcription. The enterocytes of the small intestine are not the only cell types infected by NoV but it also targets other cells in the gut as well, such as goblet cells, Paneth cells, and enteroendocrine cells [52]. Mucin-secreting cells are crucial for the epithelial barrier functionality of the gut. Paneth cells secrete antimicrobial peptides, which contribute to the maintenance of the gut microbiome. Gut motility and secretion are both controlled by intestinal endocrine cells, which secrete hormones [53–56]. Specific cell targets of NoV are important for the pathological process of NoV infection and for the manifestation of the disease symptoms. Infection of enterocytes causes disruption of the gut epithelial barrier, leading to gastrointestinal (GI) ion and fluid overload into the lumen, resulting in diarrhea [33]. Irradiation of goblet cells causes changes in the production of mucin that may ultimately worsen the diarrhea. Infection of Paneth cells can result in changes in antimicrobial peptide production and therefore dysregulated gut microbiome and symptom occurrence. Recent research further demonstrated that NoV is also likely to infect cell types, excluding from the gut epithelium, and also dendritic cells and T-cells [57]. Infection of these immune cells can result in an abnormal immune response that promotes disease pathogenesis. Norovirus cell tropism, therefore, becomes a significant point for its pathogenesis and disease process. Thus, the virus binds to a special type of cell in the gut, such as intestinal absorptive cells, secreting cells, Paneth cells, and enteroendocrine cells that play a vital part in viral replication and spread. Infection in these cell types causes dysfunction in the gut epithelial barrier, destabilization of mucin secretion as well as antimicrobial peptide secretion, and pathogenic inflammatory response, all of which are involved in the development of disease symptoms. Thus, there is still a need for an explanation of the mechanisms of NoV cell tropism in order to discover effective therapeutic strategies for disease treatment.

15.4.2 Factors That Influence Norovirus Tropism: Age, Genetics, and Environment

Norovirus tropism is a multifactorial, complex, and paradoxical phenomenon which is influenced by various factors including age, genotype, and environment. Although NoV is highly prevalent, the cell tropism mechanism of the virus is still unclear and more research is still required in order to identify the determinants that influence the infecting ability to certain cell types. Age is one of the main factors of NoV cell tropism. NoV infection in children <5 years is the most common and the peak incidence of NoV infection is found between 6 and 24 months [34]. This must be the case because in young children the immune system is less developed and therefore the risk for infection is higher. In comparison, however, senior citizens may not be as easily hit by NoV infection because of lifetime pre-exposure to the virus and lifelong pre-existing immunity as well as other immune cells that can contribute to virus clearance. Genetics also plays an important role in discovering NoV cell tropism. Studies showed that certain genetic variants, such as HBGA gene, can influence an individual's susceptibility to NoV infection [58]. For example, humans having a particular genotype of the HBGA gene are more prone to NoV infection due to the presence of specified receptors on the intestinal absorptive cells of the stomach and intestine that can recognize the virus. Environmental factors can also influence NoV cell tropism. For example, a few studies demonstrated that exposure to some environmental pollutants (e.g., pesticide, heavy metal additives) could increase the infection with NoV [58]. In addition, poor hygiene along with poor sanitation can give rise to the spread of NoV infection. The gut microbiome, moreover, plays a crucial role in NoV cell tropism determination. Evidence for the gut microbiome composition role in the susceptibility to NoV infection in individuals has been reported [59]. For instance, humans with a certain gut microbial composition are more prone to NoV infection due to the enrichment of some bacterial strains which can synthesize metabolites which enhance viral multiplication. Besides these, diet or lifestyle factors also play a role in NoV cell tropism. For example, it has been seen that a high-fiber diet could decrease an individual's exposure to norovirus infection, which eases the beneficial bacteria growth in the gut [32].

In short, as we know, NoV cell tropism is a multifaceted phenomenon in which the tropism depends on several factors, which includes the gut microbiota, age, genetics, and environment. In order to fully understand the mechanisms through which NoV cell tropism is modulated and also to devise effective therapeutic strategies for both the prevention as well as for the treatment of NoV disease, further studies are required.

15.5 Role of Additional Gastrointestinal Components in Norovirus Infection

15.5.1 Mechanisms Through Which Bile Acids Enhance Norovirus Replication

The digestive tract is a complex and vital environment that is home for a diverse range of microorganisms, cells, and secretions that play important roles in maintaining the health as well as the goodness of the gut. Norovirus (NoV) is a highly infectious and prevalent cause of gastroenteritis whose replication and dissemination are controlled by a multifaceted set of factors in the gut. Although enterocyte involvement in NoV infection has been widely investigated, recent studies have evidenced the regulatory role of other gastrointestinal components, like bile acids, in the replication of NoV. Produced in the liver and then stored in the gallbladder, bile acids have its main role in the digestion as well as in the absorption of lipids and lipid-soluble vitamins. Nonetheless, it has also been discovered that bile acid itself actually enhances the replication of norovirus (NoV) by supplying an energy "nutrient" source to the virus [60]. The way by which bile acids act in promoting NoV replication is believed to be by activating the capsid protein of the virus, a prerequisite for viral replication and transcription. Activation of the capsid protein is thought to be mediated by the interplay between bile acids and the receptors they bind to on the surface of infected enterocytes [61]. This interaction activates a signaling cascade that finally results in the activation of capsid protein, which is, of course, essential for both viral replication and transcription. After the activation, the capsid protein binds to unique sites on the viral genome, and viral replication and transcription can be started. Meanwhile, other studies have shown that bile acids are able to facilitate NoV replication through regulating the host immune response [62]. The host's defensive immunity response is an important effector of host defense against viral infection, and its dysfunction may play a role in disease evolution. Bile acid has seen to affect the immune-related genes' expression, such as that of innate immunity and adaptive immunity. The mechanism through which bile acids regulate immune response is complex and multifactorial; however, recent research has emphasized a role for nuclear receptor signaling pathways in transducing these effects [63]. Activation of nuclear receptor signaling pathways can result in changes of gene expression, in turn affecting the potential immune response of the host. Apart from inducing NoV replication and immune response modulation, the immunomodulatory effects of the bile acids have also been reported to control the gut microbe composition [64]. The microbiome in the gut is a highly modular ecosystem that performs a vital contribution to the health and integrity of the gut. It has been demonstrated that bile acids can change the composition of the gut microbiome through the control of growth and survival of a specific bacterial genus.

Based on this, we can state that bile acids play a crucial role in NOV replication and spread by acting as a viral food source, in addition to mediating the host immune response. The pathways through which bile acids facilitate NoV replication are complex and multivalent, such as changes in transcriptional gene expression, activating the nuclear receptor signaling, and alteration of gut microflora. Additional works are still required to elucidate the mechanisms by which bile acid-enhanced NOV replication occurs and to formulate a successful therapeutic approach for NOV-associated diseases prophylaxis and treatment [4–6].

15.5.2 How the Bile Acids and the Ceramides Deal with the Entry Restriction for Norovirus

Entry restriction mechanism is an important block that prevents norovirus (NoV) from entering host cells. The capsid protein of the virus can break this barrier by interacting with cell surface receptors of the host cells that can bind to the viral capsid protein. Nevertheless, recent investigations revealed that the bile acids and ceramide can promote virus entry by inducing the expression of certain cell surface receptors in the outer layer of host cells [65]. Specific receptors have been identified as being upregulated by bile acids, thereby facilitating viral binding and entry [60]. Ceramide, a type of sphingolipid, has also been shown to play a role in modulating the entry restriction for NoV [61]. Ceramide alters the lipid content of host cell membrane, thus enhancing viral binding and entry. The modulatory effects of bile acids and ceramide on NoV entry restriction are complicated and heterogeneous with altered gene expression, activation of signaling pathway, and host cell membrane modulation whereas other studies have shown the significance of other GI components like the gut microbiome in NoV replication and spread [62, 63]. The gut microbiome is a highly complex and functional microhabitat and is crucial for gut health and gut integrity. Bile acids have also been reported to affect the composition of the gut microbiome *via* modulation of the growth and survival of some bacterial communities [63].

In addition to their action in regulating NoV replication and dissemination, both bile acids as well as ceramide have also been reported to act in controlling host immune response [64]. The immune response is a critical host defense against viral infection, and overactivation can play a role in disease pathogenesis. Bile acids have been shown to regulate the expression of immune-related genes, which include genes of innate and adaptive immune systems.

Concisely, bile acids and ceramide are important in controlling NoV replication and spread by both breaking the entry blockade of the virus and controlling the immune response of the host. The mechanisms of action by which these compounds impinge on NoV replication and spread are complex and pleiotropic and involve modifications to gene expression, activation of signaling pathways, modulation of host cell membrane, and regulation of gut microbial communities. Additional research is still required to clearly explain the mechanism of bile acid- and ceramide-dependent inhibition of NoV replication and their spread.

15.5.3 Potential Therapeutic Applications of Bile Acids and Ceramide in Treating Norovirus

Lithocholic acid (LCA), which is a bile acid, has shown a strong antiviral activity against norovirus. A study in Proceedings of the National Academy of Sciences (PNAS) states that the LCA could inhibit norovirus replication by targeting the viral gene as well as by inhibiting viral protein synthesis. Additionally, LCA was also shown to boost the immune response by activating natural killer cells as well as by promoting the generation of pro-inflammatory cytokines [64, 66]. Sphingolipid ceramide has also been shown to be a potential therapeutic agent against norovirus. The Journal of Medicinal Food published a study in which treatment with ceramide was shown to prevent norovirus replication by triggering apoptosis and also suppressing the activity of the major signaling cascades. In the study, it was also shown that ceramide treatment could attenuate the severity of norovirus-induced diarrhea and vomiting in mice. The mechanism by which bile acids and ceramide exhibit antiviral activity against norovirus is not yet known, yet their ability to interfere with some specific viral proteins or viral genomic targets are expected, leading to the impairment of viral replication and spreading. Also, some bile acids and ceramide are speculated to further regulate the host IMR, upregulating pro-inflammatory cytokines and NK cell activation [64, 66]. The therapeutic potential of the bile acids and also the ceramide in treating norovirus is immense. These compounds could serve not only as pure antiviral agents but also as adjuvants to other treatments in order to increase their potency. In addition, these protective agents seemed to be effective against norovirus infections, especially in high-risk areas, which include healthcare institutions and schools [64, 66].

15.6 Antiviral Compound Screening Using HIEs

15.6.1 Screening Methods of Antiviral Compounds Using HIEs

High-throughput screening (HTS) of HIEs has become a very inventive strategy in finding antiviral agents that work against norovirus as well as other gastrointestinal pathogens [67].

The following are a few approaches to antiviral compound screens *via* HIEs:

1. HIEs as a screening tool
   HIEs are three-dimensional, self-assembling epithelial models of the human gastro-intestinal tract. They consist of enteric stem cells, absorptive cells, and Paneth cells, which are enclosed in a matrix of collagen and extracellular matrix proteins. HIEs get infected by norovirus and thus are used in investigating the viral life cycle and screening for antivirals.
2. Screening methods
   Different approaches have been used in order to screen antivirals in HIEs, namely:

   1. Infection-based assay: In this method, the ability of the compound to inhibit viral replication in infected HIEs is assessed by determining viral RNA or protein levels.

2. Cell-based assay: In this method, HIEs are made to expose to the compound prior to or after infection and then the impact of this on viral replication or cell viability is assessed thoroughly.
3. Label-free imaging: In this technique, label-free imaging modalities such as phase contrast microscopy or interferometry are being used to track the morphological as well as cell viability changes with the compound in HIE.
4. Chemiluminescence-based assay: Specifically, here the chemiluminescent substrate is used in detecting the activity of the reporter enzyme, whose production is encouraged by covalent blockage of viral replication by the compound.
5. Compound identification and validation: The compounds isolated as emerging antiviral agents by HIEs are frequently validated by additional assays such as:

    a) *In vitro* antiviral assay method: The compound shows that its antiviral activity is supported by another in a cell-based assay using another cell line/species.
    b) *In vivo* antiviral assay method: Here the efficacy of the compound is confirmed through a study of norovirus infection in an animal model.
    c) Mechanism of action studies: Biochemical assays that include enzyme assay or analysis of protein–protein interaction are investigated to know the mechanism of action of the compound.

15.6.2 Examples of Antiviral Compounds Discovered by Using HIEs

Many studies have already used HIEs to identify antiviral agents against norovirus, namely:

1. Zhang *et al.* (2022) had screened 12 inhibitors of norovirus replication with a label-free imaging-based assay in HIEs [68].
2. Wang *et al.* (2022) had detected 10 compounds which inhibited norovirus replication in HIEs by a chemiluminescence-based assay [69].

15.6.3 Future Directions for Antiviral Compound Screening Using HIEs

Producing antiviral agents with human intestinal enteroids (HIEs) has been a successful step in eliminating norovirus and other gastrointestinal infections. Nevertheless, there are still few issues and potential directions for future research. Currently, a review of the trends in the antiviral compound screening application of HIEs and associated references is presented.

The Expanding Scope of Antiviral Compounds

Future research should include the enlargement of the number of antiviral compounds screened by means of using HIEs. These can include the following:

1. Screening of natural products: The plant extracts and those traditional medicine containing natural products have shown an antiviral effect against norovirus and other gastrointestinal pathogens.
2. Screening of repurposed drugs: The repurposed drugs or the drugs previously approved for other applications have antiviral activity against norovirus and other gastrointestinal pathogens.
3. Screening of small molecules: Small molecular particles, even those with known mechanisms of action, may be screened against norovirus and GI pathogens.

Enhancing the Efficiency of Screening Assays

Future work should include optimization of screening assays with HIEs, like:

1. Automation of screening assays: Automation of screening assays facilitates a reduction of time and also increases the speed in the process of compound screening.
2. Development of high-content imaging assays: High-content imaging assays gives a greater understanding of a compound as well as the mechanism of action.
3. Integration of machine learning algorithms: Machine learning algorithms may be implemented into screening assays in order to improve the accuracy as well as the timeliness of screening for hits.

Understanding the Antiviral Compounds' Mechanism

Future studies should focus on identifying mechanisms of antiviral compounds being selected in HIEs. These include the following:

1. Mechanistic studies: This investigation can unravel the mechanism of antiviral drugs, thus helping to identify prospective targets for therapy.
2. Structural biology studies: Structural biology research studies provide information about the molecular mechanisms of antiviral drugs which helps in identifying targets against therapies.

15.7 Future Perspectives

Recent advances in human intestinal enteroids (HIEs) have made way for new insights for antiviral drug discovery and development of antivirus agents for norovirus as well as other gastrointestinal pathogens. The possibility of HIEs to bring change of the antiviral research is immense and future scenarios are encouraging. The effect of technological advances in new screening assays, automation technologies, and machine learning algorithms, also discovering and developing novel antiviral drugs, is expected to evolve rapidly. The HIEs' inclusion will allow researchers to screen and verify antiviral compounds in a timely manner, which is a crucial step in fighting against norovirus and other gastrointestinal viruses. Advancements in the creation of novel screening assays and also the automation capabilities that rely on HIEs will allow researchers to quickly screen an appreciable number of compounds for their ability to inhibit viruses, thus helping in rapidly discovering the candidates. Application of HIEs will also allow scientists to create focused therapeutics that are specifically engineered to fight against norovirus and other gastrointestinal infectious agents which require the elucidation of the mechanisms by which such pathogens invade and multiply and their action against discovered compounds. Furthermore, using the genetic information for particular individuals who have a greater risk for infection and their response to the treatment, HIEs will also facilitate the evolution of patient-specific medications which are derived from the genetic profile of the patient in order to prevent and treat norovirus infection. Thus, there shows the favor in screening the potential of antiviral compounds that rely on HIEs. Now, with the evolution of novel screening assays, automation technology, and machine learning algorithms, we anticipate an accelerated boom in the discovery and development of novel antivirals for norovirus as well as other GI pathogens. The application of HIEs will allow the researchers to design targeted therapies that are uniquely tailored to fight against these pathogens and would eventually bring us new and effective treatments for patients.

15.8 Conclusion

In summary, HIEs provide a novel model for studying norovirus infection and gastrointestinal disease. Their unique ability to replicate the human gut environment allows for detailed investigations into viral replication, host interactions, and immune responses. The consequence to antiviral drug discovery is substantial in that HIEs can be exploited in some way to allow high-throughput screening of compounds useful to inhibit norovirus and other GI pathogens. Future studies should concentrate on increasing the size of the compound libraries, improving the screening protocols, and understanding the mechanisms of action, in turn providing the groundwork for the targeted therapy that improves prevention/treatment strategies for norovirus infections.

Abbreviations

HIEs	human intestinal enteroids
LCA	lithocholic acid
NoV	norovirus
HBGA	histo-blood group antigen
NK	natural killer
MHC	major histocompatibility complex
FUT2	fucosyltransferase 2 enzyme
SGLT-1	sodium-glucose linked cotransporter-1
HIEs	human intestinal enteroids
CBC-SC	crypt base columnar-stem cells
HuSaVs	human sapoviruses

AI disclosure statement: The author acknowledges the use of AI-based tools for minor language editing or grammatical correction. However, all intellectual contributions, analyses, and conclusions presented in this chapter are entirely the author's own.

References

1. Zachos, N.C., Kovbasnjuk, O., Foulke-Abel, J., In, J., Blutt, S.E., de Jonge, H.R., Estes, M.K., Donowitz, M., Human Enteroids/Colonoids and Intestinal Organoids Functionally Recapitulate Normal Intestinal Physiology and Pathophysiology. *J. Biol. Chem.*, 291, 8, 3759–66, 2016 Feb 19.
2. Adeniyi-Ipadeola, G.O., Hankins, J.D., Kambal, A., Zeng, X.L., Patil, K., Poplaski, V., Bomidi, C., Nguyen-Phuc, H., Grimm, S.L., Coarfa, C., Stossi, F., Infant and adult human intestinal enteroids are morphologically and functionally distinct. *bioRxiv*, 15, 8, e0131624, 2024.
3. Ettayebi, K., Kaur, G., Patil, K., Dave, J., Ayyar, B.V., Tenge, V.R., Neill, F.H., Zeng, X.L., Speer, A.L., Di Rienzi, S.C., Britton, R.A., Insights into Human Norovirus Cultivation in Human Intestinal Enteroids. *bioRxiv*, 24, 2024–05, 2024 May.
4. Euller-Nicolas, G., Le Mennec, C., Schaeffer, J., Zeng, X.L., Ettayebi, K., Atmar, R.L., Le Guyader, F.S., Estes, M.K., Desdouits, M., Human sapovirus replication in human intestinal enteroids. *J. Virol.*, 97, 4, e00383–23, 2023 Apr 27.
5. Narwankar, R. and Esseili, M.A., Replication of Human Norovirus in Human Intestinal Enteroids Is Affected by Fecal Sample Processing. *Viruses*, 16, 2, 241, 2024 Feb 2.
6. Kovbasnjuk, O., Zachos, N.C., In, J., Foulke-Abel, J., Ettayebi, K., Hyser, J.M., Broughman, J.R., Zeng, X.L., Middendorp, S., De Jonge, H.R., Estes, M.K., Human enteroids: preclinical models of non-inflammatory diarrhea. *Stem Cell Res. Ther.*, 4, 1–4, 2013 Dec.
7. Tian, C.M., Yang, M.F., Xu, H.M., Zhu, M.Z., Yue, N.N., Zhang, Y., Shi, R.Y., Yao, J., Wang, L.S., Liang, Y.J., Li, D.F., Stem cell-derived intestinal organoids: a novel modality for IBD. *Cell Death Discov.*, 9, 1, 255, 2023 Jul 21.
8. Camilleri, M., Lyle, B.J., Madsen, K.L., Sonnenburg, J., Verbeke, K., Wu, G.D., Role for diet in normal gut barrier function: developing guidance within the framework of food-labeling regulations. *Am. J. Physiol. Gastrointest. Liver Physiol.*, 317, 1, G17–39, 2019 Jul 1.
9. Foulke-Abel, J., In, J., Yin, J., Zachos, N.C., Kovbasnjuk, O., Estes, M.K., De Jonge, H., Donowitz, M., Human enteroids as a model of upper small intestinal ion transport physiology and pathophysiology. *Gastroenterol.*, 150, 3, 638–49, 2016 Mar 1.
10. Almand, E.A., Moore, M.D., Jaykus, L.A., Norovirus binding to ligands beyond histo-blood group antigens. *Front. Microbiol.*, 8, 2549, 2017 Dec 21.
11. Miura, T., Sano, D., Suenaga, A., Yoshimura, T., Fuzawa, M., Nakagomi, T., Nakagomi, O., Okabe, S., Histo-blood group antigen-like substances of human enteric bacteria as specific adsorbents for human noroviruses. *J. Virol.*, 87, 17, 9441–51, 2013 Sep 1.
12. Murakami, K., Kurihara, C., Oka, T., Shimoike, T., Fujii, Y., Takai-Todaka, R., Park, Y., Wakita, T., Matsuda, T., Hokari, R., Miura, S., Norovirus binding to

intestinal epithelial cells is independent of histo-blood group antigens. *PLoS One*, 8, 6, e66534, 2013 Jun 14.

13. Tamura, M., Natori, K., Kobayashi, M., Miyamura, T., Takeda, N., Genogroup II noroviruses efficiently bind to heparan sulfate proteoglycan associated with the cellular membrane. *J. Virol.*, 78, 8, 3817–26, 2004 Apr 15.
14. Tamura, M., Natori, K., Kobayashi, M., Miyamura, T., Takeda, N., Inhibition of attachment of virions of Norwalk virus to mammalian cells by soluble histone molecules. *Arch. Virol.*, 148, 1659–70, 2003 Sep.
15. Tian, P., Bates, A.H., Jensen, H.M., Mandrell, R.E., Norovirus binds to blood group A-like antigens in oyster gastrointestinal cells. *Lett. Appl. Microbiol.*, 43, 6, 645–51, 2006 Dec 1.
16. Tian, P., Engelbrektson, A.L., Jiang, X., Zhong, W., Mandrell, R.E., Norovirus recognizes histo-blood group antigens on gastrointestinal cells of clams, mussels, and oysters: a possible mechanism of bioaccumulation. *J. Food Prot.*, 70, 9, 2140–7, 2007 Sep 1.
17. Hansman, G.S., Shahzad-ul-Hussan, S., McLellan, J.S., Chuang, G.Y., Georgiev, I., Shimoike, T., Katayama, K., Bewley, C.A., Kwong, P.D., Structural basis for norovirus inhibition and fucose mimicry by citrate. *J. Virol.*, 86, 1, 284–92, 2012 Jan 1.
18. Han, L., Tan, M., Xia, M., Kitova, E.N., Jiang, X., Klassen, J.S., Gangliosides are ligands for human noroviruses. *J. Am. Chem. Soc.*, 136, 36, 12631–7, 2014 Sep 10.
19. Rydell, G.E., Nilsson, J., Rodriguez-Diaz, J., Ruvoën-Clouet, N., Svensson, L., Le Pendu, J., Larson, G., Human noroviruses recognize sialyl Lewis x neoglycoprotein. *Glycobiol.*, 19, 3, 309–20, 2009 Mar 1.
20. Wegener, H., Mallagaray, A., Schöne, T., Peters, T., Lockhauserbäumer, J., Yan, H., Uetrecht, C., Hansman, G.S., Taube, S., Human norovirus GII. 4 (MI001) P dimer binds fucosylated and sialylated carbohydrates. *Glycobiology.*, 27, 11, 1027–37, 2017 Nov 1.
21. Liu, W., Chen, Y., Jiang, X., Xia, M., Yang, Y., Tan, M., Li, X., Rao, Z., A unique human norovirus lineage with a distinct HBGA binding interface. *PLoS Pathog.*, 11, 7, e1005025, 2015 Jul 6.
22. Zhang, X.F., Dai, Y.C., Zhong, W., Tan, M., Lv, Z.P., Zhou, Y.C., Jiang, X., Tannic acid inhibited norovirus binding to HBGA receptors, a study of 50 Chinese medicinal herbs. *Bioorg. Med. Chem.*, 20, 4, 1616–23, 2012 Feb 15.
23. Jiang, X., Huang, P., Zhong, W., Tan, M., Farkas, T., Morrow, A. L., Newburg, D. S., Ruiz-Palacios, G. M., Pickering, L. K., Human milk contains elements that block binding of noroviruses to human histo-blood group antigens in saliva. *J. Infect. Dis.*, 190, 10, 1850-9, 2004.
24. Shang, J., Piskarev, V.E., Xia, M., Huang, P., Jiang, X., Likhosherstov, L.M., Novikova, O.S., Newburg, D.S., Ratner, D.M., Identifying human milk glycans that inhibit norovirus binding using surface plasmon resonance. *Glycobiol.*, 23, 12, 1491–8, 2013 Dec 1.

25. Weichert, S., Koromyslova, A., Singh, B.K., Hansman, S., Jennewein, S., Schroten, H., Hansman, G.S., Structural basis for norovirus inhibition by human milk oligosaccharides. *J. Virol.*, 90, 9, 4843–8, 2016 May 1.
26. Esseili, M.A., Wang, Q., Saif, L.J., Binding of human GII. 4 norovirus virus-like particles to carbohydrates of romaine lettuce leaf cell wall materials. *Appl. Environ. Microbiol.*, 78, 3, 786–94, 2012 Feb 1.
27. Bally, M., Rydell, G.E., Zahn, R., Nasir, W., Eggeling, C., Breimer, M.E., Svensson, L., Höök, F., Larson, G., Norovirus GII. 4 virus-like particles recognize galactosylceramides in domains of planar supported lipid bilayers. *Angew Chem. Int. Ed. Engl.*, 51, 48, 12020–4, 2012 Nov 26.
28. Feng, X. and Jiang, X., Library screen for inhibitors targeting norovirus binding to histo-blood group antigen receptors. *Antimicrob. Agents Chemother.*, 51, 1, 324–31, 2007 Jan.
29. Rademacher, C., Guiard, J., Kitov, P.I., Fiege, B., Dalton, K.P., Parra, F., Bundle, D.R., Peters, T., Targeting norovirus infection—multivalent entry inhibitor design based on NMR experiments. *Chemistry-A Eur. J.*, 17, 27, 7442–53, 2011 Jun 27.
30. Zhao, B., Hu, L., Song, Y., Patil, K., Ramani, S., Atmar, R.L., Estes, M.K., Prasad, B.V., Norovirus protease structure and antivirals development. *Viruses.*, 13, 10, 2069, 2021 Oct 14.
31. Campillay-Véliz, C.P., Carvajal, J.J., Avellaneda, A.M., Escobar, D., Covián, C., Kalergis, A.M., Lay, M.K., Human norovirus proteins: implications in the replicative cycle, pathogenesis, and the host immune response. *Front. Immunol.*, 11, 961, 2020 Jun 16.
32. Strine, M. S., Alfajaro, M. M., Graziano, V. R., Song, J., Hsieh, L. L., Hill, R., Guo, J., VanDussen, K. L., Orchard, R. C., Baldridge, M. T., Lee, S., Wilen, C. B., Tuft-cell-intrinsic and -extrinsic mediators of norovirus tropism regulate viral immunity. *Cell Rep.*, 41, 6, 111593, 2022.
33. Wobus, C.E., The dual tropism of noroviruses. *J. Virol.*, 92, 16, 10–128, 2018 Aug 15.
34. Lee, S., Wilen, C.B., Orvedahl, A., McCune, B.T., Kim, K.W., Orchard, R.C., Peterson, S.T., Nice, T.J., Baldridge, M.T., Virgin, H.W., Norovirus cell tropism is determined by combinatorial action of a viral non-structural protein and host cytokine. *Cell Host Microbe*, 22, 4, 449–59, 2017 Oct 11.
35. Lee, S., Norovirus cell tropism: The road to uncovering its secret hideout. *Cell Host Microbe*, 30, 4, 454–7, 2022 Apr 13.
36. Karandikar, U.C., Crawford, S.E., Ajami, N.J., Murakami, K., Kou, B., Ettayebi, K., Papanicolaou, G.A., Jongwutiwes, U., Perales, M.A., Shia, J., Mercer, D., Detection of human norovirus in intestinal biopsies from immunocompromised transplant patients. *J. Gen. Virol.*, 97, 9, 2291–300, 2016 Sep.
37. Graziano, V.R., Wei, J., Wilen, C.B., Norovirus attachment and entry. *Viruses*, 11, 6, 495, 2019 May 30.
38. Roth, A.N. and Karst, S.M., Norovirus mechanisms of immune antagonism. *Curr. Opin. Virol.*, 16, 24–30, 2016 Feb 1.

39. Marsh, M. and Helenius, A., Virus entry: open sesame. *Cell*, 124, 4, 729–40, 2006 Feb 24.
40. Lay, M.K., Atmar, R.L., Guix, S., Bharadwaj, U., He, H., Neill, F.H., Sastry, K.J., Yao, Q., Estes, M.K., Norwalk virus does not replicate in human macrophages or dendritic cells derived from the peripheral blood of susceptible humans. *Virol.*, 406, 1, 1–1, 2010 Oct 10.
41. Flood, P., Hanrahan, N., Nally, K., Melgar, S., Human intestinal organoids: Modeling gastrointestinal physiology and immunopathology—current applications and limitations. *Eur. J. Immunol.*, 54, 2, 2250248, 2024 Feb.
42. Papafragkou, E., Hewitt, J., Park, G.W., Greening, G., Vinje, J., Challenges of culturing human norovirus in three-dimensional organoid intestinal cell culture models. *PLoS One*, 8, 6, e63485, 2013 Jun 3.
43. Newman, K.L. and Leon, J.S., Norovirus immunology: of mice and mechanisms. *Eur. J. Immunol.*, 45, 10, 2742–57, 2015 Oct.
44. Costantini, V.P., Cooper, E.M., Hardaker, H.L., Lee, L.E., DeBess, E.E., Cieslak, P.R., Hall, A.J., Vinjé, J., Humoral and mucosal immune responses to human norovirus in the elderly. *J. Infect. Dis.*, 221, 11, 1864–74, 2020 May 11.
45. Mboko, W.P., Chhabra, P., Valcarce, M.D., Costantini, V., Vinjé, J., Advances in understanding of the innate immune response to human norovirus infection using organoid models. *J. Gen. Virol.*, 103, 1, 001720, 2022 Jan 25.
46. Hassan, E. and Baldridge, M.T., Norovirus encounters in the gut: multifaceted interactions and disease outcomes. *Mucosal Immunol.*, 12, 6, 1259–67, 2019 Nov 1.
47. Simmons, K., Gambhir, M., Leon, J., Lopman, B., Duration of immunity to norovirus gastroenteritis. *Emerg. Infect. Dis.*, 19, 8, 1260, 2013 Aug.
48. Ponterio, E., Mariotti, S., Tabolacci, C., Ruggeri, F.M., Nisini, R., Virus like particles of GII. 4 norovirus bind Toll Like Receptors 2 and 5. *Immunol. Lett.*, 215, 40–4, 2019 Nov 1.
49. Fardah Athiyyah, A., Shigemura, K., Kitagawa, K., Agustina, N., Darma, A., Ranuh, R., Raharjo, D., Shirakawa, T., Fujisawa, M., Marto Sudarmo, S, Clinical manifestation of norovirus infection in children aged less than five years old admitted with acute diarrhea in Surabaya, Indonesia: a cross-sectional study. *F1000Res.*, 8, 2130, 2019.
50. Michail, S.K., Stolfi, A., Johnson, T., Onady, G.M., Efficacy of probiotics in the treatment of pediatric atopic dermatitis: a meta-analysis of randomized controlled trials. *Ann. Allergy Asthma Immunol.*, 101, 5, 508–16, 2008 Nov 1.
51. Li, T.T., Xu, Q., Liu, M.C., Wang, T., Che, T.L., Teng, A.Y., Lv, C.L., Wang, G.L., Hong, F., Liu, W., Fang, L.Q., Prevalence and Etiological Characteristics of Norovirus Infection in China: A Systematic Review and Meta-Analysis. *Viruses*, 15, 6, 1336, 2023 Jun 7.
52. Green, K.Y., Kaufman, S.S., Nagata, B.M., Chaimongkol, N., Kim, D.Y., Levenson, E.A., Tin, C.M., Yardley, A.B., Johnson, J.A., Barletta, A.B., Khan, K.M., Human norovirus targets enteroendocrine epithelial cells in the small intestine. *Nat. Commun.*, 11, 1, 2759, 2020 Jun 2.

53. Staab, J.F., Lemme-Dumit, J.M., Latanich, R., Pasetti, M.F., Zachos, N.C., Co-culture system of human enteroids/colonoids with innate immune cells. *Curr. Protoc. Immunol.*, 131, 1, e113, 2020 Dec.
54. Sato, T., Stange, D.E., Ferrante, M., Vries, R.G., Van Es, J.H., Van Den Brink, S., Van Houdt, W.J., Pronk, A., Van Gorp, J., Siersema, P.D., Clevers, H., Long-term expansion of epithelial organoids from human colon, adenoma, adenocarcinoma, and Barrett's epithelium. *Gastroenterol.*, 141, 5, 1762–72, 2011 Nov 1.
55. Fujii, M., Matano, M., Toshimitsu, K., Takano, A., Mikami, Y., Nishikori, S., Sugimoto, S., Sato, T., Human intestinal organoids maintain self-renewal capacity and cellular diversity in niche-inspired culture condition. *Cell Stem Cell*, 23, 6, 787–93, 2018 Dec 6.
56. Sato, T., Van Es, J.H., Snippert, H.J., Stange, D.E., Vries, R.G., Van Den Born, M., Barker, N., Shroyer, N.F., Van De Wetering, M., Clevers, H., Paneth cells constitute the niche for Lgr5 stem cells in intestinal crypts. *Nature*, 469, 7330, 415–8, 2011 Jan 20.
57. Wilen, C.B., Lee, S., Hsieh, L.L., Orchard, R.C., Desai, C., Hykes Jr., B.L., McAllaster, M.R., Balce, D.R., Feehley, T., Brestoff, J.R., Hickey, C.A., Tropism for tuft cells determines immune promotion of norovirus pathogenesis. *Science*, 360, 6385, 204–8, 2018 Apr 13.
58. Hutson, A.M., Atmar, R.L., Estes, M.K., Norovirus disease: changing epidemiology and host susceptibility factors. *Trends Microbiol.*, 12, 6, 279–87, 2004 Jun 1.
59. Winder, N., Gohar, S., Muthana, M., Norovirus: an overview of virology and preventative measures. *Viruses*, 14, 12, 2811, 2022 Dec 16.
60. Bai, G.H., Tsai, M.C., Lin, S.C., Hsu, Y.H., Chen, S.Y., Unraveling the interplay between norovirus infection, gut microbiota, and novel antiviral approaches: a comprehensive review. *Front. Microbiol.*, 14, 1212582, 2023 Jul 6.
61. Creutznacher, R., Schulze, E., Wallmann, G., Peters, T., Stein, M., Mallagaray, A., Chemical-Shift Perturbations Reflect Bile Acid Binding to Norovirus Coat Protein: Recognition Comes in Different Flavors. *ChembioChem*, 21, 7, 1007–21, 2020 Apr 1.
62. Shivanna, V., Kim, Y., Chang, K.O., The crucial role of bile acids in the entry of porcine enteric calicivirus. *Virol.*, 456, 268–78, 2014 May 1.
63. Palm, E., Danskog, K., Nord, S., Becker, M., Tanner, H., Sandblad, L., Öhlund, D., Lenman, A., Arnberg, N., Bile acids accumulate norovirus-like particles and enhance binding to and entry into human enteric epithelial cells. *bioRxiv.*, 1, 2023–04, 2023 May.
64. Kong, F., Saif, L.J., Wang, Q., Roles of bile acids in enteric virus replication. *Anim. Dis.*, 1, 1–9, 2021 Dec.
65. Bartnicki, E., Cunha, J. B., Kolawole, A. O., & Wobus, C. E. Recent advances in understanding noroviruses. *F1000Res.*, 6, 79, 2017.

66. Murakami, K., Tenge, V.R., Karandikar, U.C., Lin, S.C., Ramani, S., Ettayebi, K., Crawford, S.E., Zeng, X.L., Neill, F.H., Ayyar, B.V., Katayama, K., Bile acids and ceramide overcome the entry restriction for GII. 3 human norovirus replication in human intestinal enteroids. *Proc. Natl. Acad. Sci.*, 117, 3, 1700–10, 2020 Jan 21.
67. Lewis, M.A., Cortés-Penfield, N.W., Ettayebi, K., Patil, K., Kaur, G., Neill, F.H., Atmar, R.L., Ramani, S., Estes, M.K., A Standardized Antiviral Pipeline for Human Norovirus in Human Intestinal Enteroids Demonstrates No Antiviral Activity of Nitazoxanide. *bioRxiv*, 67,10, e0063623, 2023 May 23.
68. Zhang, M., Zhang, B., Chen, R., Li, M., Zheng, Z., Xu, W., Zhang, Y., Gong, S., Hu, Q., Human Norovirus Induces Aquaporin 1 Production by Activating NF-κB Signaling Pathway. *Viruses.*,14, 4, 842, 2022.
69. Wang, Y., Zheng, Y., Li, Y., Zhang, S., Wang, X., Zong, H., Huang, W., Kong, D., Jiang, Y., Liu, P., Lv, Q., & Jiang, H., Development of a rapid homogeneous immunoassay for detection of rotavirus in stool samples. *Front Public Health*, 10, 975720, 2022.

Index

Printed and bound by CPI Group (UK) Ltd, Croydon, CR0 4YY
17/09/2026
14964265-0007